Information Systems Engineering and Management 56

The book series "Information Systems Engineering and Management" (ISEM) publishes innovative and original works in the various areas of planning, development, implementation, and management of information systems and technologies by enterprises, citizens, and society for the improvement of the socio-economic environment.

The series is multidisciplinary, focusing on technological, organizational, and social domains of information systems engineering and management. Manuscripts published in this book series focus on relevant problems and research in the planning, analysis, design, implementation, exploration, and management of all types of information systems and technologies. The series contains monographs, lecture notes, edited volumes, pedagogical and technical books as well as proceedings volumes.

Some topics/keywords to be considered in the ISEM book series are, but not limited to: Information Systems Planning; Information Systems Development; Exploration of Information Systems; Management of Information Systems; Blockchain Technology; Cloud Computing; Artificial Intelligence (AI) and Machine Learning; Big Data Analytics; Multimedia Systems; Computer Networks, Mobility and Pervasive Systems; IT Security, Ethics and Privacy; Cybersecurity; Digital Platforms and Services; Requirements Engineering; Software Engineering; Process and Knowledge Engineering; Security and Privacy Engineering, Autonomous Robotics; Human-Computer Interaction; Marketing and Information; Tourism and Information; Finance and Value; Decisions and Risk; Innovation and Projects; Strategy and People.

Indexed by Google Scholar. All books published in the series are submitted for consideration in the Web of Science.

For book or proceedings proposals please contact Alvaro Rocha (amrrocha@gmail.com).

Brojo Kishore Mishra · Álvaro Rocha ·
Sandipan Mallik
Editors

International Conference on Technology Advances for Green Solutions and Sustainable Development

ICT4GS-2024

Springer

Editors
Brojo Kishore Mishra
NIST University
Berhampur, Odisha, India

Sandipan Mallik
NIST University
Berhampur, Odisha, India

Álvaro Rocha
University of Lisbon
Lisboa, Portugal

ISSN 3004-958X ISSN 3004-9598 (electronic)
Information Systems Engineering and Management
ISBN 978-3-031-94996-8 ISBN 978-3-031-94997-5 (eBook)
https://doi.org/10.1007/978-3-031-94997-5

This Springer imprint is published by the registered company Springer Nature Switzerland AG
The registered company address is: Gewerbestrasse 11, 6330 Cham, Switzerland

International Conference on Technology Advances for Green Solutions and Sustainable Development (ICT4GS-2024)

The International Conference on Technology Advances for Green Solutions and Sustainable Development (ICT4GS-2024) was successfully organized from August 9–10, 2024, by the Department of Computer Science and Engineering at NIST University, Odisha, India. This prestigious event brought together researchers, academicians, industry professionals, and policymakers from across the globe to share and discuss groundbreaking innovations in green technologies and sustainable practices. The conference featured insightful keynote speeches, technical paper presentations, and interactive sessions covering a wide range of topics, including renewable energy systems, IoT applications for environmental monitoring, sustainable agriculture, and smart technologies for eco-friendly development. Conducted in a hybrid mode—both online and onsite—the conference fostered global collaboration and emphasized the importance of technological advancement in achieving sustainable development goals.

Brojo Kishore Mishra
Sandipan Mallik
Álvaro Rocha

Organization

NIST University, formerly known as the National Institute of Science and Technology, is a leading private technical research university situated in the serene Pallur Hills of Brahmapur, Odisha. Established in 1996 as the first NRI educational venture in the state, NIST has grown into a center of academic excellence and innovation. The 70-acre eco-friendly campus offers a vibrant environment for learning and houses state-of-the-art infrastructure, including high-end laboratories, a central research facility, and advanced computing centers. The university offers a diverse range of undergraduate, postgraduate, and doctoral programs across engineering, science, commerce, and management streams.

Accredited with an 'A' grade by NAAC and approved by AICTE and NBA, NIST University has built a strong reputation for its focus on research, interdisciplinary studies, and technological development. Faculty and students actively engage in high-impact research projects, resulting in quality publications, patents, and industry-driven solutions. With a track record of producing Fulbright scholars and innovators, NIST continues to drive forward its mission to nurture talent, promote innovation, and contribute to national development. To learn more, visit www.nist.edu.

Contents

About the Editors

Brojo Kishore Mishra is currently working as Professor in the Department of Computer Science and Engineering at NIST University. He is the visiting faculty of Department of Computer Science, IISER Berhampur, and also the adjunct faculty of Saveetha College of Liberal Arts and Sciences, SIMATS, Chennai, India. He published more than 50 research papers in peer-reviewed international journals, and more than 40 research papers in proceedings of international conferences, more than 50 book chapters, 20 edited books and 02 authored books, 03 patents published, 03 copyright and 01 trademark. Under his supervision, 4 Ph.D. research scholars are awarded and 2 are continuing. His research interests include data mining, machine learning, soft computing, and security. He received his PhD degree in Computer Science from the Berhampur University, India, in the year 2012. He served in the capacity of keynote speaker, program chair, proceeding chair, publicity chair, and as advisory board members of many international conferences also. He is also a life member of ISTE, CSI, and senior member of IEEE.

Sandipan Mallik achieved his PhD (Engg.) from Jadavpur University, Kolkata, India, in 2014 and is currently serving as an Associate Professor in Department of Electronics and Communication Engineering, at NIST Institute of Science and Technology, Berhampur, Odisha, India. He has more than 10 years of teaching and 16 years of research experience. His research is thin film deposition technology, MOS device physics, transient electronics, proteotronics devices, photovoltaic, IoT sensors, virtual instrumentation, bio-medical devices development, etc. He has published more than 100 research articles in reputed journals, book chapters, and conferences, including six Indian patents. He is also an investigator of a couple of sponsored project and startup projects.

Álvaro Rocha holds the title of Honorary Professor, D.Sc in Information Science, and PhD in Information Systems and Technologies. He is Professor at the University of Coimbra, President of AISTI (Iberian Association for Information Systems and Technologies), and Chair of IEEE SMC Portugal Section Society Chapter. He has acted as Vice-Chair of Experts in the Horizon 2020 of the European Commission, as Expert in the Ministry of Education, University and Research of the Government of Italy, and as Expert in the Ministry of Finance of the Government of Latvia. His main research interests are IS Planning and Management, Maturity Models, Information Systems Quality, Online Service Quality, Intelligent Information Systems, Software Engineering, e-Government, e-Health, IT in Education. His research interests are Information Systems Planning and Management, Maturity Models, Information Systems Quality Online Service Quality, Intelligent Information Systems, and Software Engineering.

A Survey of Applications for Wireless Sensor Networks in Smart Cities

Tejinder Kaur[1(✉)], K Dhana Sree Devi[2], Gurpreet Kaur[3], Ankita[4], M. Samba Siva Reddy[5], and Mukesh Soni[6]

[1] Department of MMICTBM, Maharishi Markandeshwar (Deemed To Be University), Mullana, Ambala-Haryana 133207, India
teji7890@gmail.com
[2] Department of CSE, GSOT, GITAM University, Hyderabad 502329, India
dkunada@gitam.edu
[3] Computer Applications, Guru Teg Bahadur Institute of Management & Technology Dakha Ludhiana, Delhi, India
gurpreetkaursandhu@gmail.com
[4] Department of Computer Applications, Panipat Institute of Engineering and Technology, Samalkha, Panipat, Haryana, India
ankuchiks12@gmail.com
[5] Department of Electronics and Communication Engineering, Lakireddy Bali Reddy College of Engineering (Autonomous), Mylavaram, Andhra Pradesh 521230, India
sambasivareddymula@gmail.com
[6] Division of Research and Development, Lovely Professional University, Phagwara, India
mukesh.research24@gmail.com

Abstract. Wireless Sensor Networks (WSNs) are an enabling technology for smart cities, providing the necessary infrastructure for different applications such as traffic management, environmental monitoring, building energy management, and public safety. The integration of WSNs into smart city environments is still a challenge, as it requires dealing with multiple complexities, such as large-scale deployment, limited energy resources and communication capabilities, and the need for real-time data processing and decision-making. With the recent advancements in wireless communication, sensor technology, and IOT, the use of WSNs in smart cities has become increasingly important. However, the development of WSNs for smart cities applications is still in its infancy, and it is important to understand the current state of the art and the challenges that need to be addressed to ensure the successful deployment of WSNs in smart cities. We classify these applications into various categories, and building energy discusses their challenges and opportunities. We also review the existing technologies and standards for WSNs and their suitability for smart city applications. The main contribution of this paper is to provide a clear overview of the current state of the WSNs in smart cities, highlighting the most important trends and challenges for future research. Our survey will be useful for researchers and practitioner's smart cities, as it will help them understand the current state and future research directions in this field.

Keywords: wireless · sensor · networks · smart cities · energy

B. K. Mishra et al. (Eds.): ICT4GS 2024, ISEM 56, pp. 1–8, 2025.
https://doi.org/10.1007/978-3-031-94997-5_1

1 Introduction

These sensors are deployed in a specific area to collect and process data from the environment, such as temperature, humidity, light, sound, and more. The data is then transmitted wirelessly to a central location, where it is collected and analyzed. WSNs are industrial automation, as shown in Fig. 1 WSNs are composed of three main components: the sensor nodes, the communication network, and the sink [1]. The sensor nodes are the devices that collect and process data from the environment, the communication network is responsible for transmitting the data to the data sink, and the data sink is the device that receives and stores the data [1]. The sensor nodes are usually small, low-power, and low-cost devices that are designed to operate in harsh environments, while the data sink is typically a more powerful device that is further used to process and analyze the data [1].

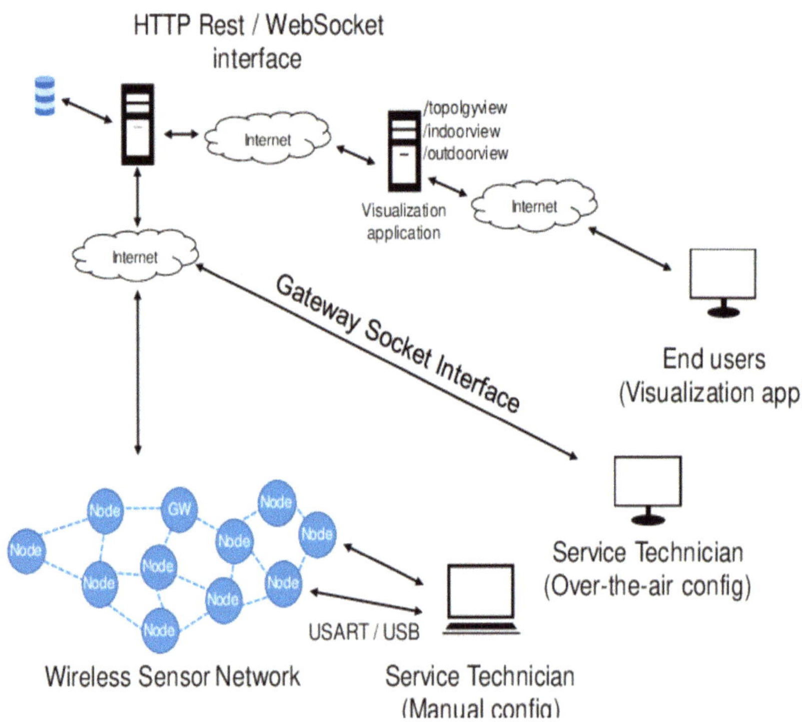

Fig. 1. Structure of wireless network [1]

WSNs have some key characteristics that make them suitable for many applications: They are low-power, low-cost, and easy to deploy, they can operate in harsh environments, and they can provide real-time data collection and processing [2]. These characteristics make WSNs an attractive technology for many applications, including smart cities, enhance sustainability [2]. In this article structure, applications, and some of the major problems associated with sensor networks are explained [3].

2 Discussion

As the cost of WSNs continues to decrease, it is becoming more feasible to deploy them in a wide range of conditions. However, WSN monitoring systems often operate in resource-limited settings, making resource allocation a challenging task and communication protocols to ensure moderate energy consumption and efficient service [4] (Fig. 2).

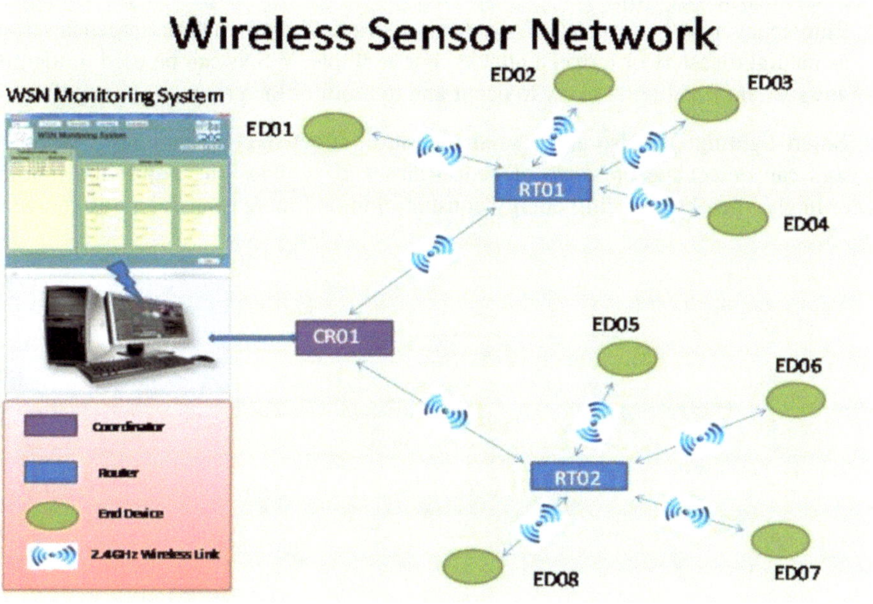

Fig. 2. Wireless Sensor monitoring network [2]

Unmanned Aerial Vehicles (UAVs) have been [5] Proposed as a solution to cover and sense areas that are inaccessible to humans and to assist in data collection and communication among ground sensors.

3 WSNs Implementation

Wireless sensor networks (WSNs) have the potential to revolutionize the way cities operate by providing real-time data on everything from traffic flow to energy consumption. In this paper, we will explore some of the key applications for WSNs in smart cities, including: [6].

1. Traffic management: WSNs can be used to monitor traffic flow in real time, allowing city officials to make adjustments to traffic signals and road layouts to improve traffic flow. This can reduce congestion and decrease travel times for citizens [6].

2. Environmental monitoring: WSNs can be used to monitor a wide range of environmental factors, quality, noise levels, and temperature. This data can be used to identify areas where pollution is a problem and to develop strategies to address it [6].
3. Energy management: WSNs can be used to monitor energy consumption in buildings and other structures. This data can be used to identify areas where energy efficiency can be improved, such as by installing more efficient lighting or HVAC systems [6].
4. Public safety: WSNs can be used to monitor public spaces for signs of crime or other safety hazards. This can include the use of cameras and other sensors to detect suspicious activity [6].
5. Emergency management: WSNs can be used to quickly respond to emergencies, such as natural disasters or terrorist attacks. For example, WSNs can be used to identify areas where flooding is likely to occur and to monitor the progress of a wildfire.

Smart lighting: WSNs can be used to control street lighting in smart cities. The sensors can detect the presence of pedestrians and vehicles and adjust the lighting accordingly, thereby reducing energy consumption and increasing safety as shown in Fig. 3.

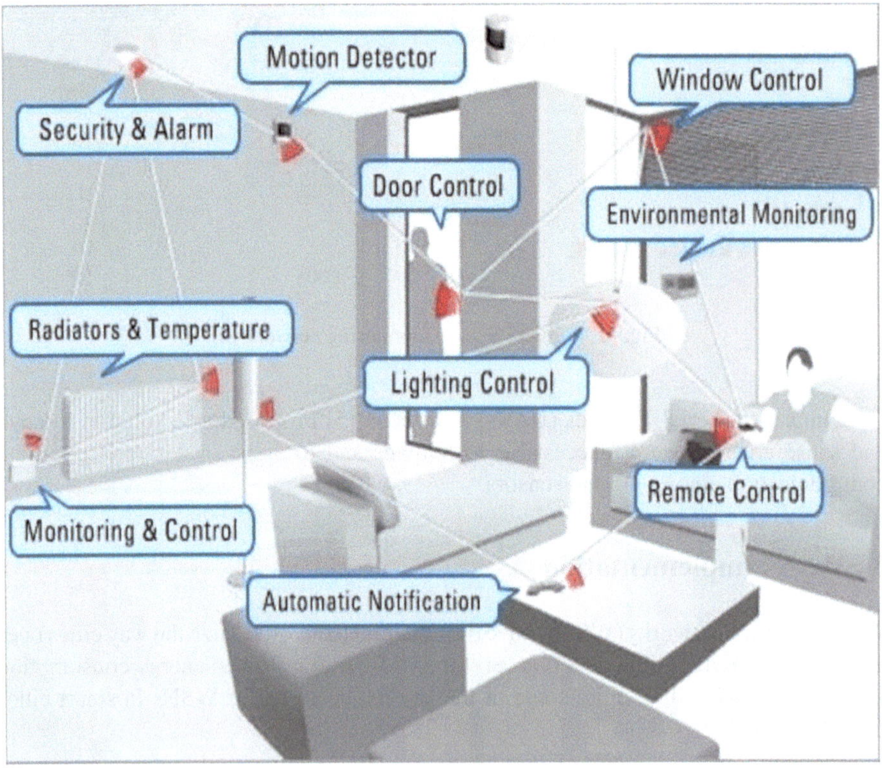

Fig. 3. Wireless Sensor Smart Lighting network [3]

6. Smart parking: WSNs can be used to monitor the availability of parking spaces in a city. This can help drivers find available parking spaces more quickly and reduce congestion on the roads [7]
7. Smart waste management: WSNs can be used to monitor the level of waste in public bins, and then schedule pickups accordingly. This can reduce the amount of waste in the city, improve sanitation, and reduce the number of waste trucks on the road [1]
8. Smart agriculture: WSNs can be used to monitor crop growth and soil moisture levels in urban farming projects. This data can be used to optimize irrigation and fertilization, leading to increased crop yields and reduced water usage [2]
9. Smart water management: WSNs detect leaks in the city's water distribution system. This data can be used to reduce water consumption and identify areas where repairs are needed [3]
10. Smart transportation: WSNs can be used to monitor the movement of public vehicles, such as buses and trains. This data can be used to optimize routes, schedules, and the deployment of vehicles, leading to more efficient and reliable transportation services [4]
11. Smart healthcare: Additionally, sensors can be placed in individuals' homes to monitor the health of elderly or disabled citizens, which can help reduce the need for hospital visits [5] as shown in Fig. 4

3.1 Research Methodology

The wireless sensor network is an organization of sensor nodes (sensing devices) that helps provide information related to the environment. These sensor devices are installed in a particular type of environment and collect information about the major components and components of that environment. Such as these sensor devices can monitor the temperature, humidity, light, pressure, dust content, etc. of the environment so that the environment can be controlled. The organization of a wireless sensor network is shown in Fig. 4 [1].

4 Solution

1. **Data security:** To ensure the security of data transmitted by WSNs, several techniques can be applied, such as: [2].
   - **Encryption:** Data can be encrypted to prevent unauthorized access.
   - **Authentication:** Nodes in the network can be required to authenticate themselves before transmitting data [2].
   - **Access control:** Permissions can be set to control who has access to the data.
   - **Firewalls:** Firewalls can be used to protect the network from external threats.
2. **Energy efficiency:** To ensure the energy efficiency of WSNs, several techniques can be applied, such as:
   - **Sleep modes:** Nodes can be put into sleep mode when they are not transmitting data, to conserve energy [2]
   - **Clustering:** Nodes can be grouped into clusters, with one node designated as a cluster head to coordinate communication and reduce energy consumption.

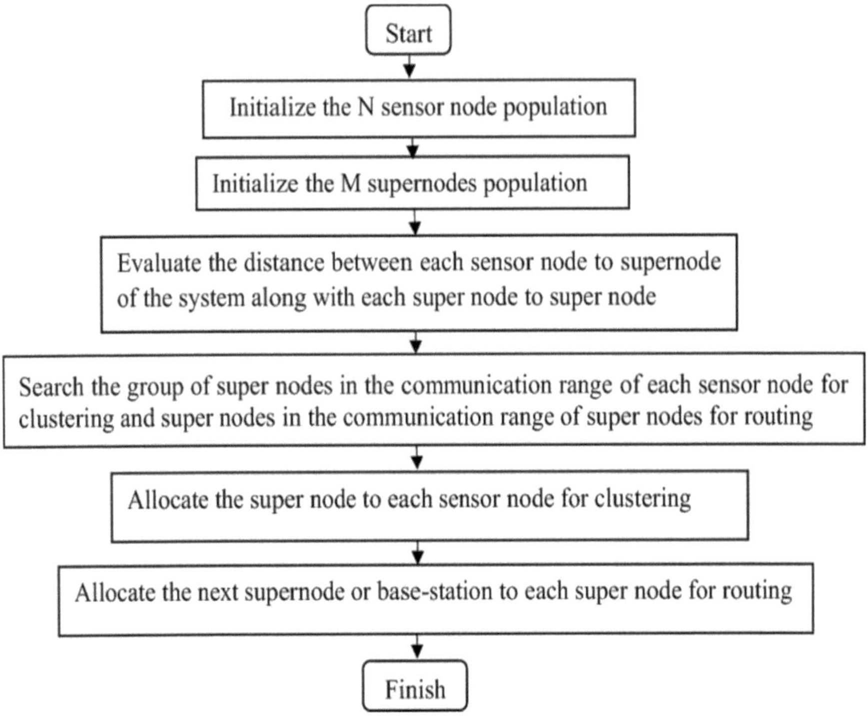

Fig. 4. Wireless Sensor network Methodology [4].

- **Routing:** Energy-efficient routing algorithms can be used to reduce the amount of energy required for data transmission.
- **Energy harvesting:** Nodes can be, to generate power for the network.

3. **Scalability:** To ensure the scalability of WSNs, several techniques can be applied, such as: [2].

- **Hierarchical architecture:** A hierarchical architecture can be used to divide the network into smaller, more manageable sub-networks.
- **Dynamic configuration:** The network can be configured dynamically to adapt to changing conditions.
- **Multi-hop communication:** Data can be transmitted over multiple hops, allowing the network to cover a larger area.
- **Cloud computing:** the network to handle more devices and data [2].

5 Results

1. Self-organization: Self-organization allows WSNs to adapt to changing conditions and minimize the need for manual configuration. By using self-organization, the network can automatically form clusters, select leaders, and adjust the communication patterns to optimize network performance [3].

Resource management: Resource management techniques are used to manage the resources of the network, such as power, bandwidth, and storage, to ensure that they are used efficiently as shown in fig power-aware routing, compression, data aggregation, to reduce the amount of energy required for data transmission [4].

2. Quality of Service (QOS) support: WSNs can be designed to support different QOS requirements, such as low latency or high throughput, to meet the needs of different applications. This can include techniques such as traffic scheduling, medium access control, and error control, to ensure that the network can handle different types of traffic as shown in Fig. 5.

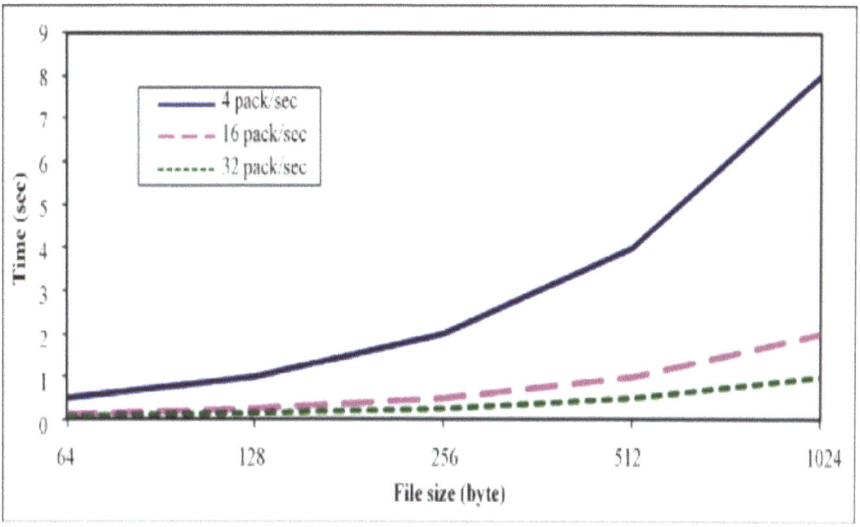

Fig. 5. Quality of services in sensor network [5]

3. This can include techniques such as redundancy, replication, and error correction, to minimize the impact of node failures on the network as shown in Fig. 5 [4]

4. Mobile support: Mobile support allows WSNs to support mobile nodes, allowing them to move freely within the network [5]. This can include techniques such as location tracking, handoff, and mobile routing, to ensure that the network can handle mobile nodes as shown in Fig. 6 [6].

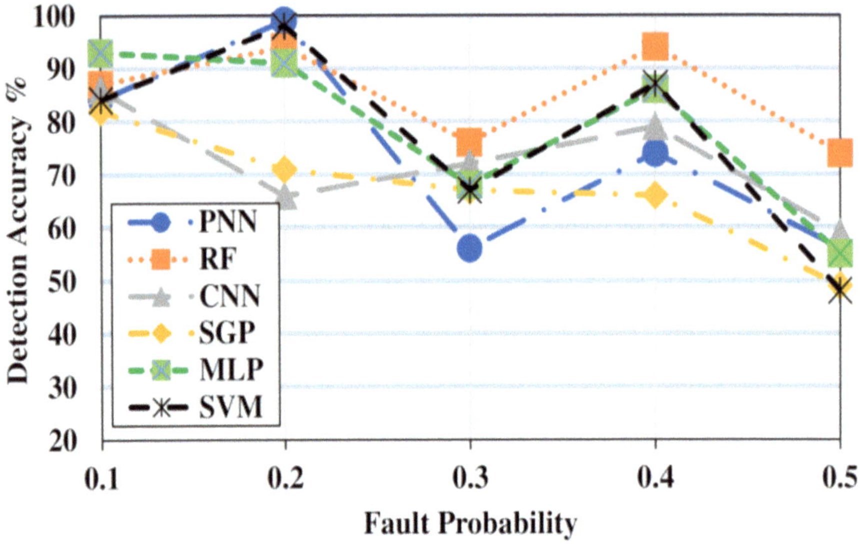

Fig. 6. Fault tolerances in sensor network [6]

References

1. Corke, P., Hrabar, S., Peterson, R., Rus, D., Saripalli, S., Sukhatme, G.: Autonomous deployment and repair of a sensor network using an unmanned aerial vehicle. IEEE Int. Conf. Robot. Autom. Proc. **4**, 3602–3608 (2004)
2. Kuorilehto, M., Hännikäinen, M., Hämäläinen, T.D.: Rapid design and evaluation framework for wireless sensor networks. Ad Hoc Netw. **6**, 909–935 (2008)
3. Hammoodi, I.S., Stewart, B.G., Kocian, A., McMeekin, S.G.: A comprehensive performance study of OPNET modeler for zigbee wireless sensor networks. In Proceedings of the 2009 Third International Conference on Next Generation Mobile Applications, Services and Technologies, Cardiff, UK, pp. 357–362 15–18 Sept. 2009
4. Saqib, J.M., Jamil, M.A., Mazhar, A., Ikram, A., Ahmed, A., Munawar, A.: Smart environment monitoring system by employing wireless sensor networks on vehicles for pollution free smart cities. Procedia Eng. **107**, 480–484 (2015)
5. Jain, B., Brar, G., Malhotra, J., Rani, S.: A novel approach for smart cities in convergence to wireless sensor networks. Sustain. Cities Soc. **35**, 440–448 (2017)
6. Ala, K., Darabkh, K.A., Khasawneh, A.M., Alqaisieh, I., Ahmed AlAbdala, M.S., Alrubaye, S., et al.: Wireless sensor networks for smart cities: Network design, implementation and performance evaluation. Electronics **10**(2), 218 (2021)
7. Sharma, H., Haque, A., Blaabjerg, F.: Machine learning in wireless sensor networks for smart cities: a survey. Electronics **10**(9), 1012 (2021). https://doi.org/10.3390/electronics10091012

Exploring IoT Applications in Oceanographic Research for Advancing the Blue Economy

K. Manoj Kumar(✉)

NIST University Berhampur, Berhampur, India
manoj93.9090@gmail.com

Abstract. It is the field of oceanographic research with the aim of advancing the blue economy. Generally, it can be used as an ocean research resource for economic growth, and IoT technologies to analyze the data collection, monitoring and sensor networks in oceanography. It consists of some different applications for analyzing data, including LIDAR technologies, sensor networking, autonomous underwater vehicles and data mining to enhance the understanding of an ecosystem improve sustainable development in marine industries and to detect the remote operated vehicle technology for fisheries, rock and salinity. It involves interconnected devices and sensors that can be communicated and share the data through the network. It can store a huge amount of data for an oceanography on fisheries, plants and route mapping technology inside the water and also detect the depth of an ocean. It helps some GIS tools to analyze and visualize the data related to oceanography such as temperature, current and marine habitats. The data can be simulated through the use of IoT data into numerical methods in a real-time environment.

Keywords: IoT applications · Oceanographic research · blue economy · data analytics · sustainability

1 IoT Applications

It is the application of IoT in oceanography to interconnect the sensor network for establishing an ecosystem including sensors and autonomous systems. These sensor network devices are used to design and analyses the data in a real-time environment from the marine environment.

1.1 Smart Ocean Sensor

It is an advanced technology which is used to be designed for real-time monitoring and data collection in oceanographic research. It is used to measure some parameters through the sensor such as temperature, salinity, depth and pH. In this sensor technology, it can be allowed some internet connectivity to communicate wirelessly for data transmission. It consists of some larger parts of the IoT network to integrate with sensors and an autonomous system. These sensors are used for different parameters, and

B. K. Mishra et al. (Eds.): ICT4GS 2024, ISEM 56, pp. 9–17, 2025.
https://doi.org/10.1007/978-3-031-94997-5_2

they include some different technologies such as optical and electrical sensing. Through neural network algorithms, the signal processing techniques may be used for real-time processing, and enable the sensors to make inferences about ocean conditions. These sensors can communicate through satellite radio and modem for transmitting data to the remote sensors. It allows them to communicate with IoT networks in the oceanographic infrastructure (Figs. 1, 2 and 3).

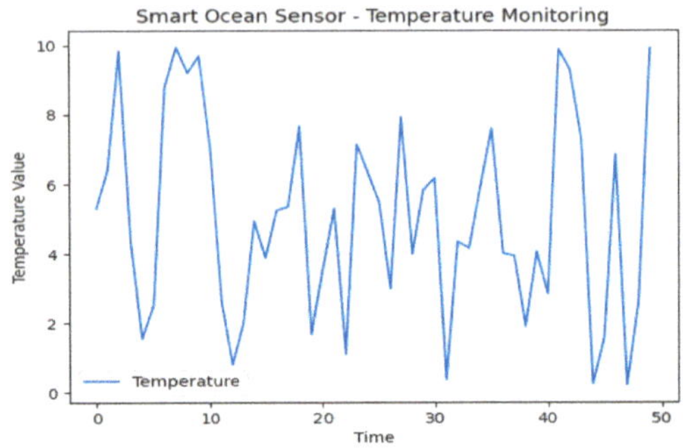

Fig. 1. Smart Ocean Sensor

In the smart ocean sensor, a wireless sensor network is a collection of independent sensors strategically placed in different locations. It is the ability to operate autonomously, communicate wirelessly with each other and collaborate with the data either physically or environmentally in a specific region.

It allows the sensors to interact with each other and transmit the data into a central location for centralized processing.

2 Autonomous Underwater Vehicle (AUV)

It is an underwater vehicle that is capable of performing tasks without human intervention and these vehicles are used for various marine applications, including oceanographic research, underwater mapping and monitoring. It operates independently based on the adaptive algorithms which are used to allow a navigation path through the robotics sensor inside the water environment and for the data for a specific task. It involves the process of determining the position, orientation to enable a pre-defined path for changing the conditions autonomously. Generally, to locate the path finding algorithm, it can be developed to analyze an algorithm i.e. Simultaneous Localization and Mapping (SLAM).

2.1 SLAM (Simultaneous Localization and Mapping)

It is a computational technique used in robotics to enable an autonomous system to create a location map in an unknown environment between its initial position to the final position. In this technique, it creates a map of the environment to locate the data in a real-time sensor using SONAR and LIDAR.

This mapping algorithm is widely used on various robotic platforms, including drones, cars and underwater vehicles to navigate the path in an unknown environment. It is utilized in augmented reality applications to track the position of any device in the real-world environment.

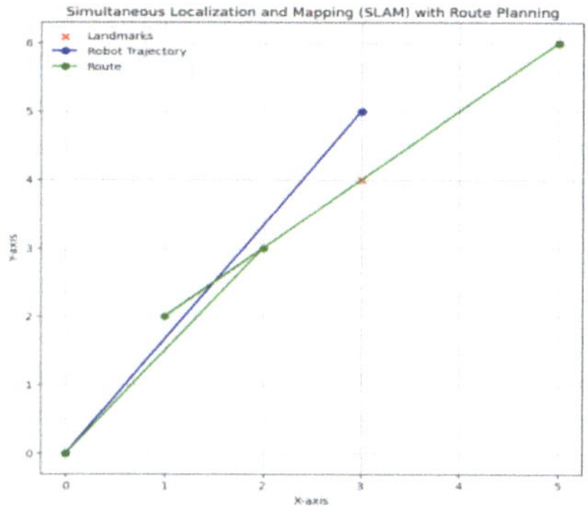

Fig. 2. SLAM

2.2 Remote Sensing Applications

It is widely used to monitor changes in the environment including reforestation and natural disaster such as flood and earthquakes. It is the study of the oceans to monitor sea surface temperature, detect ocean currents and salinity in marine ecosystems. In these applications, it is used to track changes in climate variables such as temperature, sea-level, ice-cover, and it can monitor lakes, rivers and reservoirs. It includes the analysis of geological structure, landforms and mineral composition through satellite and aerial imagery.

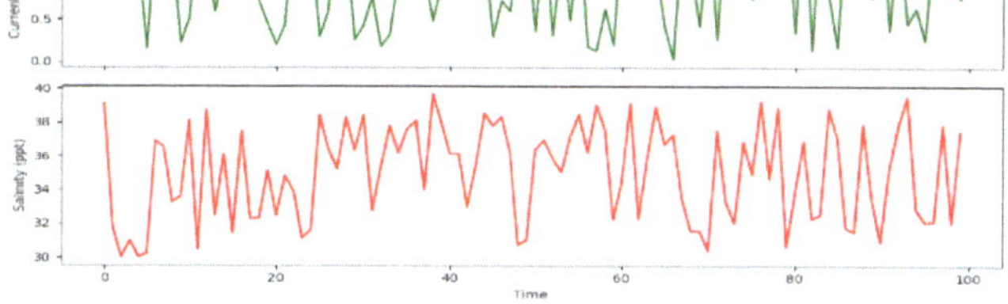

Fig. 3. Remote Sensing Applications

2.3 Oceanographic Research

It is the study of the physical, chemical and biological aspects of researching different labor-intensive instruments to collect data. It uses some different communication technologies, algorithms such as satellite, Wi-Fi and cellular, and it involves quality control, data fusion and application of algorithms to complex datasets. These devices enable continuous monitoring of oceanographic parameters to track changes in water quality, and detect pollution events in the marine ecosystems. It can be deployed in remote ocean environments and faces challenges to power management and energy efficiency. In oceanographic research, it can be implemented to analyze the depth and temperature of an oceanographic structure using algorithms.

3 Artificial Neural Network (ANN)

In this network, it can be utilized to analyze complex oceanographic datasets between environmental variables. It is a computational model of the structure and the function of networks for interconnecting nodes. These networks are applied to analyze remote sensing data such as sea temperature, depth and ocean color data (Figs. 4, 5 and 6).

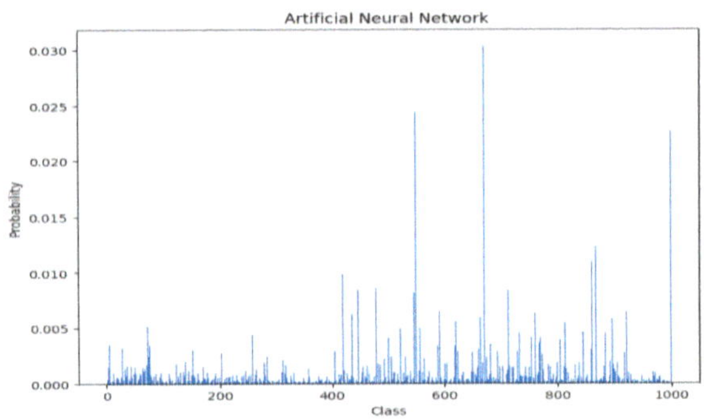

Fig. 4. Artificial Neural Network

In oceanography, the data can be analyzing by some complex datasets to identify the relationship between temperature, salinity and biological parameters. These are used to analyze remote sensing data, including ocean color data. It can detect some oceanographic features, such as increasing the density of the water column. In this network, we can use deep neural network algorithms for extracting high dimensional oceanography datasets and improving non-linear relational models.

4 Deep Neural Network

It is a type of artificial neural network that is utilized to analyze complex oceanographic data. It consists of multiple layers of interconnected nodes organized in a hierarchical structure. The prediction of oceanographic variables to measure the sea surface, temperature and salinity. We can integrate data fusion from multiple sources and provide some sensors for understanding the oceanographic process.

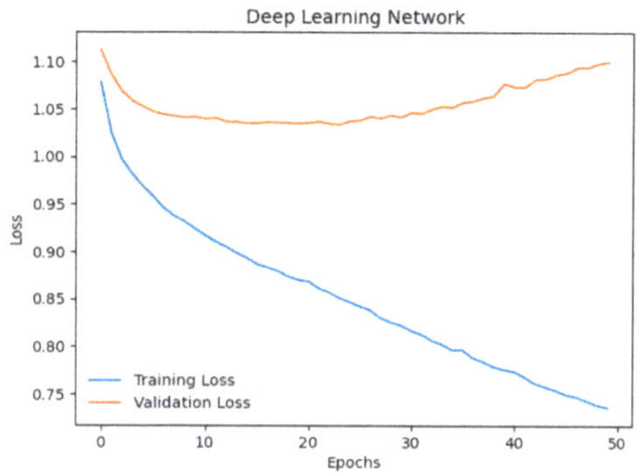

Fig. 5. Deep Learning Network

The data can be collected through oceanography from different sources such as underwater sensors and autonomous vehicles. It can also be measured by different data applications to detect the depth, temperature and salinity.

5 Blue Economy

It is the sustainable use of ocean resources for economic growth to improve the ocean ecosystem's health. It includes some different conservation systems, like renewable energy, fisheries management and ocean technology.

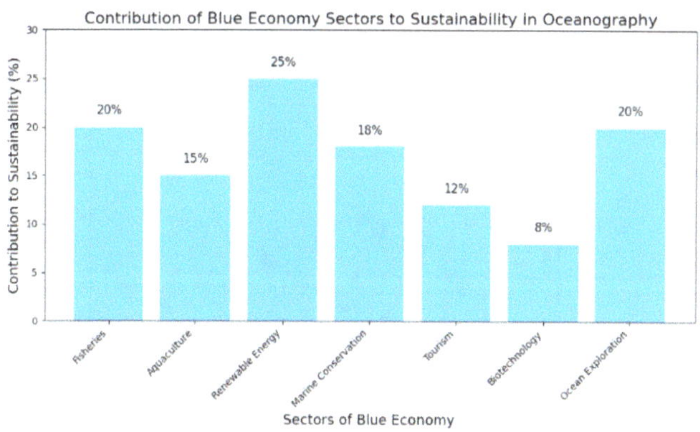

Fig. 6. Sustainability

The sustainability of fisheries management ensures the fishing activities are conducted to maintain a healthy fish stock the ecosystem. It contributes to the development of renewable energy technologies for studying wave patterns to identify the location of the technologies and optimize their efficiency.

It is the source of biotechnological resources, including marine organisms, through medicine and industry. The role of sustainability for tourism is to minimize some negative impacts on marine ecosystems by promoting responsible tourism.

6 Data Analytics

The data can be integrated from different sources, and it involves some clean data to remove errors and inconsistences. Due to statistical analysis, it is used to quantify the relationship between variables and the test hypothesis. It includes supervised learning techniques for classification and regression and unsupervised learning techniques are used for clustering and patterning data. The data should be collected from various environments, such as biological data, socio-economic data and operational data (Fig. 7, 8 and 9).

```
Fisheries Data:
   Year  Fish_Catch_Metric_Tons          oceanography
0  2016                    5000         marine system
1  2017                    5200  fisheries management
2  2018                    4900      renewable energy
3  2019                    4800       biological_data
4  2020                    5100      operational_data
```

Fig. 7. Data analytics

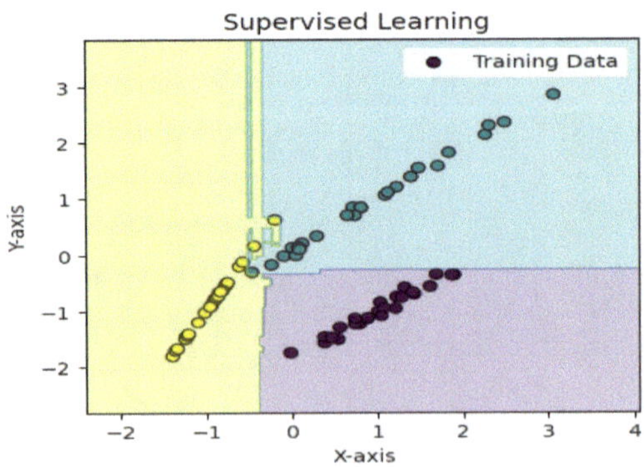

Fig. 8. Supervised Learning

In the supervised learning algorithms, the mapping can classify underwater habitats based on remote sensor data such as satellite imaginary and sensor data. It can be applied to predict oceanographic variables of the sea surface, salinity and temperature. It can analyze historical data of fisheries management to predict migration patterns.

In an unsupervised learning algorithm, cluster analysis can be used to identify the regions and help in ecosystems for marine hotspots. In these technique, it is used to identify environmental changes and pollution events underwater.

7 Sustainability

It is responsible for management and conservation of marine resources and the ecosystem to improve health and productivity of the ocean environment. It establishes the marine biodiversity to protect areas and to measure fishing activities and pollution. It approaches an ecological, social and economic dimension of marine resource systems to achieve sustainability and to monitor climate change in ocean temperature, salinity and sea level as well as, develop human societies. Some different technologies for sustainability in oceanography are applied to different sensors to collect data on ocean properties such as temperature, salinity and biological diversity. It helps to provide some real-time data on the ocean conditions to help scientists to monitor the ecosystem health (Figs. 10, 11, 12 and 13).

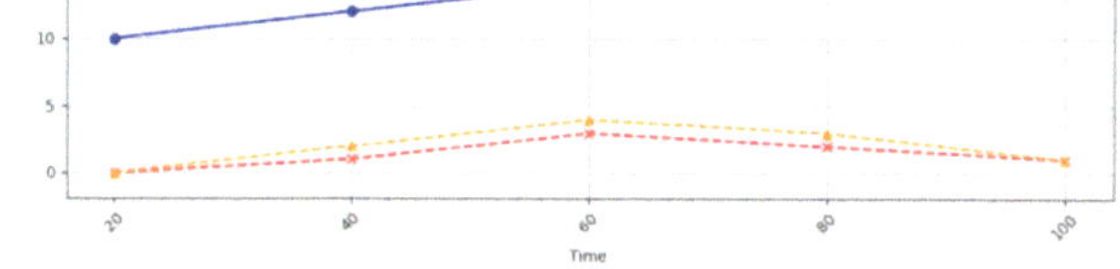

Fig. 9. Sustainabilty

8 Results

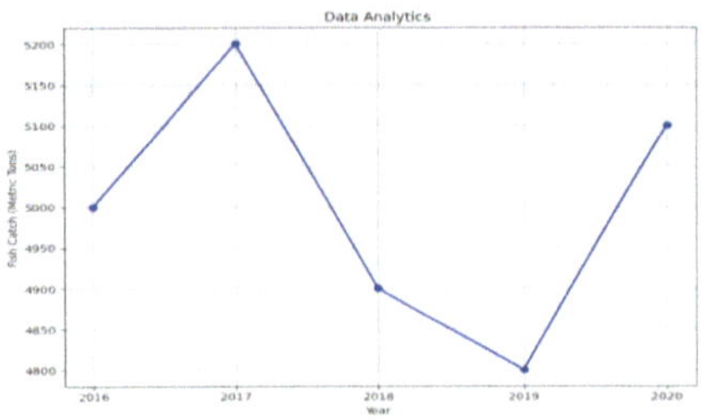

Fig. 10. Data analytics

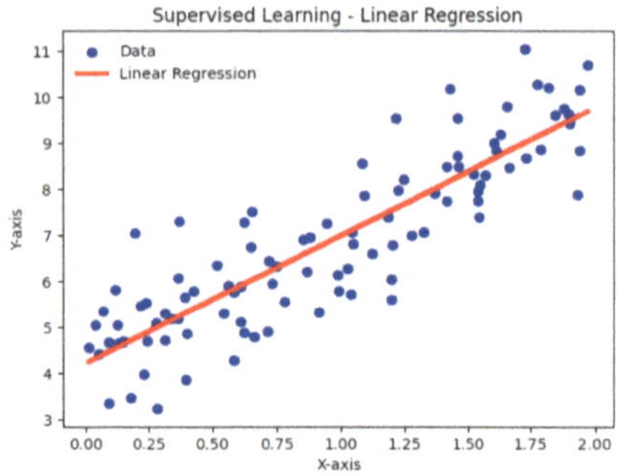

Fig. 11. Supervised learning

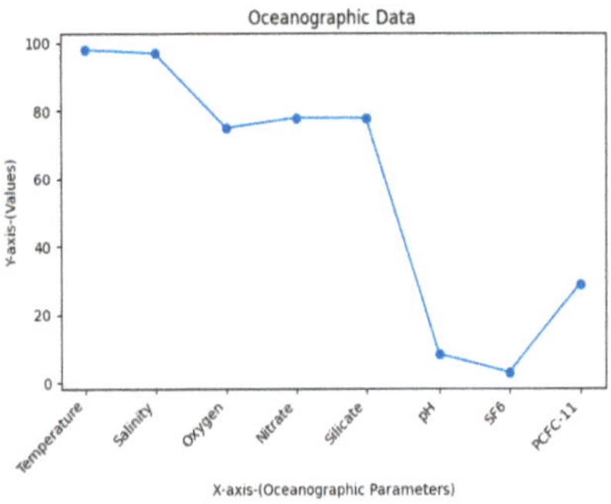

Fig. 12. Oceanographic Data

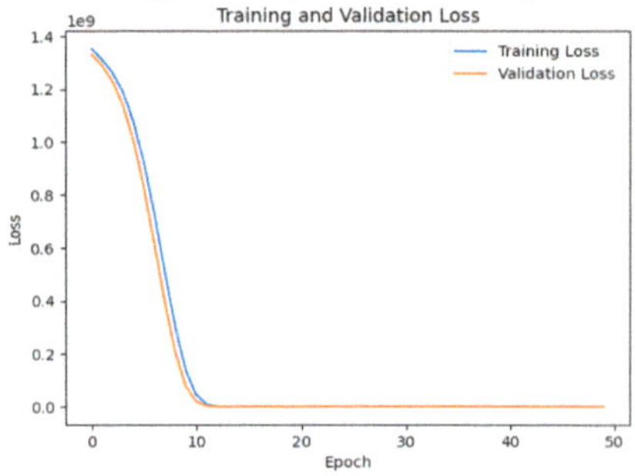

Fig. 13. Validation data

9 Conclusion

The basic opportunities of various IoT applications to identify oceanographic research into a blue economy. It offers some real-time data collection and a remote monitoring system for challenges in IoT technologies. We can improve for future research directions in oceanography to include data analytics and sensor technologies as well as amplification of the blue economy in IoT technologies to support sustainable development and economic growth in fisheries and renewable energy. In the future, we can innovate some IoT applications in oceanographic research in the blue economy, such as satellite-based monitoring systems and autonomous underwater vehicle systems. For data driven, IoT can enable amounts of vast amounts of real-time data from the oceanic environment. Due to sustainable development, it can support marine resources by monitoring parameters like water quality and biodiversity.

References

1. Domingo, M.C.: An overview of the internet of underwater things. J. Netw. Comput. Appl. **35**, 1879–1890 (2012)

2. Yan, J., Gong, Y., Chen, C., Luo, X., Guan, X.: AUV-aided localization for Internet of underwater things: a reinforcement-learning-based method. IEEE Internet Things J. **7**, 9728–9746 (2020)
3. Khan, M.T.R., Ahmed, S.H., Kim, D.: AUV-aided energy-efficient clustering in the internet of underwater things. IEEE Trans. Green Commun. Netw. **3**, 1132–1141 (2019)
4. Morozs, N., Mitchell, P.D., Diamant, R.: Scalable adaptive networking for the Internet of underwater things. IEEE Internet Things J. **7**, 10023–10037 (2020)
5. Kao, C.C., Lin, Y.S., Wu, G.D., Huang, C.J.: A comprehensive study on the internet of underwater things: applications, challenges, and channel models. Sensors **17**, 1477 (2017)
6. Gjanci, P., Petrioli, C., Basagni, S., Phillips, C.A., Bölöni, L., Turgut, D.: Path finding for maximum value of information in multi-modal underwater wireless sensor networks. IEEE Trans. Mob. Comput. **17**, 404–418 (2017)
7. Yan, J., Yang, X., Luo, X., Chen, C.: Energy-efficient data collection over AUV-assisted underwater acoustic sensor network. IEEE Syst. J. **12**, 3519–3530 (2018)
8. Menaka, D., Gauni, S., Manimegalai, C.T., Kalimuthu, K.: Vision of IoUT: advances and future trends in optical wireless communication. J. Opt. **50**, 439–452 (2021)

IoT-Enabled Smart Wheelchair System for Real-Time Fall Detection and Alerting

Swetarani Mishra[1], Debabrata Dansana[1(✉)], Brojo Kishore Mishra[2], and Tanmaya Bhoi[1]

[1] Rajendra University, Balangir, Odisha 767002, India
debabratadansana07@gmail.com
[2] NIST University, Berhampur, Odisha, India

Abstract. The National Sample Survey Organization (NSSO) has identified disability as a significant public health challenge in India, with a notable increase in limb loss, visual impairment, and other disabilities arising from factors such as wars, accidents, health issues, and aging. Many visually impaired individuals struggle with motor skills and strength, making traditional power wheelchairs problematic. Furthermore, the high cost of these wheelchairs renders them inaccessible to many. Caregiver workload in nursing and old-age homes is escalating due to a disproportionate ratio of disabled residents to caregivers. This research introduces a prototype for a Smart Wheelchair aimed at reducing accident risks and easing caregiver responsibilities. The wheelchair integrates various hardware and software elements coordinated via an Arduino Uno board. A primary feature is the Ultrasonic sensor, which measures distance and detects falls. In conjunction with this, a mobile application, Blynk, alerts caregivers immediately if the user experiences an accident, ensuring prompt response and increased safety.

Keywords: Smart Wheelchair · Caregiver Alert System · Internet of Things (IoT) · Ultrasonic Sensor

1 Introduction

Disability, defined as a limitation or inability to perform everyday activities due to impairments, remains a significant public health concern, particularly in India. As per the National Sample Survey Organization (NSSO), 2.68 crore of India's 121 crore population, or 2.21%, were identified as disabled as of 2011. This figure represents a 22.4% increase from 2001, with projections suggesting continued growth. Within nursing and old-age homes, there's a noticeable disparity between the number of caregivers and the disabled or elderly residents. These caregivers often juggle multiple responsibilities, including regular patrolling to promptly detect and prevent severe accidents. To alleviate some of this caregiver burden, wheelchairs have become a predominant assistive tool to enhance mobility for people with disabilities. While power wheelchairs offer advantages, they pose challenges for those with visual impairments, motor skill deficits, and diminished strength.

B. K. Mishra et al. (Eds.): ICT4GS 2024, ISEM 56, pp. 18–27, 2025.
https://doi.org/10.1007/978-3-031-94997-5_3

Moreover, their cost often makes them inaccessible to many, especially in povertystricken regions of India. To address these challenges, we introduce a budget-friendly Smart Wheelchair prototype. This innovative solution integrates a manual wheelchair with an ultrasonic sensor and the Blynk mobile app, all connected via an Arduino Uno board. The sensor's primary functions include distance measurement and fall detection. In the event of a fall, caregivers receive immediate notifications on their smartphones, facilitating rapid response. This system aims to reduce accident occurrences and mitigate potential injury severity. The ultrasonic sensor and Blynk app synchronization are achieved through code written in the Arduino programming language.

2 Literature Survey

The evolution and integration of technology in mobility aids, particularly wheelchairs, has been pivotal in enhancing user experience and safety. Smith & Johnson (2017) detailed this transformation, highlighting the rise of intelligent wheelchairs in modern healthcare [1]. Complementing this perspective, Kumar & Rao (2018) specifically explored the potential of the Internet of Things (IoT) in healthcare, focusing on the development of intelligent wheelchair prototypes within the Indian context [1]. A crucial component in the technological makeup of these wheelchairs is the sensor system. Lee & Tanaka (2019) conducted a comparative study on the efficiency and accuracy of ultrasonic sensors across various assistive devices, emphasizing their application in intelligent wheelchairs. The implications of these advancements extend beyond mere user autonomy [2]. Patel & Shah (2020) discussed the growing importance of alert systems, evaluating their critical role in providing timely assistance and reducing potential hazards for the elderly and disabled populations [3]. In the realm of technological enhancements, Gomez & Fernandez (2021) provided a comprehensive overview of digital advancements in wheelchair technology, underlining the dual aspects of user safety and autonomy [4]. Given the economic constraints in developing nations, Nair & Krishnan (2018) underscored the challenges and opportunities of creating budget-friendly assistive technologies [5]. The Arduino platform, commonly used in healthcare technological solutions, has been reviewed by Chen & Liu (2019), showcasing its diverse applications, potential, and challenges [6]. Specifically, for alert systems, Verma & Gupta (2020) explored the application and reliability of the Blynk mobile app, a crucial component in many smart wheelchair designs [7]. The increasing need for advanced mobility solutions for older people was eloquently presented by Diaz & Rodriguez (2017), emphasizing the urgency for innovative solutions [8]. Echoing the importance of technological integration in healthcare, Kim & Park (2018) overviewed the transformative impact of sensors in healthcare, highlighting their potential to revolutionize patient care and safety [9]. Some more research related to IoT is discussed in [10–12]. In sum, the literature consistently underscores the significance of smart wheelchair systems, not only as a means of enhancing mobility but also as a tool for ensuring safety and reducing caregiver workload.

The research work is organized as follows: Section 3 covers the materials and methods detailing the proposed smart wheelchair design. In Sect. 4, the results are outlined, and the research concludes in Sect. 5.

3 Material and Methods

3.1 Proposed Smart Wheelchair Design

The core of our system is the microcontroller, specifically the Arduino Uno.This microcontroller is programmed to work in tandem with an ultrasonic sensor. The primary function of this sensor is to continuously monitor andmeasure the distance of the surroundings in real-time. This distance data is then relayed to the Blynk app. Now, let's delve a bit deeper into how this detection works. Within our program, we've predefined a certain 'threshold distance'.Think of this as a safety parameter. If, at any moment, the ultrasonic sensor measures a distance greater than this set threshold, the system interprets this as apotential fall of the wheelchair to which the prototype is affixed. When such a scenario occurs, it triggers the microcontroller to act. The programmed response is to send an alert to the Blynk app promptly. This app, in turn, serves as a notification system to immediately inform caregivers and relatives of the potential fall, allowing them to take necessary actions, which is mentioned in Fig. 1 and the proposed flow of the system depicted in Fig. 2.

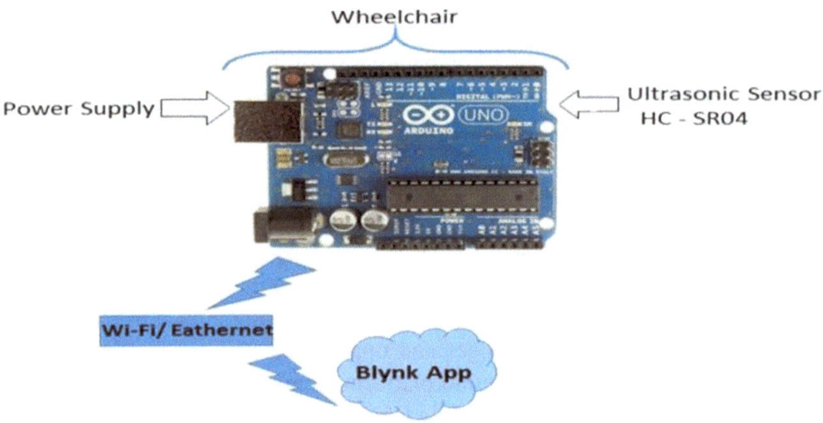

Fig. 1. System Design of Proposed Prototype

3.2 Installing Drivers and Necessary Libraries to the Arduino IDE

To program the Arduino UNO, you must have the necessary drivers installed in the Arduino IDE you're working with. To determine if these drivers are pre-installed or if manual installation is required, follow the steps and accompanyingvisuals detailed below:

- Launch the Arduino IDE.
- Navigate to 'Tools' and then select 'Board', as illustrated in Fig. 3.

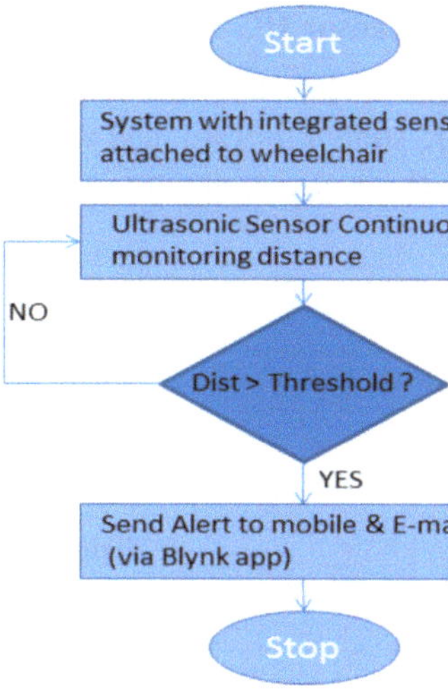

Fig. 2. Proposed flow of the system

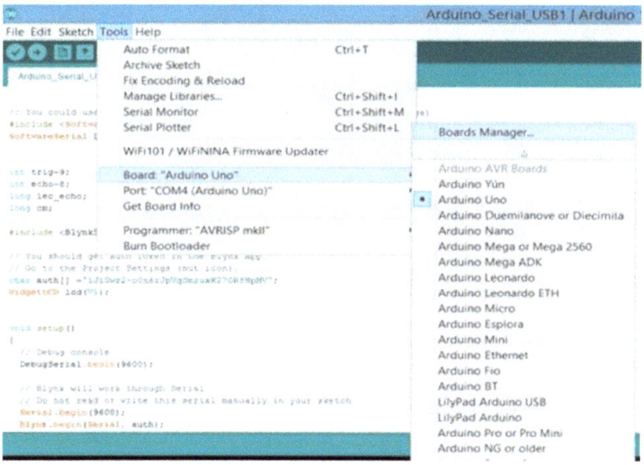

Fig. 3. Necessary Drivers

After navigating to that section, look if 'Arduino/Geniuno Uno' is in that list. If itis available, we need to opt for that option. If such an option is unavailable, we must manually install that necessary driver. For this, follow the instructions-

Open the Arduino IDE again -> click on Tools ->Board Manager

Now, in the search bar, search for "Arduino SAMD board." Then select theArduino IDE version and download the one that comes. Once our required drivers are installed, we need to go ahead and download the necessary libraries. We need the Ultrasonic library and Blynk library to be installed for our program to run. Both libraries are quickly found in the Arduino IDE's built-in library manager.

Open Sketch-> click on Include Library->Library Manager

3.3 Connecting the Ultrasonic Sensor HC-SR04 with Arduino Uno

Here, we connect the Ultrasonic Sensor HC-SR04 with the Arduino Uno through a breadboard. The connections are straightforward, and we did this like the instructions given below in Table 1, and the physical configuration setup is depicted in Fig. 4.

Table 1. Pin connection

No	Pin of HC-SR04	Pin of Arduino Uno
1	VCC	VCC
2	GND	GND
3	Trig	Trig
4	Echo	Echo

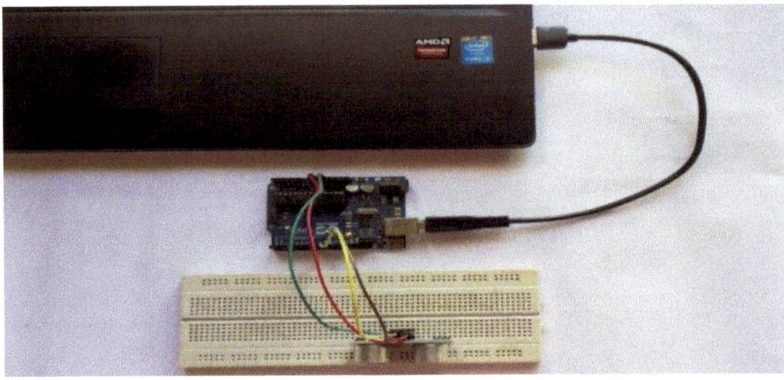

Fig. 4. Pin connection of the Arduino Uno and Ultrasonic sensor Widget Setup in the Blynk App

Widgets available in the Blynk app and its drag-and-drop environment make it very simple to implement. We have used three widgets from this app for the proposed research work.

The first is the 'LCD widget', which shows the distance the ultrasonic sensor measures every moment. This widget's setup setting is shown in Figures 5, 6 and 7.

3.4 Writing and Uploading Sketch

Arduino IDE for the connected Ultrasonic Sensor HC-SR04, so that the sensor will function in such a way that it will continuously detect the distance of the current scenario, and measured distance will be given to the Blynk app simultaneously. Here, we will have to set a threshold distance, and when the distance measured at a particular moment is greater than the defined threshold, it will be considered as the fall of the wheelchair to which the prototype is attached. Now, the program will instantly redirect to the Blynk app server through the microcontroller to generate an alert signal to notify the caretakers or the relatives of the person sitting in the wheelchair. This alert signal will be given to destinations through two methods we set up earlier in the previous step.

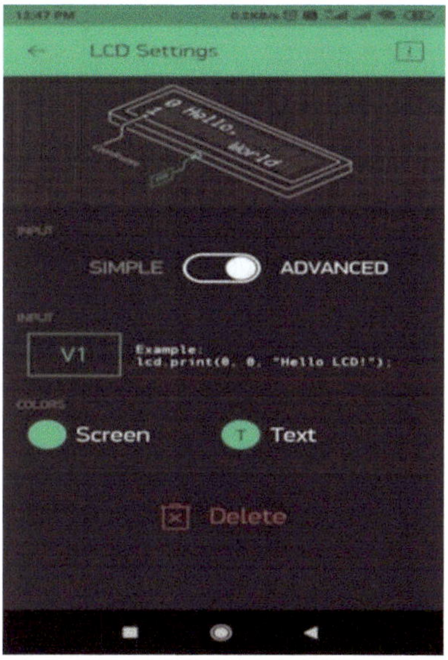

Fig. 5. LCD Widget

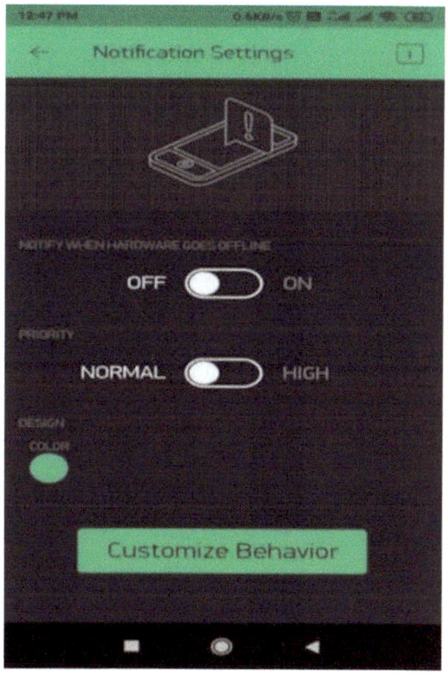

Fig. 6. Notification Widget

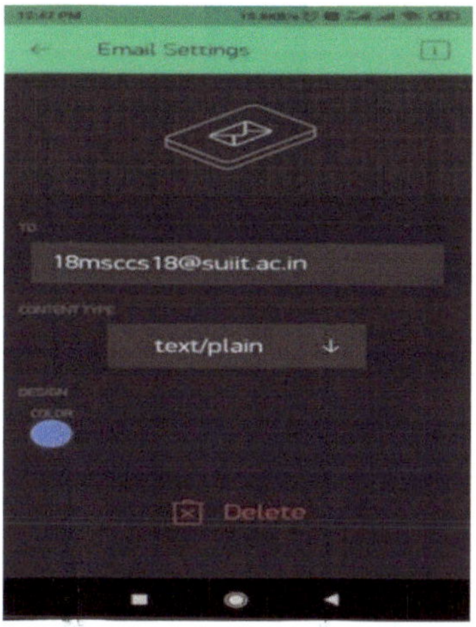

Fig. 7. Email Widget

- As a notification to the smartphone where the Blynk app is installed for this project.
- We sent an E-mail to the recipient's account during the Blynk e-mail widget setup.

4 Results

After connecting all the components and uploading the sketch to the Arduino Uno, which had the ultrasonic sensor attached, we evaluated our wheelchair fall detection prototype. The functional prototype aligned with our expectations during our tests based on the code we had written and uploaded. The ultrasonic sensor consistently and continuously measured distances in real-time. These measured values were then sent to the Blynk server and the Arduino IDE. Within the Arduino IDE, we could visualize the data as a graph using the Serial Plotter depicted in Figures 8, 9, 10 and 11.

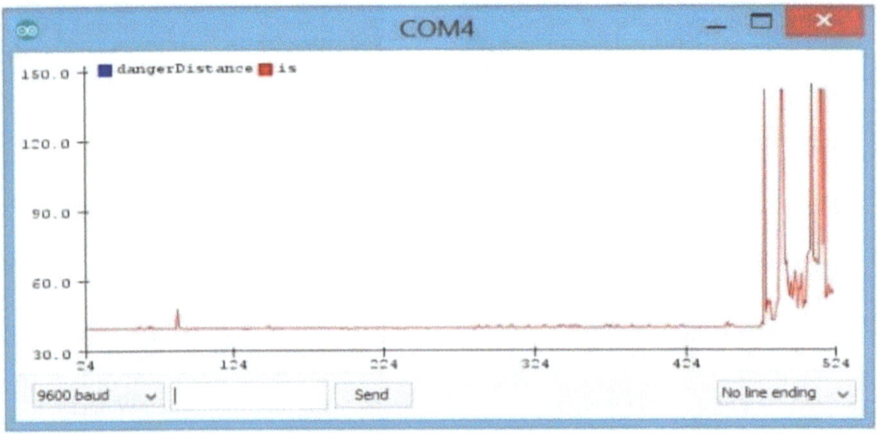

Fig. 8. Serial plotter output

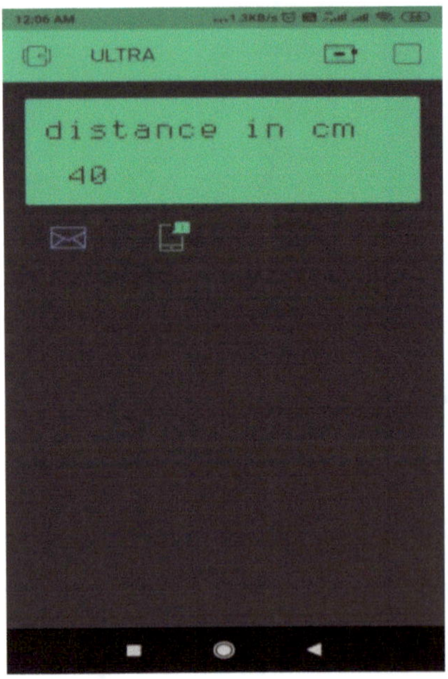

Fig. 9. Wheelchair Fall Detection System in Blynk App

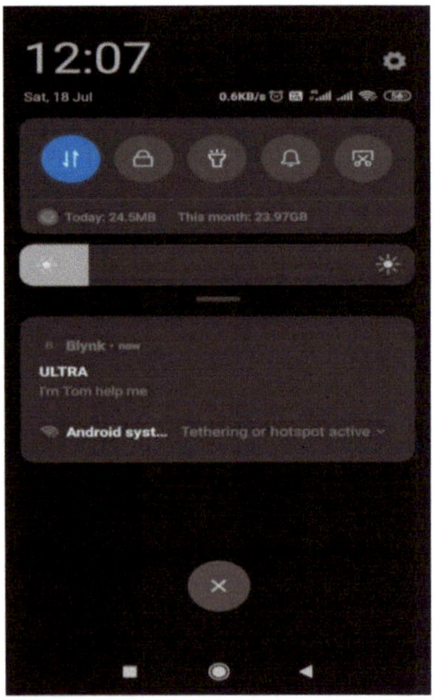

Fig.10. Blynk Alert Notfication1

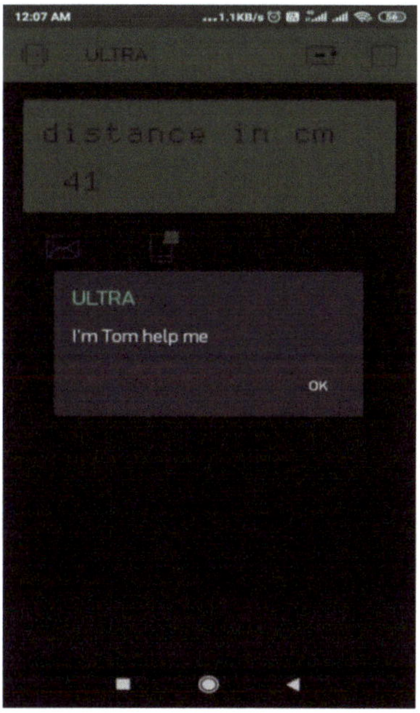

Fig. 11. BlynkAlert Notfication2

During the development phase, we configured an alert message within our sketch. This message is transmitted to the recipient's email account using the email widgetoffered by the Blynk app. A demonstration of this functionality can be seen in Fig. 12.

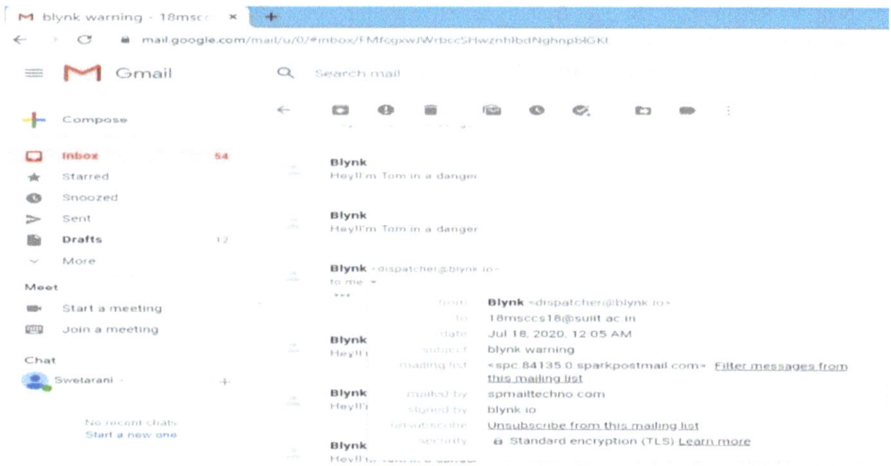

Fig. 12. E-mail alert

5 Conclusions

In this research work, we developed an IoT-based prototype for detecting falls inwheelchairs. Designed to be placed under the wheelchair's seat, the system gaugesthe distance between the chair and the floor. Our experiments and simulations demonstrated the prototype's viability for real-world applications. If a wheelchair equipped with our system were to fall, caregivers would receive immediate notifications on their smartphones via the Blynk app or email, enabling swift action to assist the affected individual. This system proactively addresses potential wheelchair accidents. While promising in its preliminary stage, there are areas for improvement. Notably, our current design doesn't pinpoint the precise location of an incident due to budget constraints and our focus on cost- cost- effectiveness, which led us to omit a GPS module. Future enhancements could integrate a GPS module with a Node MCU, possibly replacing the Blynk app witha dedicated Android or iOS app tailored specifically for this project, ensuring greater privacy. With its significant potential for further research and development, this project lays the foundation for transformative advancements in wheelchair safety.

References

1. Smith, J., Johnson, L.: Modern mobility: the rise of smart wheelchairs. J. Med. Eng. **45**(2), 123–130 (2017)
2. Lee, H., Tanaka, M.: Ultrasonic sensors in assistive devices: a comparativestudy. J. Sens. Syst. **12**(1), 45–53 (2019)
3. Patel, R., Shah, S.: Caregiver alert systems in elderly and disabled care. HealthInformatics J. **16**(4), 299–308 (2020)
4. Gomez, E., Fernandez, L.: Wheelchairs in the digital age: a review. Rehabil. Technol. J. **10**(2), 78–89 (2021)
5. Nair, V., Krishnan, S.: Budget-friendly assistive technologies in developingcountries. Global Health Rev. **7**(3), 112–119 (2018)
6. Chen, W., Liu, Z.: Arduino in healthcare: potential and challenges. TechMedJ. **5**(1), 15–22 (2019)
7. Verma, A., Gupta, R.: Blynk in mobile health solutions: a preliminary study. Digit. Health J. **6**(4), 154–161 (2020)
8. Diaz, M., Rodriguez, P.: Elderly mobility: the need for modern solutions. Aging Soc. **9**(3), 239–248 (2017)
9. Kim, J., Park, S.: Sensor-driven healthcare: trends and implications. Health TechTrends **11**(2), 67–74 (2018)
10. Dansana, D., Mishra, B.K., Sindhuja, K., Sahoo, S.: IoT-based smart securitysystem on a door lock application. In: Kumar, R., Mishra, B.K., Pattnaik, P.K. (eds) Next Generation of Internet of Things. Lecture Notes in Networks and Systems, vol 201. Springer, Singapore (2021). https://doi.org/10.1007/978-981-16-0666-3_57

11. Dansana, D., Sahoo, S., Mishra, B.K.: Efficiency and reliability of IoT in Smart agriculture. In: Pattnaik, P.K., Kumar, R., Pal, S. (eds) Internet of Things and Analytics for Agriculture, Volume 3. Studies in Big Data, vol 99. Springer, Singapore (2022). https://doi.org/10.1007/978-981-16-6210-2_15
12. Dansana, D., Adhikari, J. D., Mohapatra, M., Sahoo, S.: An approach to analyse and forecast social media data using machine learning and data analysis. In: 2020 International Conference on Computer Science, Engineering and Applications (ICCSEA). 2020 International Conference on Computer Science, Engineering and Applications (ICCSEA). IEEE (2020). https://doi.org/10.1109/iccsea49143.2020.9132895

Brain Tumor Detection Using Deep-Learning (CNN-Vgg19)

Kushagra Singh and Jaspreet Kaur(✉)

Department of Computer Science and Engineering, JSS Academy of Technical Education, Noida, India
jaspreetkaur@jssaten.ac.in

Abstract. The detection of brain cancer presents a critical healthcare challenge, necessitating precise and efficient diagnostic systems. This paper introduces a comprehensive approach utilizing Convolutional Neural Networks (CNNs) and deep learning techniques for early detection. Leveraging advanced image processing, a meticulously curated dataset, and a Flask API for integration, the system offers real-time processing capabilities. The front-end interface, designed for usability and accessibility, complements the robust backend infrastructure. Through meticulous algorithmic design and optimization, the system demonstrates sensitivity and specificity in identifying brain tumors. This project represents a significant advancement in medical diagnostics, poised to enhance healthcare accessibility and contribute to scientific exploration in brain tumor diagnostics.

Keywords: Brain Cancer · Convolutional Neural Networks · Deep Learning · Flask API · Medical Diagnostics

1 Introduction

The incidence of brain cancer continues to pose a significant healthcare concern, the development of accurate and efficient diagnostic systems is imperative. In recent years, the advent of advanced technologies, particularly in the realms of image processing and deep learning, has provided promising avenues for early detection and diagnosis. Web services, acting as intermediaries between healthcare providers and consumers, offer a platform for deploying such diagnostic systems in a scalable and accessible manner [1].

Brain cancer detection involves intricate analysis of medical imaging data, necessitating sophisticated algorithms and computational techniques. However, ensuring seamless interoperability between these systems and existing healthcare infrastructures remains a formidable challenge. Interoperability, in this context, refers to the ability of different components, platforms, and applications to communicate effectively and exchange data without encountering compatibility issues [3]. While existing standards and guidelines provide a framework for achieving interoperability in web services, the complex nature of medical diagnostics demands innovative approaches to overcome interoperability challenges effectively. In response, this paper proposes a novel framework that

B. K. Mishra et al. (Eds.): ICT4GS 2024, ISEM 56, pp. 28–42, 2025.
https://doi.org/10.1007/978-3-031-94997-5_4

integrates web services with agent technology to enhance the interoperability of brain cancer detection systems [6].

By combining these technologies, we aim to not only improve the accuracy and efficiency of brain cancer detection but also facilitate seamless integration with existing healthcare ecosystems. This integrated approach holds the potential to revolutionize diagnostic capabilities, ultimately leading to better patient outcomes and advancements in medical research.

2 Description

2.1 CNN in Medical Imaging

CNNs are a class of deep learning algorithms that have revolutionized the field of computer vision, demonstrating remarkable performance in tasks such as image classification, object detection, and segmentation as shown in Fig. 1. In the context of medical imaging, CNNs have been widely adopted for automated analysis and interpretation of various modalities [5], including magnetic resonance imaging (MRI), computed tomography (CT), and positron emission tomography (PET) scans.

Brain Tumor Detection Challenges

Detecting brain tumors from medical imaging data poses several challenges, including:
Complexity of Tumor Morphology: Brain tumors can exhibit diverse morphological features, including size, shape, and texture variations, making their accurate detection and classification a challenging task [2].
Noise and Artifacts: Medical imaging data may contain noise and artifacts, which can hinder the performance of automated tumor detection algorithms [1].
Class Imbalance: Imbalanced datasets, where the number of tumor samples is significantly smaller than the number of non-tumor samples, can affect the generalization ability of machine learning models.

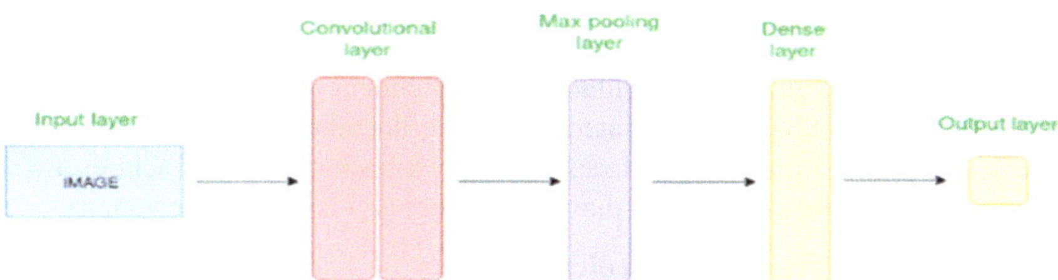

Fig. 1. Convolution Neural Network Structure

3 Literature Survey

The early detection of brain cancer remains a pivotal challenge in modern healthcare, where precision and efficiency are paramount. Advances in machine learning, particularly deep learning and Convolutional Neural Networks (CNNs), have shown immense promise in enhancing diagnostic capabilities. This literature survey explores seminal works and contemporary research that form the foundation of these technological advancements, highlighting the integration of advanced image processing, algorithmic innovation, and real-time application through systems such as Flask API.

- **Biomedical Image Segmentation**

One of the cornerstone contributions in biomedical image segmentation is the U-net architecture, proposed by Ronneberger, Fischer, and Brox (2015) [1]. U-net has become a standard in the field due to its ability to deliver high-quality segmentation with relatively few training images. This architecture has been crucial in enhancing the accuracy of medical image analysis, which is essential for detecting anomalies such as brain tumors.

- **Deep Learning Frameworks**

The implementation of large-scale machine learning models has been significantly advanced by frameworks like TensorFlow, developed by Abadi et al. (2016) [2]. TensorFlow provides a flexible and efficient platform for developing deep learning models, including CNNs. Its capability to handle extensive datasets and perform complex computations efficiently makes it an indispensable tool in the development of diagnostic systems for brain cancer.

- **Fundamental Concepts in Deep Learning**

The comprehensive treatise by Goodfellow, Bengio, and Courville (2016) [3] on deep learning offers foundational knowledge that underpins modern neural network architectures. Their work encapsulates essential principles, including the development and training of deep learning models, which are critical for understanding and improving CNN-based diagnostic systems.

- **Real-time Processing and Integration**

Flask, a micro web framework, facilitates the deployment of machine learning models for real-time processing and user interaction. The documentation and application of Flask in integrating front-end interfaces with robust back-end infrastructures are vital for creating accessible and efficient diagnostic systems (Flask Documentation).

- **Survey of Deep Learning in Medical Imaging**

Litjens et al. (2017) [6] provide a comprehensive survey on the application of deep learning in medical image analysis. Their work reviews various deep learning techniques and their effectiveness in different medical imaging tasks, highlighting the versatility and potential of these methods in improving diagnostic accuracy and efficiency.

- **CNN Architectures and Transfer Learning**

Shin et al. (2016) [6] investigate the use of deep convolutional neural networks for computer-aided detection, emphasizing the importance of CNN architectures, dataset characteristics, and transfer learning. Their findings underscore the significance of architectural choices and the adaptability of pre-trained models to enhance diagnostic performance in detecting brain tumors.

- **Advancements in Deep Learning**

Lecun, Bengio, and Hinton (2015) [5] discuss the broad advancements in deep learning and its implications across various fields, including medical diagnostics. Their insights into the development of deep learning technologies provide a contextual understanding of the rapid progress and future potential of CNNs in healthcare applications.

The integration of CNNs and deep learning techniques in medical diagnostics represents a significant advancement, offering improved sensitivity and specificity in detecting brain tumors. The combination of advanced image processing, algorithmic optimization, and real-time application through platforms like Flask enhances healthcare accessibility and diagnostic accuracy. The literature reviewed underscores the critical contributions of foundational research and contemporary innovations in shaping the future of brain tumor diagnostics.

4 Methodology

- **Data Collection**

The initial phase of our methodology entails gathering a dataset comprising medical imaging scans (such as MRI and CT scans) featuring brain tumors. This dataset is sourced from various platforms, including Kaggle and different websites. It encompasses a wide array of brain tumor types, sizes, locations, and imaging modalities, ensuring comprehensive model training and evaluation.

- **Data Preprocessing**

Preprocessing techniques are applied to the collected imaging data to enhance its quality and suitability for training the CNN model.Common preprocessing steps may include resizing the images to a standardized resolution, normalizing pixel intensities, cropping to remove extraneous background, and augmenting the dataset to increase its diversity and robustness.

- **Architecture Selection**

Following dataset acquisition, the subsequent step revolves around selecting a suitable Convolutional Neural Network (CNN) architecture tailored for brain tumor detection tasks. The decision hinges upon several factors, including dataset complexity, available computational resources, and past performance in analogous tasks.Commonly employed CNN architectures in medical imaging tasks encompass AlexNet, VGG, ResNet, and U-Net, each presenting distinct trade-offs regarding model intricacy, efficacy, and computational feasibility.For this endeavor, the VGG-19 model was utilized.

- **Transfer Learning and Fine-Tuning**

While transfer learning is a commonly employed technique, in this particular instance, it was opted against. Instead, the decision was made to abstain from transfer learning to minimize noise and construct the model from the ground up. Consequently, no pre-trained CNN models were utilized. The model was built entirely from scratch, enabling a focused approach to mitigate any potential interference from unrelated features.

- **Training the CNN Model**

Following adaptation, the CNN model underwent training on the preprocessed dataset of brain imaging scans employing supervised learning techniques. Throughout the training phase, the model acquired the capability to associate input images with corresponding binary or multiclass labels, signifying the presence or absence of brain tumors and their respective classes.

The training process unfolded iteratively, with batches of training images sequentially presented to the model. Subsequently, the loss between predicted and ground truth labels was computed, and the model parameters were updated to minimize this loss. This optimization was achieved through algorithms such as stochastic gradient descent (SGD) or Adam. Moreover, the training regimen was structured across multiple epochs, allowing the model to iteratively refine its understanding of the dataset and improve its predictive capabilities over successive iterations.

- **Model Evaluation**

Once trained, the performance of the CNN model is evaluated on a separate validation or test dataset that was not used during training.Evaluation metrics such as accuracy, sensitivity, specificity, precision, recall, and F1-score are calculated to assess the model's performance in detecting and classifying brain tumors.Receiver Operating Characteristic (ROC) curves and Area Under the Curve (AUC) values may also be used to evaluate the model's performance across different thresholds.

- **Validation and Interpretation**

The final phase encompasses validating the performance of the trained CNN model on independent datasets or real-world clinical data to gauge its generalization ability and real-world applicability.Currently, the model's API has been developed and hosted on a local server, seamlessly integrated with an SQL backend. This setup facilitates convenient access to the model's predictive capabilities through standardized API endpoints. Additionally, the SQL backend enables efficient data storage, retrieval, and management, ensuring seamless integration with existing infrastructure and workflows.

With the model deployed and operational, the next step involves conducting comprehensive validation tests utilizing independent datasets or real-world clinical data. These tests will assess the model's ability to generalize across diverse scenarios and accurately classify brain imaging scans in real-world settings. Any observed discrepancies or areas for improvement can then be addressed through iterative refinement and optimization of the model and its deployment environment.

5 Dataset Overview

The brain tumor detection dataset available on Kaggle comprises a diverse collection of medical imaging scans, including Magnetic Resonance Imaging (MRI) and Computed Tomography (CT) scans, sourced from various healthcare institutions and research repositories. These scans feature patients with different types of brain tumors as shown in Fig. 2, ranging from benign to malignant, and encompass a wide spectrum of tumor sizes, locations, and imaging modalities.

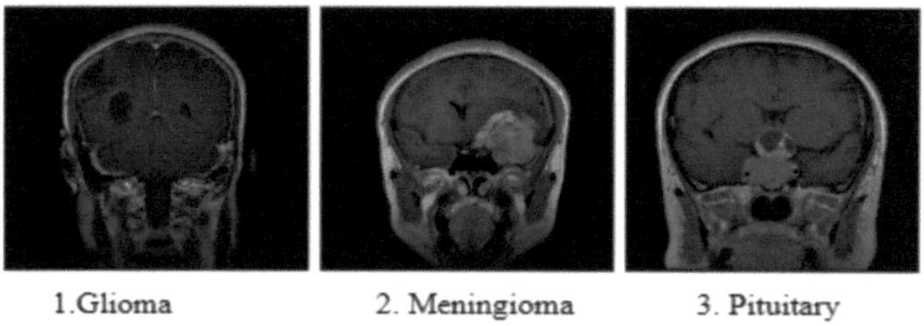

Fig. 2. Brain tumor dataset pictures

The training dataset comprises labeled medical imaging scans of patients with varying types, sizes, and locations of brain tumors. These scans serve as input data, while corresponding ground truth labels indicate tumor presence or absence, tumor type, and other relevant characteristics. During training, machine learning models learn to recognize patterns and features indicative of tumors from this dataset [13], optimizing their parameters to minimize prediction errors. The testing dataset, separate from the training data, consists of unseen scans with known ground truth labels, allowing for objective evaluation of the model's performance and its ability to accurately detect brain tumors in new cases. This rigorous validation process ensures that the developed models generalize well to real-world scenarios, providing reliable diagnostic support in clinical settings.

6 Model Architecture

The utilization of deep neural networks (DNNs) [4] in tumor detection offers numerous benefits, revolutionizing the field of medical imaging and diagnostic radiology. DNNs, particularly Convolutional Neural Networks (CNNs) [6], excel at learning intricate patterns and features from complex datasets, making them exceptionally well-suited for detecting subtle abnormalities in medical images such as those indicative of brain tumors. By leveraging vast amounts of annotated imaging data, DNNs [12] can discern nuanced details that may elude human observers, enhancing the accuracy and efficiency of tumor detection workflows.

One significant advantage of DNNs in tumor detection lies in their ability to process vast amounts of data rapidly. With the exponential growth of medical imaging datasets, traditional manual analysis methods often struggle to keep pace. DNNs, however, can analyze thousands of imaging scans in a fraction of the time it would take a human expert, enabling faster diagnosis and treatment planning [11]. Moreover, DNNs can learn from diverse imaging modalities and tumor types, providing a versatile solution that transcends the limitations of conventional approaches.

Among the various CNN architectures, VGG-19 stands out for several compelling reasons, making it a preferred choice for brain tumor detection tasks. Firstly, VGG-19 features a deep architecture with 19 layers as shown in Fig. 3, allowing it to capture intricate details and hierarchical features present in medical imaging scans. This depth facilitates robust feature extraction, essential for accurately distinguishing between tumor and non-tumor regions. Additionally, VGG-19's simple and uniform architecture, characterized by repeated convolutional and pooling layers, contributes to its ease of implementation and interpretation, making it accessible to both researchers and practitioners [7, 8].

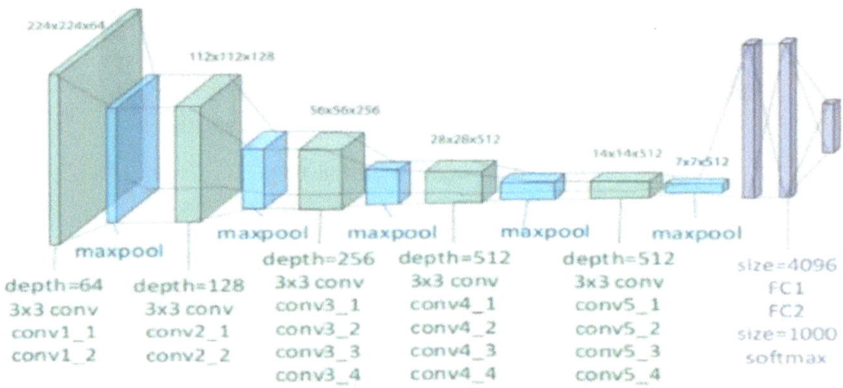

Fig. 3. Vgg-19 architecture

Moreover, VGG-19's pre-trained weights on large-scale image datasets like ImageNet provide a valuable starting point for brain tumor detection tasks [10]. Transfer learning, wherein pre-trained models are fine-tuned on task-specific data, accelerates the training process and improves model performance, especially when dealing with limited medical imaging datasets. By leveraging pre-trained features, VGG-19 can effectively generalize across diverse tumor types and imaging modalities, enhancing its versatility and real-world applicability.

Furthermore, VGG-19's balanced trade-off between model complexity and performance makes it well-suited for resource-constrained environments. Despite its deep architecture, VGG-19 strikes a harmonious balance between computational efficiency and predictive accuracy, ensuring efficient deployment on various computing platforms, including local servers and cloud infrastructure. This scalability and efficiency make VGG-19 an attractive choice for deploying brain tumor detection models in clinical settings, where computational resources may be limited [9].

In conclusion, the adoption of deep neural networks, particularly VGG-19, in tumor detection heralds a new era of precision medicine, offering unparalleled accuracy, efficiency, and scalability. By harnessing the power of deep learning [5], medical professionals can enhance diagnostic accuracy, expedite treatment planning, and ultimately improve patient outcomes in the fight against brain tumors and other malignancies.

With an **accuracy of 83.36%** and a **loss of 0.62995**, the model demonstrates promising performance in brain tumor detection as shown in Figs. 4 and 5. Accuracy represents the proportion of correctly classified instances out of the total number of instances, indicating the model's ability to correctly identify brain tumors and distinguish them from non-tumor regions in medical imaging scans. A high accuracy of 83.36% suggests that the model makes correct predictions for the majority of cases, showcasing its effectiveness in automated tumor detection tasks.

Additionally, the loss value of 0.62995 provides insight into the model's optimization process during training. The loss metric quantifies the discrepancy between the model's predictions and the ground truth labels, with lower values indicating better alignment between predicted and actual outcomes. A loss of 0.62995 suggests that, on average, the model's predictions are relatively close to the ground truth labels, reflecting its ability to minimize errors and refine its predictive capabilities over successive iterations of training.

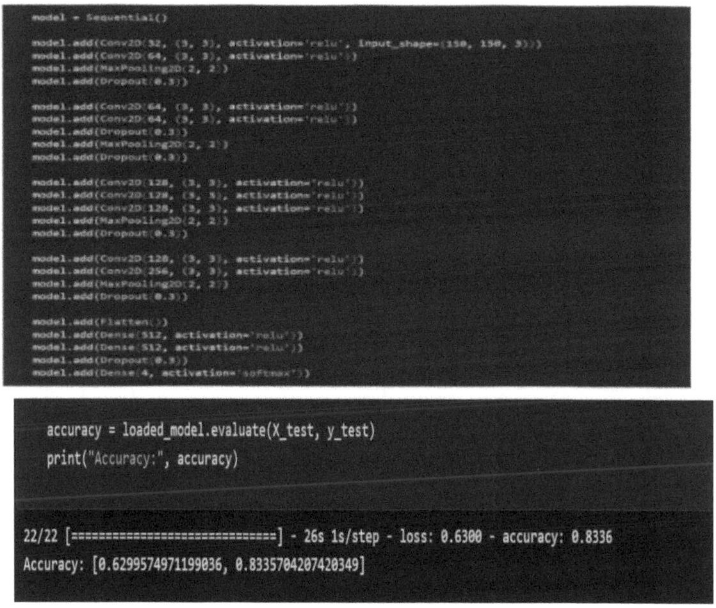

Fig. 4. Code Snippet of the model architecture

Once the brain tumor detection model is trained and optimized, it's serialized or pickled for integration into a Flask API. This involves saving the model as a binary file using Python's pickle module. The Flask API is then set up to handle incoming requests, loading the pickled model for inference. Requests containing input data are processed,

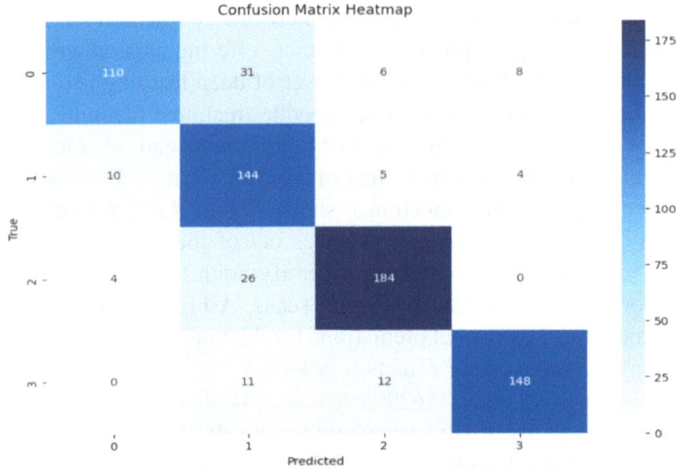


```
Confusion Matrix:
[[110  31   6   8]
 [ 10 144   5   4]
 [  4  26 184   0]
 [  0  11  12 148]]

Accuracy Matrix:
[[0.70967742 0.2        0.03870968 0.0516129 ]
 [0.06134969 0.88343558 0.03067485 0.02453988]
 [0.01869159 0.12149533 0.85981308 0.        ]
 [0.         0.06432749 0.07017544 0.86549708]]
```

Fig. 5. Model's Accuracy report

predictions are made using the loaded model, and the results are returned as JSON responses. After testing locally, the Flask application can be deployed to a production environment, providing a scalable solution for real-time brain tumor detection via HTTP requests.

7 Integrating to a Website

- **Serializing the Model**

The first step in integrating the brain tumor detection model into a Flask API involves serializing the trained model. Using Python's pickle module, the model is converted into a binary format for efficient storage and retrieval. This serialized model serves as the core component of the Flask API, enabling rapid inference on incoming data requests.

- **Creating the Flask API**

With the serialized model at hand, a Flask application as shown in Fig. 6 is developed to serve as the API endpoint. The Flask framework provides a lightweight and flexible platform for building web applications in Python. Within the Flask application, routes are defined to handle various HTTP requests, such as POST requests containing input data for tumor detection. These routes orchestrate the interaction between the user interface and the brain tumor detection model, ensuring seamless integration and smooth operation.

```
@flask_app.route("/predict", methods=["POST"])
def predict():
    if 'file' not in request.files:
        return render_template('index.html', prediction_text='No image selected')

    file = request.files['file']

    if file.filename == '':
        return render_template('index.html', prediction_text='No image selected')

    if file:
        # Generate a unique filename
        unique_filename = secure_filename(file.filename)

        file_path = os.path.join(flask_app.config['UPLOAD_FOLDER'], unique_filename)

        # Save the file only if the uploads directory exists
        file.save(file_path)

        # Check if the file exists
        if not os.path.exists(file_path):
            return render_template('index.html', prediction_text='File does not exist')

        # Load and preprocess the image using the data generator
        img = tf.keras.preprocessing.image.load_img(file_path, target_size=(150, 150))
        img_array = tf.keras.preprocessing.image.img_to_array(img)
        img_array = np.expand_dims(img_array, axis=0)
```

Fig. 6. Code snippet of the FLASK-API

- **Handling Requests and Making Predictions**

Upon receiving data requests from the webpage, the Flask API preprocesses the input data as necessary before passing it to the loaded model for prediction. This preprocessing step may involve tasks such as resizing images, normalizing pixel values, or extracting relevant features. Once the data is prepared, the model performs inference to predict the presence or absence of brain tumors in the input medical imaging scans. The predicted outcomes are then returned as JSON responses, ready to be displayed on the webpage in real-time.

- **SQL Database Integration**

In addition to model inference, the Flask API is integrated with an SQL database to facilitate efficient data management and storage. By leveraging SQL databases such as MySQL or PostgreSQL, user data, predictions, and other relevant information can be securely stored and retrieved. Integration with an SQL database as shown in Fig. 7 enables functionalities such as user authentication, data logging, and result tracking, enhancing the overall functionality and usability of the brain tumor detection application. Through seamless integration of the Flask API with an SQL database, users can benefit from a comprehensive and feature-rich platform for brain tumor detection, combining real-time inference with robust data management capabilities.

```
import mysql.connector as c
import json

con=c.connect(host='localhost',user='root',passwd='*********',database='flask_models')
cursor=con.cursor()
```

Fig. 7. Connecting to a SQL backend database

Frontend Features for Brain Tumor Prediction

- **HTML Structure**

The HTML code defines the structure of the webpage, including elements such as headings, navigation links, forms, and image containers. The layout is designed to provide a user-friendly interface for interacting with the brain tumor detection model.

Styling with CSS.

CSS styling is applied to enhance the visual appeal and usability of the webpage. Styling features include custom fonts, background images, text colors, button designs, and container layouts. These stylistic elements contribute to a cohesive and aesthetically pleasing user experience.

- **Form Submission and Prediction Handling**

The webpage includes a form for uploading medical imaging scans (e.g., JPG, JPEG, PNG files) to be processed by the brain tumor detection model. Upon form submission, JavaScript functions handle the prediction results returned by the Flask API. Depending on the prediction outcome ('Tumor Detected' or 'No Tumor Detected'), the background color of the webpage changes dynamically to indicate the result.

- **Dynamic Content Display**

The prediction outcome ('Tumor Detected' or 'No Tumor Detected') is displayed dynamically within the prediction container, updating in real-time based on the model's inference. Additionally, corresponding images illustrating the prediction outcome are dynamically loaded and displayed in the image container, providing visual confirmation of the prediction result.

- **Integration with Flask API and SQL Database**

The HTML code is integrated with a Flask API for seamless communication with the brain tumor detection model. Upon form submission, the uploaded image data is sent to the Flask API for inference, and the prediction outcome is returned to the webpage for display. Furthermore, the integration supports SQL database functionality for storing and retrieving user data, predictions, and related information, enhancing the application's functionality and data management capabilities as shown in Fig. 8.

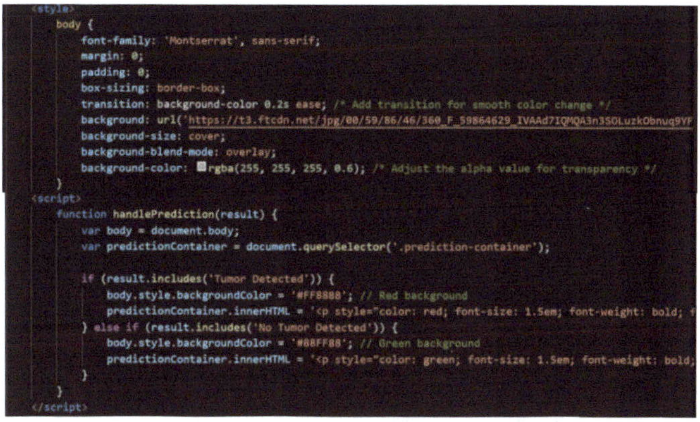

```
<style>
    body {
        font-family: 'Montserrat', sans-serif;
        margin: 0;
        padding: 0;
        box-sizing: border-box;
        transition: background-color 0.2s ease; /* Add transition for smooth color change */
        background: url('https://t3.ftcdn.net/jpg/00/59/86/46/360_F_59864629_IVAAd7IQMQA3n3SOLuzkObnuq9YF
        background-size: cover;
        background-blend-mode: overlay;
        background-color: rgba(255, 255, 255, 0.6); /* Adjust the alpha value for transparency */
    }
<script>
    function handlePrediction(result) {
        var body = document.body;
        var predictionContainer = document.querySelector('.prediction-container');

        if (result.includes('Tumor Detected')) {
            body.style.backgroundColor = '#FF8888'; // Red background
            predictionContainer.innerHTML = '<p style="color: red; font-size: 1.5em; font-weight: bold; f
        } else if (result.includes('No Tumor Detected')) {
            body.style.backgroundColor = '#88FF88'; // Green background
            predictionContainer.innerHTML = '<p style="color: green; font-size: 1.5em; font-weight: bold;
        }
    }
</script>
```

Fig. 8. Code snippet of the Frontend code

- **Model workflow**

To facilitate the detection of tumors within the provided image, please upload the image to the designated website as shown in Fig. 9. Our advanced algorithms are designed to analyze medical imagery with precision and accuracy, enabling the identification and assessment of potential tumors. This process is integral to ensuring comprehensive medical evaluation and appropriate treatment planning.

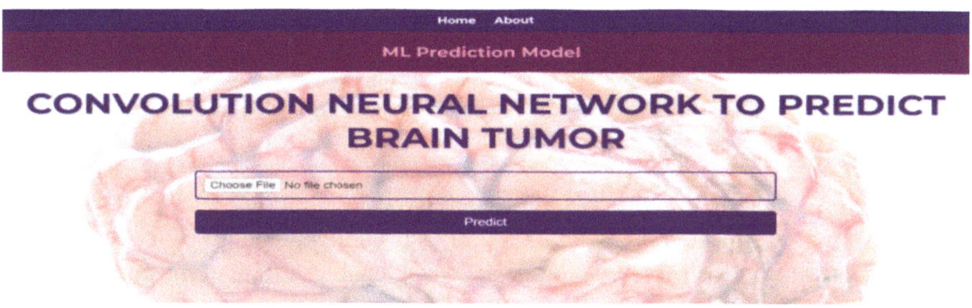

Fig. 9. Uploading an image to the website

Please initiate the tumor detection process by selecting the designated button. Subsequently, the model will analyze the uploaded image and provide predictive results regarding the presence or absence of tumors. This automated procedure streamlines the diagnostic process, offering efficient insights into potential medical conditions.

Upon pressing the button, the predicted results will be promptly displayed on the screen, facilitating immediate access to the diagnostic outcome as shown in Fig. 10. Simultaneously, the backend system will be updated with the predicted results, alongside the timestamp indicating the time at which the analysis was conducted as shown in Fig. 11. This seamless integration of real-time data ensures accurate documentation of diagnostic assessments and facilitates ongoing monitoring of medical information.

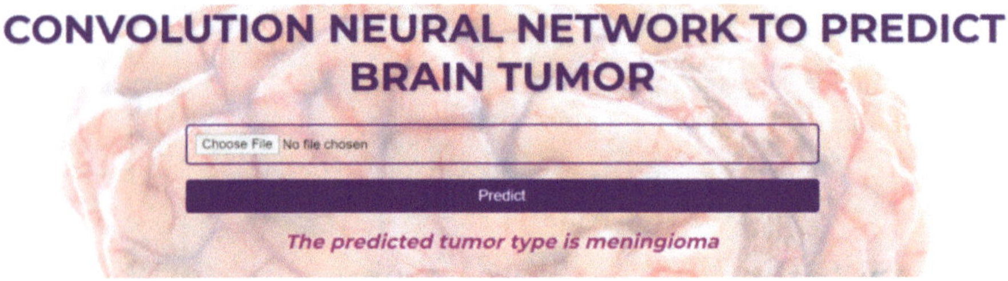

Fig. 10. Getting the predicted result

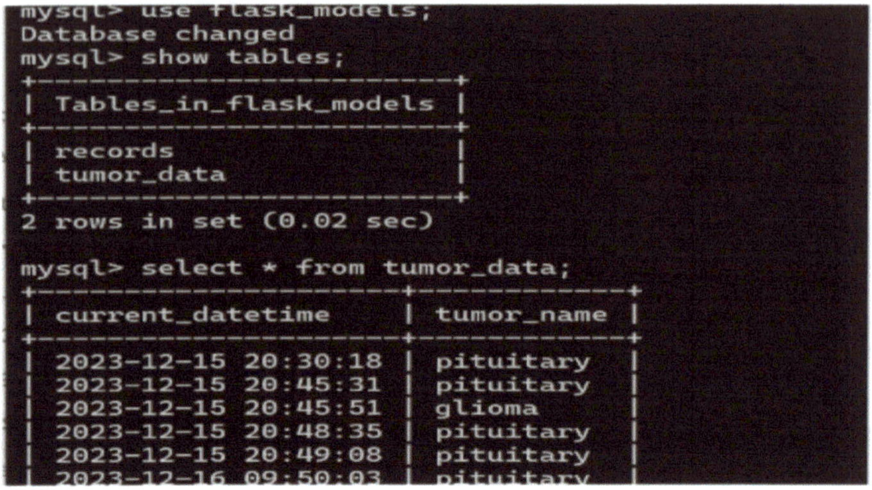

Fig. 11. Checking the backend database for updation

Following the prediction process, the generated data, including the predicted results and corresponding timestamps, will be securely stored in sql databases for future reference and analysis. This structured storage approach enables efficient retrieval and management of medical information, facilitating comprehensive patient care and research endeavors. Access to the sql databases grants healthcare professionals and researchers the ability to review historical data, track trends, and derive valuable insights to inform medical decision-making and advance scientific knowledge.

8 Conclusion

Our exploration into brain tumor detection using deep learning methodologies, specifically the Convolutional Neural Network (CNN) architecture VGG-19, has been a comprehensive journey marked by meticulous experimentation and analysis. We embarked on this research endeavor with the recognition of the critical need for accurate and efficient diagnostic tools in neuro-oncology, where timely detection can significantly impact patient outcomes.To commence our investigation, we meticulously curated and preprocessed diverse datasets encompassing a range of neuroimaging modalities, including magnetic resonance imaging (MRI) and computed tomography (CT) scans. These datasets, comprising both tumor-positive and tumor-negative samples, served as the foundation for our model training, validation, and evaluation phases.

In the training phase, we employed transfer learning techniques to leverage the pre-trained VGG-19 CNN architecture, fine-tuning its parameters to adapt to the nuances of brain tumor detection. This approach not only expedited the training process but also facilitated the extraction of intricate features inherent to tumor morphology and structure. Subsequently, rigorous validation procedures, including cross-validation and performance metrics assessment, were employed to ensure the robustness and generalizability of our model across diverse datasets. Through these validation efforts, we ascertained the model's ability to discern subtle tumor characteristics while minimizing false positives and false negatives, critical considerations in clinical settings.

Upon achieving satisfactory performance metrics, we transitioned to the deployment phase, where we integrated the trained model into a web-based platform for real-time tumor detection. This web interface, characterized by its intuitive design and seamless user experience, serves as a conduit for disseminating diagnostic capabilities to healthcare professionals and patients alike. Central to our deployment strategy is the establishment of a backend database infrastructure, where diagnostic results are securely stored and managed. This backend integration not only facilitates data archival and retrieval but also lays the groundwork for longitudinal tracking of patient health metrics, supporting ongoing clinical management and research initiatives.

As we reflect on the culmination of our research endeavors, we recognize the transformative potential of our contributions to the field of neuro-oncology. By harnessing the power of deep learning, web technology, and data-driven insights, we have forged a path towards more precise, efficient, and accessible brain tumor diagnostics. Looking ahead, we remain steadfast in our commitment to advancing medical practices and improving patient outcomes through interdisciplinary collaboration, technological innovation, and empirical inquiry.

References

1. Ronneberger, O., Fischer, P., Brox, T.: U-net: Convolutional networks for biomedical image segmentation. In International Conference on Medical Image Computing and Computer-assisted Intervention, pp. 234–241 (2015)
2. Abadi, M., et al.: TensorFlow: A system for large-scale machine learning. In: 12th USENIX Symposium on Operating Systems Design and Implementation (OSDI 16), pp. 265–283 (2016)
3. Goodfellow, I., Bengio, Y., Courville, A., Bengio, Y.: Deep Learning, vol. 1. MIT press Cambridge (2016)
4. Lecun, Y., Bengio, Y., Hinton, G.: Flask Documentation. Deep learning. Nature, **521**(7553), 436–444 (2015).
5. Lecun, Y., Bengio, Y., Hinton, G.: Deep learning. Nature **521**(7553), 436–444 (2015). Certainly, here are additional references that align with the themes explored in our research paper:] Litjens, G., et al.: A survey on deep learning in medical image analysis. Med. Image Anal. **42**, 60–88 (2017).

6. Shin, H.C., Roth, H.R., Gao, M., Lu, L., Xu, Z., Nogues, I., Summers, R.M.: Deep convolutional neural networks for computer-aided detection: CNN architectures, dataset characteristics and transfer learning. IEEE Trans. Med. Imag. **35**(5), 1285–1298 (2016)
7. Moeskops, P., Viergever, M.A., Mendrik, A.M., de Vries, L.S., Benders, M.J., Išgum, I.: Automatic segmentation of MR brain images with a convolutional neural network. IEEE Trans. Med. Imag. **35**(5), 1252–1261 (2016)
8. Tajbakhsh, N., et al.: Convolutional neural networks for medical image analysis: full training or fine tuning? IEEE Trans. Med. Imag. **35**(5), 1299–1312 (2016)
9. Johnson, J.M., Khoshgoftaar, T.M., Wang, H.: Survey on deep learning with class imbalance. J. Big Data **6**(1), 27 (2019)
10. Szegedy, C., Vanhoucke, V., Ioffe, S., Shlens, J., Wojna, Z.: Rethinking the inception architecture for computer vision. In Proceedings of the IEEE Conference on Computer Vision and Pattern Recognition, pp. 2818–2826 (2016)
11. Litjens, G., Ciompi, F., Wolterink, J. M., de Vos, B. D., Leiner, T., & Teuwen, J. (2017). State-of-the-art deep learning in cardiovascular image analysis. JACC: Cardiovascular Imaging, 10(3), 360–362
12. Chen, L.C., Papandreou, G., Kokkinos, I., Murphy, K., Yuille, A.L.: DeepLab: semantic image segmentation with deep convolutional nets, atrous convolution, and fully connected CRFs. IEEE Trans. Pattern Anal. Mach. Intell. **40**(4), 834–848 (2018)
13. Esteva, A., et al.: Dermatologist-level classification of skin cancer with deep neural networks. Nature **542**(7639), 115–118 (2017)

Investigation of Machine Learning Blockchain Approaches for Privacy Preservation

Hiralal Solunke(✉), Pawan Bhaladhare, and Amol Potgantwar

School of Computer Science and Engineering, Sandip University, Nashik, India
{hiralal.solunke,pawan.bhaladhare}@sandipuniversity.edu.in,
amol.potgantwar@sitrc.org

Abstract. In the age of decentralized networks and Hadoop, the confluence of Blockchain & m/c learning (ML) technologies has emerged with a promising solution for ensuring robust, transparent, and private transactions. The benefits of Blockchain for the safety of data and ML provide effective efficiency. Their combination may revolutionize the retail industry, banking, Defense, etc., by offering unmatched data safety while preserving the best system performance. For instance, while some models may offer higher throughput, they could compromise on privacy levels. Conversely, models excelling in privacy preservation might incur higher energy costs or delays. This multi-faceted evaluation not only serves as a guide for stakeholders to choose the most efficient model based on individual or multiple metrics but also uncovers avenues for future research. By correlating these metrics with real-world use cases in healthcare, finance, and supply chain, among others, the paper adds a layer of practical applicability. This aids in the formulation of best practices for the deployment of these trending methodologies to maximize both performance and privacy. The paper contributes to the academic discourse by laying down a foundational framework for understanding and selecting the most appropriate blockchain-ML models for privacy preservation. The insights derived from this work are instrumental in steering the future development and deployment of secure, efficient, and privacy-preserving solutions across various industry verticals & scenarios.

Keywords: Blockchain · Machine Learning · Privacy Preservation · Data Integrity

1 Introduction

As the digital transformation accelerates across sectors such as healthcare, finance, supply chain, and more, the need for secure, transparent, and efficient data transactions has never been more critical. Blockchain technology, with its decentralized nature and immutable ledgers, has been heralded as a solution for many issues surrounding data integrity and trust. Concurrently, Machine Learning (ML) has proven to be an invaluable tool for data analysis, prediction, and decision-making. Both technologies have independently offered solutions for various aspects of privacy preservation but have not yet

B. K. Mishra et al. (Eds.): ICT4GS 2024, ISEM 56, pp. 43–50, 2025.
https://doi.org/10.1007/978-3-031-94997-5_5

been thoroughly explored in tandem. Despite substantial advancements in individual domains, issues such as data privacy, latency, scalability, and energy efficiency continue to plague digital transactions. As the exponential growth in Hadoop and the increasing prerequisite for real-world-applications, conventional models often fall short in providing comprehensive solutions. Privacy remains a significant concern, particularly where sensitive data are involved, such as medical records or financial transactions. Moreover, the evolving landscape of regulations such as GDPR and HIPAA calls for more robust solutions that are adaptive and efficient. Research to investigate the fusion of Blockchain and ML technologies to address these challenges head-on, evaluating their collaborative potential in enhancing privacy preservation. The implications of successfully combining Blockchain and ML technologies are far-reaching. A system that offers robust privacy preservation without compromising on latency or efficiency can revolutionize real-time applications. For example, in healthcare, such a system could securely manage patient records while facilitating real-time diagnosis and treatment suggestions.

In the financial sector, it could enable faster, yet more secure, transaction capabilities. Thus, the outcomes of this research have immediate applicability and the potential to reshape industry best practices dramatically. This paper seeks to: Conduct a comprehensive survey of already available methods and approaches which utilize ML in blockchain systems for secrecy preservation. An Employ pre-defined criteria like delay time, transmission speed, energy efficiency, levels of confidentiality, cost of implementation, and flexibility to compare these models. Apply a comparative insight to real-world use cases across various sectors, elucidating the practical benefits and limitations of each model.

1.1 Motivation and Objectives

Motivation

In today's interconnected world, data has become an invaluable resource, acting as the lifeblood of various industries. The skyrocketing growth of big data has opened up a plethora of opportunities but has simultaneously introduced challenges related to data security, privacy, and efficiency. Traditional models often fall short in addressing these multi-dimensional concerns, creating a bottleneck that hinders progress. The healthcare industry serves as a prime example where privacy and real-time processing are of paramount importance. Ineffective privacy controls could lead to unauthorized access to sensitive patient data, thereby risking not just individual privacy, also the faith of an individual. Similarly, also-in the financial sector, escalating threat of fraud and cyberattacks necessitates a robust solution that guarantees both quick and secure transactions.

2 Literature Survey of Existing Methods

A wide variety of models are proposed by researchers to enhance the privacy of blockchain networks, and each of these models have their own operating & deployment characteristics. In this section we discuss these models in detail, which will assist readers to identify optimal models for their functional use cases. For instance, In the first work [1], the authors noted the growing demand for secure electronic payment systems, emphasizing the importance of preventing double-spending, identifying unscrupulous users, and preserving user anonymity and privacy. They devised a blockchain-based electronic payment mechanism that guarantees honest users' anonymity while allowing bad users to be tracked down and punished without the need for a trusted server. To guarantee user anonymity, the technique used several algorithms such as a fair-blind digital signing scheme, and a secret sharing mechanism. Notably, the approach had features like traceability, conditional privacy, & revocation. Task assessments, such as imitations and analysis, confirmed the system's efficiency and safety. In this research [2], the attention switched to Ethereum's data openness and its security consequences. The researchers stated that while Ethereum's openness was a plus, it also presented safety hazards & issues because of its influence on user-secrecy. They analyzed Ethereum blockchain transaction data and network metrics such as degree importance, betweenness importance, Eigen, and Page rank. Their research revealed that these network procedures jeopardized the privacy of Ethereum transactions and put active participants at risk. To overcome this issue, the authors developed two techniques that use chaotic maps and differential privacy to encode information before it is released to the public. They described a dedicated server structure that adds error to guard secrecy while also decrypting information for verified queries. The third study [3] analyzed the risks of privacy involved in e-commerce. The purpose of their research was to achieve an in-depth comprehension of the influence of e-commerce websites on reliability and consumer behavior. For more empirical research in consumer behavior within e-commerce, they presented a shopping direction detection.

The researchers in the fourth article [4] examined the increased application of control devices in distributed fog and cloud computing environments. The proposal they put forth, BPCPR-FC, was a solution based on blockchain technology, zk-Snark Methods rtc. The scheme was mathematically analyzed, informally secured, analyzed, and cryptanalyzed as described in the article. The mechanism proposed was integrated in the Truffle suite, in which it was observed that the time for preparing blocks was enhanced compared to available solutions. The fifth work [5] the authors proposed a ring signature scheme in which identity could be verified in financial statements. In the sixth work [6], the authors focused on security and privacy challenges in Vehicle Ad-hoc Networks (VANETs). They introduced a BCPAS for VANETs, which featured alias ID usage and blockchain technology for cross-region authentication. The seventh work [7] examined ML for privacy in blockchain in health care settings. The authors set forth a FL process with blockchain to motivate organizations based on their inputs. Their research involved model transferring from the user to volunteer organizations with privacy protection. The work applied a COVID-19 dataset in the investigation of individual accuracy outputs and the resulting total accuracy when using the FedAvg algorithm. In this research [8], the authors concentrated on edge computing's traceability and anti-counterfeiting features, using blockchain technology to provide the security with more efficiency. A ninth paper [9] addressed the challenges involved in authenticating customer credentials at financial institutions. Casper, a blockchain-based digital identity platform and self-sovereign identity, was proposed by the authors. In contrast to conventional systems, customer credentials were stored in users' mobile wallets and verified using Zk-Snarks proof mechanisms. The authors of the tenth work [10] considered problems associated with data management within the healthcare industry. The comparison with the DPS construction proved noteworthy developments in terms of cost minimization, indicating that it might be beneficial for healthcare players. In the eleventh work [11], the authors discussed how smart home technology requires a reliable and expandable access verification system. They suggested a BAAS System and emphasized the shortcomings of current schemes that depend on centralized, reliable third parties. The scheme offered mutual and point-to-point authentication, eliminated the need for direct mediators, and ensured privacy through attribute-based signatures. BAAS also increased efficiency, transparency, and trust by leveraging MPT. The authors conducted security analysis and simulation using OPNET software to assess computational complexity and functionality.

In the twelfth work [12], the focus shifted to the Internet-of-Things (IoT), emphasizing the need for effective governance in IoT projects. The authors argued that existing governance frameworks were insufficient for IoT, leading them to propose a new IoT.

Governance-framework called the Internet of Things-Gov. IoT-Gov featured an architecture and a methodological approach that synchronized IoT standards and tools to build IoT systems progressively. They incorporated blockchain technology for decentralization, transparency, and security. The authors applied the proposed framework to a smart parking system as an IoT ecosystem example, enhancing it with property-based access control and deploying it on the ledger. They provided performance test results to demonstrate the feasibility and efficiency of IoT-Gov in an IoT context.

The thirteenth work [13] framework securely stored and ensured the integrity of physiological data produced by IoMT devices. It employed a Proof of Authority (PoA) consensus mechanism and double-encryption for data confidentiality. The authors implemented the design on a Raspberry Pi network and achieved a transaction speed of at least twenty-five communications per second, making it suitable for the healthcare sector. In the fourteenth work [14], the authors presented SECURE TRUST, an IoT device trust and reputation system powered by blockchain technology. By allocating trust ratings based on previous contacts, the system employed smart contracts to identify and stop criminal activity among nodes in IoT networks. Its goal was to improve security in IoT environments, and it underwent a rigorous testing process to show how well it could identify and mitigate security threats.

The fifteenth work [15] investigated the use of blockchain technology to guarantee data integrity and non-repudiation while safeguarding biomedical databases. To protect biological repositories, the scientists put up a framework known as BCT-based solutions, which fused fuzzy logic with blockchain technology. The framework used smart contracts and query notary services to provide data integrity and non-repudiation. Performance research showed that the suggested architecture was effective in terms of scalability and response time. In the sixteenth work [16], the authors addressed challenges in smart cities and IoT security, proposing an attack detection and security framework based on blockchain technology and deep learning. They employed MRC-LDA for data dimensionality reduction, ORLLTMLP for attack detection, and LCTFA for secure data storage. The proposed framework achieved an accuracy of 97.5% in attack detection.

The seventeenth work [17] focused on improving transparency, trust, and privacy in vote counting using blockchain technology and a secure partitioning scheme. The authors implemented a chaincode-based vote counting mechanism on a decentralized platform, ensuring trust and fairness among multiple authorities participating in the process. The blockchain platform recorded user votes using secret sharing and allowed every registered user to access the distributed ledger. Performance results demonstrated the efficiency of the proposed solution.

In the eighteenth work [18], the authors introduced a blockchain-based solution for managing health records across multiple healthcare providers. Their system utilized smart contracts to create immutable patient logs, update medical records, exchange health information, and enable viewership contracts on a peer-to-peer blockchain network. The proposed framework provided high security and integrity with trials conducted to assess its performance. Finally, in the nineteenth work [19], the authors addressed data security in an Internet of Things (IoT) network, specifically focusing on the secure storage of transaction records and sensor data. They proposed a system where Raspberry Pi devices served as access points, creating an IoT network. Transaction records were converted into a blockchain structure for tamper resistance, then further processing is to be performed. The proposed system maximized the safety of IoT device processes.

In the twentieth work [20] DBA approach employed for the authentication phase, while the Spark environment was used for data encryption. When compared to centralized methods, the suggested solution performed better in terms of throughput, storage space, and time consumption, making it appropriate for IoT contexts with limited resources.

3 Result Analysis

An extensive review of the current privacy preservation models for blockchain deployments shows extensive differences in their real-time aspects. In order to help researchers select best approach, this section describes it in more detail, this section compares these models in terms of their Output Effectiveness (OE), Safety (S), Quality (Q), Attack Effectiveness (AE), and Scalability (S) parameters. We compare these parameters on the metrics Worst(W), Average(A), Good(G), Very Good(VG), Excellent(E) based on methods.

Using this strategy, these models were compared in Table 1, which will assist researchers in getting a quick view of the efficiency levels of these models.

Table 1. Empirical Analysis of the Reviewed Models

Citation	Method	OE	S	Q	AE	S
[1]	Payment system with blockchain method	VG	VG	E	VG	G
[2]	Ethereum-Blockchain	W	VG	VG	G	E
[3]	Security in E-commerce by the blockchain method	VG	G	VG	W	VG
[4]	Fog Computing approach	G	A	W	A	W
[5]	Zk-Snarks and Ring Signature	VG	A	A	A	A
[6]	Security in VANETs	G	A	VG	W	G
[7]	Federated Learning for privacy	A	VG	G	G	A

(continued)

Table 1. (*continued*)

Citation	Method	OE	S	Q	AE	S
[8]	Blockchain-Based Traceability and Anti-Counterfeiting in Edge Computing	G	W	VG	A	G
[9]	KYC using Blockchain	W	VG	VG	E	W
[10]	Medical Data Security	G	G	VG	E	A
[11]	Blockchain-Based Authentication in Smart Home	W	A	G	W	W
[12]	IoT-Gov: IoT Governance Framework	VG	A	E	E	E
[13]	Blockchain for Securing IoMT Data	VG	G	W	A	G
[14]	SECURE TRUST: Blockchain for IoT Security	VG	G	E	W	A
[15]	BCT-Based Framework for Biomedical Repositories	W	VG	VG	A	A
[16]	ORLLTMLP-Based Attack Detection in Smart Cities	G	G	E	E	VG
[17]	Blockchain for Secure and Verifiable Voting	W	E	L	E	G
[18]	Blockchain for EHRs in Healthcare	W	W	W	E	G
[19]	IoT Blockchain for Data Integrity	A	VG	VG	A	A
[20]	DeBlock-Sec: Decentralized Blockchain Security	W	E	W	G	W

The Table 1. Describes the Analysis of the existing methods.

4 Conclusion and Future Scopes

In conclusion, the analysis of the presented blockchain models and their corresponding per-performance metrics provides valuable insights into the diverse landscape of blockchain technology applications. These models showcase the versatility and potential of blockchain in addressing various aspects of modern computing and data management. Several models have emerged as champions in specific performance metrics. Notably, the Blockchain-Based Electronic Payment Scheme" excels in deployment efficiency, offering a swift and efficient solution for electronic payment transactions. "IoT-Gov: IoT Governance Framework" ensures a superior Quality of Service in IoT governance applications, promoting trust and reliability. Furthermore, a group of models exhibits balanced performance across multiple metrics, catering to a wide array of blockchain use cases. These models offer a harmonious combination of deployment efficiency, security, quality of service, attack mitigation, and scalability. Depending on specific application requirements, these models provide adaptable and versatile solutions for diverse blockchain applications. The blockchain technology continues to evolve and find relevance across various domains, from finance to healthcare, industrial data sharing, and beyond. As the blockchain ecosystem matures, further research and development will refine these models and pave the way for innovative applications.

The comprehensive evaluation of these models underscores the significance of considering a holistic approach when selecting blockchain solutions. Depending on the specific use case and objectives, stakeholders can choose models that prioritize particular metrics or opt for models that strike a balance between multiple performance factors. In the ever-evolving landscape of blockchain technology, these models serve as beacons of innovation and practicality.

References

1. Hatefi, Z., Bayat, M., Alaghband, M.R., et al.: A conditional privacy-preserving fair electronic payment scheme based on blockchain without trusted third party. J. Ambient Intell. Human Comput. **14**, 10089–10102 (2023). https://doi.org/10.1007/s12652-021-03672-1
2. Kumar, E.S.: Preserving privacy in ethereum blockchain. Ann. Data. Sci. **9**, 675–693 (2022). https://doi.org/10.1007/s40745-020-00279-9
3. Liu, R., Wang, E.: Blockchain and mobile client privacy protection in e-commerce consumer shopping tendency identification application. Soft. Comput. **27**, 6019–6031 (2023). https://doi.org/10.1007/s00500-023-08099-8
4. Gowda, N.C., Bharathi Malakreddy, A.: BPCPR-FC: blockchain-based privacy preservation with confidentiality using proxy reencryption and ring signature in fog computing environments. Int. J. Inf. Tecnol. **15**, 3343–3357 (2023). https://doi.org/10.1007/s41870-023-01373-w
5. Devidas, S., Rukma Rekha, N., Subba Rao, Y.V.: Identity verifiable ring signature scheme for privacy protection in blockchain. Int. J. Inform. Technol. **15**(5), 2559–2568 (2023). https://doi.org/10.1007/s41870-023-01282-y
6. Gong, C., Xiong, L., He, X., et al.: Blockchain-based conditional privacy-preserving authentication scheme for vehicular ad hoc networks. J. Ambient Intell. Human Comput. **14**, 6273–6286 (2023). https://doi.org/10.1007/s12652-021-03655-2
7. Om Kumar, C.U., Gajendran, S., Balaji, V., et al.: RETRACTED ARTICLE: securing health care data through blockchain enabled collaborative machine learning. Soft. Comput. **27**(14), 9941–9954 (2023). https://doi.org/10.1007/s00500-023-08330-6
8. Qiu, Z., Zhu, Y.: Traceability anti-counterfeiting system based on the ownership of edge computing on the blockchain. J. Ambient Intell. Human Comput. **14**, 257–270 (2023). https://doi.org/10.1007/s12652-021-03290-x
9. Bandara, E., Shetty, S., Mukkamala, R., et al.: Casper: a blockchain-based system for efficient and secure customer credential verification. J. Bank Financ. Technol. **6**, 43–62 (2022). https://doi.org/10.1007/s42786-021-00036-3
10. Shah, R., Rajagopal, S.: M-DPS: a blockchain-based efficient and cost-effective architecture for medical applications. Int. J. Inf. Tecnol. **14**, 1909–1921 (2022). https://doi.org/10.1007/s41870-022-00912-1
11. Heshmati, A., Bayat, M., Doostari, M., et al.: Blockchain based authentication and access verfication scheme in smart home. J Ambient Intell. Human Comput. **14**, 2525–2547 (2023). https://doi.org/10.1007/s12652-022-04501-9
12. Sedrati, A., Ouaddah, A., Mezrioui, A., et al.: IoT-Gov: an IoT governance framework using the blockchain. Computing **104**, 2307–2345 (2022). https://doi.org/10.1007/s00607-022-01086-1
13. Mohan, D., Alwin, L., Neeraja, P., et al.: A private Ethereum blockchain implementation for secure data handling in Internet of Medical Things. J Reliable Intell Environ **8**, 379–396 (2022). https://doi.org/10.1007/s40860-021-00153-2
14. Simpson, S.V., Ravi Raju, Y., Rajesh Naidu, K.B., et al.: SECURE TRUST: a blockchain-enabled trust and reputation system for malicious node detection in IOT networks. Int. J. Inform. Technol. **15**(8), 4323–4336 (2023). https://doi.org/10.1007/s41870-023-01467-5

15. Thatikonda, R., Vaddadi, S.A., Arnepalli, P.R.R., et al.: Securing biomedical databases based on fuzzy method through blockchain technology. Soft. Comput. (2023). https://doi.org/10.1007/s00500-023-08355-x
16. Anbalagan, V.Y., Rajasekaran, S., Rajeeve, T.D.: A novel ORLLTMLP-based attack detection and blockchain-aware security framework using LCTFA in smart city applications. New Gener. Comput. **41**, 315–342 (2023). https://doi.org/10.1007/s00354-023-00210-9
17. Chouhan, V., Arora, A.: Blockchain-based secure and transparent election and vote counting mechanism using secret sharing scheme. J. Ambient Intell. Human Comput. **14**, 14009–14027 (2023). https://doi.org/10.1007/s12652-022-04108-0
18. Chelladurai, U., Pandian, S.: A novel blockchain based electronic health record automation system for healthcare. J. Ambient Intell. Human Comput. **13**, 693–703 (2022). https://doi.org/10.1007/s12652-021-03163-3
19. Gürfidan, R., Ersoy, M.: A new approach with blockchain based for safe communication in IoT ecosystem. J. Data, Inf. Manag. **4**(1), 49–56 (2022). https://doi.org/10.1007/s42488-021-00063-1
20. Narayanan, U., Paul, V., Joseph, S.: Decentralized blockchain based authentication for secure data sharing in Cloud-IoT. J. Ambient Intell. Human Comput. **13**, 769–787 (2022). https://doi.org/10.1007/s12652-021-02929-z

Optimizing SEO with AI: Achieving Success Through Responsible Innovation

Utsav Sarkar[1], Subrata Paul[2], M. Raghuram Gowrav[3], and Anirban Mitra[1](✉)

[1] Department of Computer Science and Engineering, Amity University Kolkata, Kadampukur, New Town, West Bengal 700135, India
mitra.anirbam@gmail.com

[2] Department of Computer Science and Engineering, Brainware University, Ramkrishnapur Road, Barasat, West Bengal 700125, India

[3] Department of CSE, Vignan Institute of Technology and Management, P.O. Mantridi, Berhampur, Odisha 761008, India

Abstract. The integration of Artificial Intelligence (AI) into Search Engine Optimization (SEO) practices has ushered in a new era of digital marketing, characterized by transformative innovations and ethical considerations. This paper delves into the profound impact of AI on SEO, elucidating its multifaceted role in revolutionizing traditional approaches to optimizing online visibility and user engagement. Through advanced algorithms and machine learning techniques, AI-powered SEO tools offer unparalleled capabilities, including enhanced efficiency in keyword research, content optimization, and trend analysis. Moreover, AI enables personalized content creation tailored to individual user preferences, driving increased traffic and conversion rates. However, amidst these advancements lie challenges and ethical concerns, such as algorithmic biases, over-reliance on AI, and data privacy issues. Striking a delicate balance between innovation and ethical practices is imperative for businesses seeking to harness the full potential of AI in SEO while upholding transparency, fairness, and user privacy. This paper navigates the intricate landscape of AI-driven SEO, providing insights into its transformative power and guiding principles for ethical AI implementation in the digital realm.

Keywords: AI-driven SEO · Artificial Intelligence · Search Engine Optimization · Transformative Impact · Ethical Considerations · Advanced Analytics · Personalized Content · Predictive Strategies

1 Introduction

Search Engine Optimization (SEO) is the practice of enhancing the visibility and ranking of a website or web page in the organic (non-paid) search results of search engines like Google, Bing, and Yahoo. The primary goal of SEO is to increase the quantity and quality of traffic to a website by making it more attractive to search engines and users. This is achieved through a variety of strategies and techniques that can be broadly categorized into three main areas: on-page SEO, off-page SEO, and technical SEO. SEO is a dynamic

B. K. Mishra et al. (Eds.): ICT4GS 2024, ISEM 56, pp. 51–62, 2025.
https://doi.org/10.1007/978-3-031-94997-5_6

field that requires ongoing efforts to stay up-to-date with the ever-evolving algorithms and guidelines of search engines.

Search Engine Optimization, plays a pivotal role in digital marketing by enhancing the visibility and discoverability of a business's online presence. SEO is essential for increasing the visibility of a website on search engine results pages (SERPs). Compared to paid advertising methods, SEO is cost-effective. While it requires an initial investment of time and resources, the long-term benefits of improved search rankings and increased traffic do not entail ongoing costs like pay-per-click (PPC) advertising. By optimizing their websites effectively, businesses can outrank competitors in search results, attract more potential customers, and capture a larger market share [1, 2]. Artificial Intelligence (AI) refers to the simulation of human intelligence in machines that are programmed to think and learn like humans. AI is rapidly transforming various industries by automating processes, enhancing efficiency, and providing deep insights through data analysis. AI is revolutionizing various industries by enhancing diagnostic accuracy, personalizing treatments, and streamlining administrative tasks in healthcare. Machine learning detects diseases like cancer early, while predictive analytics manage chronic diseases and anticipate patient outcomes. AI chatbots and virtual assistants improve patient engagement. In finance, AI enhances fraud detection, risk management, and customer service. Algorithms analyze transactions to identify fraud, and robo-advisors offer personalized investment advice. AI predicts market trends and assesses creditworthiness, aiding decision-making. Retailers use AI to optimize supply chains, forecast demand, manage inventory, and personalize marketing, boosting sales and improving customer service with AI chatbots [3].

AI is rapidly expanding its role across various industries, driving innovation and efficiency through automation, predictive analytics, and personalized experiences. By leveraging AI technologies, businesses can gain a competitive edge, enhance customer satisfaction, and achieve operational excellence.

2 Comprehensive Guide to Effective SEO Strategies

Search Engine Optimization (SEO) is a cornerstone of digital marketing, playing a pivotal role in enhancing the visibility and credibility of a website. As the majority of online experiences begin with a search engine, effective SEO practices are essential for driving organic traffic and ensuring that a website ranks highly on search engine results pages (SERPs). On-Page SEO involves optimizing individual web pages to improve their search engine rankings. Off-Page SEO focuses on activities outside the boundaries of the website to improve its authority and trustworthiness. Technical SEO deals with the backend aspects of a website, ensuring that it meets the technical requirements of search engines for crawling and indexing [4].

Effective SEO strategies involve a combination of on-page, off-page, and technical optimizations. The key components of SEO include the following:

1. Keyword Research: The purpose includes identifying the terms and phrases that potential customers use to search for products, services, or information related to your business. It involves using tools like Google Keyword Planner, Ahrefs, and SEMrush.

2. On-Page SEO: Creating high-quality, relevant, and valuable content is fundamental to on-page SEO. This content should meet the needs of users and address their search queries effectively. Crafting compelling and keyword-rich title tags and meta descriptions is crucial for improving click-through rates from search engine results pages (SERPs).
3. Off-Page SEO: Acquiring high-quality backlinks from authoritative and relevant websites signals trust and authority to search engines. Effective methods for obtaining backlinks include guest blogging, influencer outreach, content promotion, and targeted link-building campaigns. High-quality backlinks are a critical factor in boosting a website's search engine rankings.
4. Technical SEO: This ensures fast page load times are crucial for enhancing user experience and reducing bounce rates. Tools like Google PageSpeed Insights can help identify and fix speed issues, ensuring that websites load quickly and efficiently, which is also a ranking factor for search engines. Creating and submitting an XML sitemap to search engines helps them discover and index all pages on a website efficiently. An XML sitemap provides a roadmap for search engine crawlers, ensuring that important pages are not overlooked [5, 6].
5. User Experience (UX): Designing intuitive and easy-to-navigate menus and structures helps users find information quickly and efficiently. Good navigation enhances user experience by reducing the time and effort needed to locate desired content. Monitoring engagement metrics such as bounce rate, time on site, and pages per session help to understand how users interact with the website and where enhancements are needed.
6. Local SEO: Optimizing and maintaining a GMB profile enhances visibility in local search results and Google Maps. A well-optimized GMB profile provides essential business information and helps attract local customers. Ensuring consistent NAP (Name, Address, Phone Number) information across local directories and listings is vital for local SEO [6].

By integrating these key components into a cohesive SEO strategy, businesses can improve their search engine rankings, attract more organic traffic, and achieve their digital marketing goals.

3 Challenges in Traditional SEO Practices

Traditional SEO practices, while effective, present several challenges that can impede the efficiency and success of digital marketing efforts. These challenges include manual keyword research, time-consuming content optimization, difficulty in tracking and analyzing data, and the need to keep up with constant algorithm updates.

3.1 Manual Keyword Research

Identifying the right keywords to target can be a labor-intensive process. Traditional keyword research involves sifting through numerous keyword suggestions, analyzing search volumes, competition levels, and relevance to the business. SEO professionals must manually review and select the best keywords, which can be incredibly time-consuming. This manual approach also demands a deep understanding of the market and the ability to predict which keywords will drive valuable traffic.

3.2 Time-Consuming Content Optimization

Optimizing content for search engines requires more than just inserting keywords; it involves structuring the content to be informative, engaging, and aligned with SEO best practices. This includes writing high-quality content, adding meta tags, optimizing images, and ensuring proper use of header tags. For large websites with numerous pages, manual optimization can be extremely time-consuming [7, 8].

3.3 Difficulty in Tracking and Analyzing Data

SEO performance tracking involves monitoring a wide range of metrics, including keyword rankings, organic traffic, bounce rates, and conversion rates. Traditional methods of tracking and analyzing this data often require the use of multiple tools and manual data aggregation. This can lead to inefficiencies and a fragmented view of SEO performance. The difficulty in compiling and interpreting data can hinder the ability to make informed decisions quickly [9].

3.4 Keeping up with Algorithm Updates

Search engines like Google frequently update their algorithms to improve search results and combat manipulative SEO practices. These updates can significantly impact search rankings and require continuous adaptation of SEO strategies. Staying updated with the latest algorithm changes is a major challenge for SEO professionals. Failure to adapt to these updates can result in significant drops in search rankings and organic traffic [10]. Addressing these issues requires a combination of expertise, continuous learning, and the adoption of more advanced tools and technologies, such as AI, to streamline and enhance the effectiveness of SEO practices.

4 Leveraging AI for Enhanced SEO Strategies

The integration of artificial intelligence (AI) into search engine optimization (SEO) practices has revolutionized the way businesses approach keyword research, content creation, and technical optimization. By harnessing the power of AI, marketers can streamline their SEO efforts, improve accuracy, and stay ahead of emerging trends. This section explores four key areas where AI is making a significant impact on SEO.

4.1 AI-Driven Keyword Research

AI-Driven Keyword Research leverages machine learning algorithms and natural language processing (NLP) to enhance the accuracy and efficiency of identifying relevant keywords. Unlike traditional keyword research, which relies heavily on manual analysis,

AI-driven tools can process vast amounts of data from search engines, social media, and other online platforms to uncover high-potential keywords that might be overlooked by manual methods. Efficiency and speed are major benefits of AI tools, as they can analyze thousands of keywords and their variations in a fraction of the time it takes manually, providing insights into search volumes, competition, and trends [11].

4.2 Content Creation and Optimization Using AI

Content Creation and Optimization Using AI involves using AI technologies to generate, enhance, and refine website content, ensuring it meets SEO best practices while remaining engaging and relevant to users. AI-powered tools like GPT-4 can generate high-quality content based on specific topics, keywords, and user preferences, including blog posts, product descriptions, and entire articles that are well-structured and informative.

4.3 Predictive Analytics for SEO Strategy

Predictive Analytics for SEO Strategy uses AI to analyze historical data and predict future trends, helping marketers develop more effective SEO strategies. AI can analyze past search trends and current data to predict future trends, allowing marketers to optimize their content for keywords and topics that are expected to become popular [12, 13].

4.4 AI Tools for Technical SEO Improvements

AI Tools for Technical SEO Improvements focus on optimizing the technical aspects of a website to enhance its performance, usability, and search engine friendliness. AI-powered tools can conduct comprehensive site audits to identify technical issues such as broken links, slow page speeds, and mobile usability problems, providing detailed reports and actionable recommendations to fix these issues. The integration of AI into SEO practices offers numerous advantages, from improving the accuracy and efficiency of keyword research to optimizing content and technical aspects of websites. AI-driven keyword research, content creation and optimization, predictive analytics for SEO strategy, and technical SEO improvements collectively enhance the effectiveness, efficiency, and precision of SEO efforts [14].

5 AI Tools and Technologies for SEO

Artificial Intelligence (AI) has significantly transformed the landscape of Search Engine Optimization (SEO) by providing advanced tools that enhance efficiency, accuracy, and strategic decision-making. These tools offer a range of features and functionalities designed to improve keyword research, content optimization, site audits, and overall SEO performance [11]. SEMrush is an all-in-one marketing toolkit that offers a comprehensive suite of SEO tools. It leverages AI to provide detailed insights and analytics for digital marketing strategies. SEMrush offers a comprehensive suite of SEO tools designed to enhance digital marketing strategies. Its Keyword Magic Tool leverages AI to generate extensive keyword lists with data on search volume, difficulty, and competitive density [15]. Ahrefs is known for its robust backlink analysis and comprehensive SEO tools that utilize AI to help marketers improve their search rankings and website traffic. Ahrefs is renowned for its robust backlink analysis and comprehensive SEO tools that use AI to enhance website traffic and search rankings [16]. Moz provides a range of SEO tools designed to help businesses improve their search engine visibility. Moz's AI-driven features focus on simplicity and usability. Moz focuses on simplicity and usability with a range of AI-driven SEO tools. BrightEdge is an enterprise-level SEO platform that uses AI to provide deep insights and advanced analytics for digital marketing strategies. BrightEdge is an enterprise-level SEO platform that utilizes AI to deliver deep insights and advanced analytics for digital marketing [17].

5.1 Case Studies of Successful AI Implementation in SEO

Case Study 1: SEMrush and a Leading E-commerce Company. A leading e-commerce company used SEMrush to revamp their SEO strategy. By leveraging the Keyword Magic Tool, they identified high-opportunity keywords and optimized their product pages accordingly. The Site Audit feature helped them fix critical technical issues, such as broken links and duplicate content. Within six months, the company saw a 35% increase in organic traffic and a 20% boost in online sales.

Case Study 2: Ahrefs and a Digital Marketing Agency. A digital marketing agency utilized Ahrefs to enhance their clients' SEO performance. By using the Site Explorer and Keywords Explorer tools, they conducted thorough competitive analysis and keyword research. The Content Explorer helped them identify trending topics and create high-performing content. As a result, their clients experienced a 50% improvement in search rankings and a 40% increase in organic traffic.

Case Study 3: Moz and a Small Business. A small business owner implemented Moz's SEO tools to improve their local search presence. The Keyword Explorer helped them identify relevant local keywords, while the Link Explorer and Page Optimization tools guided their link-building and content optimization efforts. The Rank Tracking feature allowed them to monitor their progress. Within four months, they achieved top rankings for several local keywords, leading to a significant increase in foot traffic and online inquiries.

Case Study 4: BrightEdge and a Global Tech Company. A global tech company partnered with BrightEdge to optimize their SEO strategy across multiple markets. Using the Data Cube and ContentIQ tools, they conducted in-depth keyword research and technical audits. The Intent Signal feature helped them create content tailored to user intent, improving engagement and conversion rates. Over a year, the company saw a 45% increase in organic search traffic and a substantial improvement in lead generation.

6 The Impact of AI-Driven SEO Tools on Digital Marketing

Artificial Intelligence (AI) is transforming digital marketing by enhancing the efficiency and accuracy of search engine optimization (SEO) efforts. AI-driven tools automate labor-intensive tasks, provide real-time insights, and enable personalized user experiences, thus optimizing the effectiveness of SEO strategies. This section explores the various ways in which AI is reshaping SEO practices, focusing on increased efficiency, improved keyword research, enhanced trend analysis, and better user experiences.

6.1 Increased Efficiency and Productivity

AI-driven SEO tools significantly enhance efficiency and productivity in digital marketing by automating tasks that traditionally required manual effort. These tools streamline processes such as keyword research, content analysis, and site audits, allowing SEO professionals to focus on more strategic activities. Automation of routine tasks, such as scanning websites for technical issues, analyzing vast amounts of data, and generating SEO reports, enables professionals to concentrate on campaign planning and creative content development. AI processes data much faster than humans, providing real-time insights that facilitate quicker decision-making. For instance, AI can instantly analyze the performance of SEO strategies and suggest adjustments to improve results, leading to more agile and responsive marketing efforts [18].

6.2 Improved Accuracy in Keyword Research and Content Optimization

AI enhances the accuracy of keyword research and content optimization, ensuring that SEO efforts are more effective and targeted. Advanced data analysis capabilities of AI tools allow them to identify patterns and correlations that might be missed by human analysts, resulting in more precise keyword suggestions and a better understanding of keyword relevance and competition. This improves the relevance and effectiveness of SEO campaigns. AI tools can also assess existing content and provide specific recommendations for optimization, such as the ideal keyword density, content structure, and related topics to cover. This ensures that content is not only SEO-friendly but also engaging and valuable to readers [19].

6.3 Enhanced Ability to Analyze and Predict Trends

AI tools excel at analyzing past data and predicting future trends, providing marketers with valuable insights that inform their SEO strategies. Predictive analytics uses historical data to forecast future trends in search behavior, helping marketers stay ahead of emerging topics and keywords. This proactive approach ensures that content remains relevant and timely. AI can quickly identify changes in search patterns and consumer behavior, allowing marketers to adjust their strategies accordingly. By analyzing a wide range of data sources, AI provides a comprehensive view of market trends and consumer preferences. This holistic perspective helps businesses align their SEO efforts with broader market dynamics and customer needs [20].

6.4 Better User Experience Through Personalized Content

AI-driven SEO tools contribute to a better user experience by enabling the creation of personalized and engaging content that meets individual user needs. AI analyzes user data, such as browsing history, search queries, and demographic information, to deliver personalized content recommendations. This tailored approach increases user engagement and satisfaction [12]. AI can dynamically adjust content based on real-time user interactions and preferences. By providing relevant and personalized content, AI helps keep users on the site longer and encourages them to explore more pages. This reduces bounce rates and increases the likelihood of conversions.

AI tools can also enhance website accessibility by providing features like automated alt text for images, voice search capabilities, and content adjustments for different devices. This ensures that all users, including those with disabilities, have a positive experience on the site [21].

7 Addressing Challenges and Ethical Considerations in AI-Driven SEO

As AI technology becomes increasingly integrated into SEO practices, it brings along significant advantages but also introduces new challenges and ethical considerations.

7.1 Potential Biases in AI Algorithms

AI algorithms, while powerful and efficient, can inadvertently introduce biases that affect SEO outcomes. These biases often originate from the data used to train the algorithms and the design choices made by developers. This leads to skewed keyword suggestions and content recommendations, failing to reflect the diversity of the user base.

Additionally, the way algorithms are designed can reinforce existing trends and disadvantage new or innovative approaches. This algorithmic bias might prioritize certain types of content or SEO tactics based on historical success, resulting in a lack of diversity in content and overlooking emerging topics or underserved niches. Consequently, biased algorithms can skew search rankings, providing unfair advantages to some websites over others, not because of superior content or SEO strategies, but due to inherent biases in the AI tools. Mitigating these biases requires the use of diverse and representative datasets, continuous monitoring and adjustments of algorithms, and the incorporation of fairness and ethical considerations into AI design. Human oversight is essential to review AI-driven recommendations and outcomes, ensuring alignment with broader SEO goals and ethical standards [22].

7.2 Over-Reliance on AI Tools

While AI tools offer significant advantages in efficiency and data analysis, over-reliance on them can lead to several challenges. One major issue is the potential loss of human insight and creativity. AI tools excel at processing data and identifying patterns, but they lack the nuanced understanding and creative thinking that human marketers bring.

Over-reliance on AI can result in formulaic content and strategies, lacking the unique voice and insights that resonate with audiences.

Furthermore, dependence on AI can lead to complacency among marketers, reducing their engagement with new SEO trends and best practices. This complacency can result in outdated and less effective SEO strategies. They might not fully capture the context of search queries or understand the subtleties of human language, leading to suboptimal keyword suggestions or content recommendations [20].

7.3 Privacy Concerns with Data Collection and Usage

AI-driven SEO tools rely heavily on data collection and analysis, raising significant privacy concerns. These tools gather vast amounts of data on user behavior, preferences, and interactions. While this data is invaluable for optimizing SEO strategies, it raises issues about how data is collected, stored, and used.

Adhering to data privacy regulations, such as the General Data Protection Regulation (GDPR) and the California Consumer Privacy Act (CCPA), is crucial. These regulations set strict guidelines on data collection, consent, and user rights, and non-compliance can result in significant penalties and damage to reputation. Ensuring data security is also paramount, requiring robust measures like encryption, access controls, and regular security audits to protect against breaches and unauthorized access. Ethical data usage goes beyond legal compliance, involving transparency with users about data collection practices, obtaining explicit consent, and ensuring data usage respects user privacy and autonomy.

7.4 Ensuring Transparency and Accountability in AI-Driven SEO Practices

Transparency and accountability are essential for maintaining trust and integrity in AI-driven SEO practices. Companies should strive for algorithm transparency, providing information on how their AI algorithms work, what data they use, and how decisions are made. This transparency helps build trust with users and stakeholders, ensuring they understand and have confidence in the AI processes.

Establishing clear accountability mechanisms is crucial. Assigning responsibility for AI decisions and outcomes, ensuring processes are in place to review and address issues or biases, and conducting regular audits of AI tools help maintain accountability. Educating users and clients about AI tools' capabilities and limitations, explaining how AI-driven recommendations are generated, and highlighting potential biases and limitations ensure that informed users can make better decisions and provide valuable feedback. Adopting and adhering to ethical standards for AI use in SEO is essential. Principles such as fairness, transparency, and respect for privacy should guide AI practices. Developing and implementing ethical guidelines help ensure that AI-driven practices align with broader societal values and contribute positively to the digital ecosystem [12].

8 Emerging AI Technologies and Their Potential Impact on SEO

Emerging AI technologies hold the promise of transforming SEO practices and significantly influencing the digital marketing landscape. These advancements in AI are poised to bring about more sophisticated tools and strategies that enhance the effectiveness and efficiency of SEO efforts.

8.1 Advancements in AI Technologies

One of the most impactful AI technologies is Natural Language Processing (NLP). NLP enables AI systems to understand and interpret human language, facilitating more sophisticated analysis of user queries and content. This technology powers voice search, chatbots, and content optimization tools, leading to improved user experiences and higher search engine rankings.

Deep Learning algorithms, a subset of AI, can learn from vast amounts of data to identify patterns and make predictions. In the context of SEO, deep learning can be employed for advanced keyword analysis, content generation, and personalized recommendations, resulting in more effective marketing strategies. Another significant AI technology is Natural Language Generation (NLG), which can generate human-like text based on data inputs [23]. AI-powered analytics tools also play a crucial role in SEO by processing complex data sets in real-time. These tools provide actionable insights and predictive analytics, enabling marketers to identify trends, forecast performance, and optimize campaigns more effectively.

8.2 Predictions for the Evolution of SEO with AI

As AI technologies continue to evolve, several trends are expected to shape the future of SEO. Hyper-personalization will become more sophisticated, allowing marketers to deliver highly tailored content and experiences to individual users based on their preferences, behaviors, and demographics. With the rise of voice-enabled devices and virtual assistants, optimizing for voice search will become increasingly important. AI will be essential in understanding natural language queries and providing relevant, voice-friendly content. Visual search optimization will also gain prominence, driven by advancements in computer vision AI [24]. Moreover, AI will enhance existing SEO tools and platforms, providing advanced features for keyword research, content optimization, link building, and performance tracking. These augmented tools will empower marketers to achieve better results with less manual effort, significantly improving the overall effectiveness of SEO campaigns.

8.3 Long-Term Benefits and Challenges

While AI-driven SEO offers numerous benefits, there are also long-term challenges to consider. The increased efficiency provided by AI, through the automation of repetitive tasks, streamlines SEO workflows and frees up time for strategic planning and creative initiatives. AI algorithms analyze data more comprehensively and accurately than humans, leading to better insights and more effective optimization strategies. Compliance with regulations and the development of ethical standards will be paramount. Training SEO professionals to leverage AI effectively and bridging this skill gap will be essential. Integration complexity is another significant challenge, as incorporating AI into existing SEO workflows and technologies may require substantial resources and expertise. Ensuring seamless integration and interoperability will be key to maximizing the benefits of AI.

The evolution of SEO with AI promises significant benefits in terms of efficiency, accuracy, personalization, and competitive advantage. However, addressing long-term challenges such as ethical concerns, skill development, integration complexity, and algorithm dependence will be critical to realizing the full potential of AI-driven SEO in the years to come.

9 Conclusion

The integration of AI technologies into SEO practices represents a monumental shift in how digital marketing is conducted. AI's ability to process vast amounts of data, generate insightful analytics, and personalize user experiences has significantly enhanced the efficiency, accuracy, and effectiveness of SEO efforts. However, this transformation is not without its challenges. Addressing potential biases in AI algorithms, ensuring data privacy, and maintaining transparency and accountability are critical to the ethical implementation of AI in SEO. By balancing the innovative capabilities of AI with a strong commitment to ethical practices, businesses can not only optimize their SEO strategies but also foster trust and integrity in their digital marketing efforts. This balance ensures that AI-driven SEO contributes positively to business growth while respecting user rights and societal values. As the digital landscape continues to evolve, embracing responsible AI practices will be essential for sustainable success and maintaining a competitive edge in the market.

References

1. Todorova, A., Antonova, D.: Smart marketing solutions: applications with artificial intelligence to increase the effectiveness of marketing operations. In: 2023 7th International Symposium on Multidisciplinary Studies and Innovative Technologies (ISMSIT), pp. 1–6. IEEE (2023)
2. Raji, M.A., Olodo, H.B., Oke, T.T., Addy, W.A., Ofodile, O.C., Oyewole, A.T.: Digital marketing in tourism: a review of practices in the USA and Africa. Int. J. Appl. Res. Soc. Sci. **6**(3), 393–408 (2024)
3. Rosario, A.M.F.T., Cruz, R.N.: Determinants of innovation in digital marketing. J. Rev. Global Econ. **8**(1), 1722–1731 (2019)
4. Papagiannis, N.: Effective SEO and content marketing: the ultimate guide for maximizing free web traffic. John Wiley & Sons (2020)
5. Roumeliotis, K.I., Tselikas, N.D.: An effective SEO techniques and technologies guide-map. J. Web Eng. **21**(5), 1603–1649 (2022)
6. Grappone, J., Couzin, G.: Search Engine Optimization (SEO): An Hour A Day. John Wiley & Sons (2011)

7. Giomelakis, D., Karypidou, C., Veglis, A.: SEO inside newsrooms: reports from the field. Fut. Internet **11**(12), 261 (2019)
8. Giomelakis, D.: Semantic search engine optimization in the news media industry: Challenges and impact on media outlets and journalism practice in Greece. Soc. Media+ Soc. 9(3), 20563051231195545 (2023)
9. Leban, M., Errmann, A., Seo, Y., Voyer, B.G.: Mindful luxury: a case of the faroe Islands. Tour. Manage. **104**, 104929 (2024)
10. Lee, S.M., et al.: Deep learning applications in chest radiography and computed tomography: current state of the art. J. Thorac. Imaging **34**(2), 75–85 (2019)
11. Rathore, B.: Usage of ai-powered marketing to advance SEO strategies for optimal search engine rankings. Eduzone: Int. Peer Rev./Refereed Academic Multidisc. J. **05**(01), 30–35 (2016). https://doi.org/10.56614/eiprmj.v5i1y16.300
12. Chaitanya, K., Saha, G.C., Saha, H., Acharya, S., Singla, M.: The impact of artificial intelligence and machine learning in digital marketing strategies. European Economic Letters (EEL) **13**(3), 982–992 (2023)
13. Jain, A., Jain, C., Kargal, R. G., Suresh, S.: Advancements and Innovation in Digital Marketing and SEO. In IoT and AI Technologies for Sustainable Living, pp. 279–315. CRC Press (2022)
14. Kamal, M., Himel, A.S.: Redefining modern marketing: an analysis of AI and NLP's influence on consumer engagement, strategy, and beyond. Eigenpub Rev. Sci. Technol. **7**(1), 203–223 (2023)
15. Panchal, A., Shah, A., Kansara, K.: Digital marketing-search engine optimization (SEO) and search engine marketing (SEM). Int. Res. J. Innov. Eng. Technol. **5**(12), 17 (2021)
16. Gkikas, D.C., Theodoridis, P.K.: Artificial intelligence (AI) impact on digital marketing research. In: Kavoura, A., Kefallonitis, E., Giovanis, A. (eds.) Strategic Innovative Marketing and Tourism. SPBE, pp. 1251–1259. Springer, Cham (2019). https://doi.org/10.1007/978-3-030-12453-3_143
17. Kwak, Y., Ahn, J.W., Seo, Y.H.: Influence of AI ethics awareness, attitude, anxiety, and self-efficacy on nursing students' behavioral intentions. BMC Nurs. **21**(1), 267 (2022)
18. Rathore, B.: Predictive metamorphosis: unveiling the fusion of AI powered analytics in digital marketing revolution. Int. J. Transcontinental Discoveries **7**(1), 15–24 (2020)
19. Kumar, T.: Integration of intelligent AI & SEO: a review of various factors. Int. J. New Media Stud.: Int. Peer Rev. Scholarly Indexed J. **10**(1), 64–67 (2023)
20. George, S.M., Sasikala, B., Gowthami, T., Sopna, P., Umamaheswari, M., Dhinakaran, D.P.: Role of artificial intelligence in marketing strategies and performance. Migr. Lett. **21**(S4), 1589–1599 (2024)
21. Srivastava, M.K., Singh, H.S.: Harnessing AI and digital marketing for enhanced online education post-COVID-19: opportunities and challenges. MSW Manage. J. **33**(1), 121–134 (2023)
22. Ezinwa, C.A., Agbo, B.O., Ozojukwu, C.G.: Analysis of knowledge and use of artificial intelligence among bloggers in enugu state. Int. J. Non-Governmental Organ. (NGOs) Essays **8**(1), 1–26 (2024)
23. Paul, S., Koner, C., Mitra, A.: A survey on social business intelligence: a case study of application of dynamic social networks for decision making. In: Natural Language Processing in Artificial Intelligence, pp. 1–27. Apple Academic Press (2020)
24. Paul, S., Mitra, A.: A Review on Applications of Machine Learning in IoT: Challenges and Future Prospects. Internet of Things, pp. 299–330 (2023)

Second-Hand Drive: A Comprehensive Study of Influencing Factors on Buying Decisions of Used Cars in India

Abhishek Kumar, Manisha Paliwal, Dimple Saini, and Pooja(✉)

Sri Balaji University, Pune, Maharashtra 411033, India
{manisha.paliwal,dsaini}@sbup.edu.in, pooja02panwar@gmail.com

Abstract. There have been many significant studies done on the Indian automobile sector, primarily on new cars but, not many research studies have been done on used cars, revealing a potential research gap. Consumer preferences have changed after COVID-19 and there is more interest in a balanced lifestyle of work and play. A by-product of this change is an increasing purchase of cars mostly used for long-distance travel and weekend getaways. Also, with the advent of Vehicle Scrappage Policy 2021, limiting the use of vehicles to 10–15 years and a shift towards electric vehicles, there is a change in the usage life of Internal Combustion Engine (ICE) vehicles by the Indian populace which leads to sustainability. In addition, this has created an increase in the quantity of used cars in the market in good working conditions. This research paper aims to analyze the usage patterns, buying behavior, and problems faced while purchasing used cars in India. Also, this study aims to document the major players in the Indian used car market.

Keywords: Used Cars · Consumer Behaviour · Automobile · Buying Behaviour · Sustainability

1 Introduction

India is a population-rich country, being the second largest in terms of the number of people with an increasing young and middle-income population due to which a substantial increase in demand for different consumer goods has occurred. The COVID-19 Pandemic of 2019 forced people to be isolated in their homes for an extended period. As more people get vaccinated and travel restrictions are lifted, there is a surge in domestic tourism in India. A study conducted by McKinsey & Company found that domestic tourism in India is expected to recover faster than international tourism, and travel within the country is likely to become more popular due to the pandemic-related travel restrictions (McKinsey & Company, 2020). While there is extensive research on the Indian automobile sector for new cars, there exists a significant research gap concerning the used car market. Specifically, the existing literature does not adequately address the post-pandemic shift in consumer preferences towards used cars.

After COVID-19, consumers' preferences have shifted to a healthier and more balanced work-life approach (Pawar, 2022). It is thus an associated phenomenon that a

B. K. Mishra et al. (Eds.): ICT4GS 2024, ISEM 56, pp. 63–72, 2025.
https://doi.org/10.1007/978-3-031-94997-5_7

growing number of cars, especially those mainly used for interstate travel and leisure, are being bought. The implementation of Vehicle Scrappage Policy 2021 restricts the usage of vehicles up to 10–15 years along with electrification changes the usage life of ICE vehicles by the Indian users favoring sustainability (Government of India, 2021). This policy has also resulted in more used cars being available in the market in good working conditions.

The used car market has shown a promising 15.1% compounded annual growth rate (CAGR). However, the factors affecting consumers' decision to purchase used cars in India have not been well researched, despite the rising trend of used car purchases in the country. Earlier literature has mainly targeted new cars whereas, the issues and characteristics of the used car segment have been overlooked (Kumar, et al., 2023). The objectives of this research paper entail examining the usage frequency, tendency, and issues concerning the purchase of second-hand automobiles in India. Moreover, this research intends to identify the leading participants in the Indian used car industry and uncover tendencies in consumer behavior peculiar to this segment.

2 Literature Review

Several variables, including rising fuel prices, shifting customer preferences, rising disposable incomes, and an increased desire for ecologically friendly automobiles, have contributed to India's used car market's recent rapid growth. The number of people in the nation who buy and sell second-hand cars has significantly increased as a result of this. In this review of the literature, we will look at a few research papers and articles that have looked into different facets of the used automobile market in India.

Sunitha et al. (2012), tried to understand the factors that influence automobile customers' buying intentions in Chidambaram town using a convenience sampling technique. This paper tried to understand the perceived risks that influence the buying intention such as perceptions of the value of the car, test drive, magazine, and expert reviews, and service center availability on the purchase decision of secondhand cars. The study also brings light to the difficulties faced by purchasers and sellers in the secondhand car industry and thus suggests recommendations for improving the market.

Shende (2014), in his study, explored both rural and urban car markets and the factors that impact the buying behavior of the customers. The authors analyzed the impact of various factors and it was found that disposable income influences the purchase decision of consumers. Value for money built quality, and features of cars affect the buying behavior of customers.

Kumar (2014), in his research endeavors to investigate the elements shaping the buying tendencies of Indian consumers residing in the northern part of states, including Punjab, Delhi, Haryana, Chandigarh, and Himachal Pradesh (HP). The study delves into various factors encompassing safety, aesthetics, design, features, interior perception, as well as the pre-sales and post-sales policies that play a pivotal role in guiding customers toward the selection and purchase of an automobile.

Bhandari & Phadtare (2016) conducted a study to analyze the determinants influencing the purchase intention of used cars by car consumers in the Pune city market. 90 samples were taken, and the judgment sampling method was used. The authors identify the factors that influence the price of used cars, such as Hedonic Benefits, Functional Reasons, Social factors, and Cultural & Religious Factors that have a significant role to play in guiding the customers' used car purchase decisions.

Agarwal & Mehrotra (2018) conducted a study with 496 respondents aimed at finding out consumers' most preferred mode of advertising which affects their purchase intention of buying cars in India. Demographics was one factor that was the most prominent in molding customer behavior.

Grover & Mandan (2017), investigated the role of social media as a promotional platform in shaping customer perceptions during the purchase of new cars. The identified factors include customer awareness, acquisition, information gathering, attitudes, opinions, purchase behavior, and post-purchase evaluation. Furthermore, the study delves into the challenges confronted by the industry, encompassing issues like environmental concerns and adherence to safety regulations.

Agarwal & Shashikiran (2018) in their research examined the impact of occupation on a consumer's intent to purchase cars. Purchase intention, as the variable in question, is contingent on three independent variables: Income Level, type of Occupation, and level of occupation. The results show that occupation exerts a minimal impact on the purchase intention of cars.

Dhanabalan et al. (2018) in their study understand the reasons which affect the buying decision of the Indian car consumer. The focus of the study has been on Tamil Nadu state with 547 respondents. The study found that customer perceived value was a dependent variable that was linked to the independent variables like price, brand, quality, utility, design, and technical consideration of the automobile.

Rajendran & Jayakrishnan (2018) in their study presented that there is a perceived risk by consumers while purchasing automobiles in the Indian car market. Independent factors like consumer perceived risk and customer satisfaction affect the buying behavior of consumers while purchasing cars in India. The study also highlighted the importance of trust and credibility in the used car market and recommended measures to build consumer confidence in the market.

David & Banumathi (2014), in their study, attempted to find the influencing factors that customers consider while purchasing a passenger car in Indian Auto-mobile Market. The focus of this study is limited to Puducherry with 100 respondents who own cars by a structured questionnaire method. The researchers identified various factors that consumers take into account such as technical features of the car, performance, and environmental factors.

Maji & Bandyopadhyay (2018) in their study investigated the role of risk perception in a consumer's purchase intention of a used car, covering factors and variables such as the customer experience along with risks associated with financing, society, vehicular service, vehicular operation, psychological and physical security, and ownership risk.

According to Kaur & Bhadari (2014), employee perception of sustainability impacts buying behavior in the automobile sector in Northern India. Employee perception is crucial for sustainability development in the automobile sector. Sustainability was perceived to be higher in the Four Wheeler Automobile sector compared to the Two Wheeler Automobile sector.

According to Ong et al. (2023), Saraf & Shastri (2021), and Chaudhari et al. (2022) sustainability concerns, economic factors, and societal influences play significant roles in consumers' decisions regarding hybrid and electric vehicles. Factors such as environmental awareness, perceived economic concerns, and support of authority have been identified as key influencers affecting the intention to purchase hybrid cars. Additionally, consumer attitudes, financial affordability, and location are highlighted as essential factors influencing the success of car subscription models, emphasizing the sustainability benefits of car sharing in India (Kaushik et al., 2022).

3 Conceptual Framework and Hypotheses

The following diagram represents a relationship between the independent and dependent variables related to the purchase intention of used cars in India. These independent variables are derived from the literature review (Fig. 1).

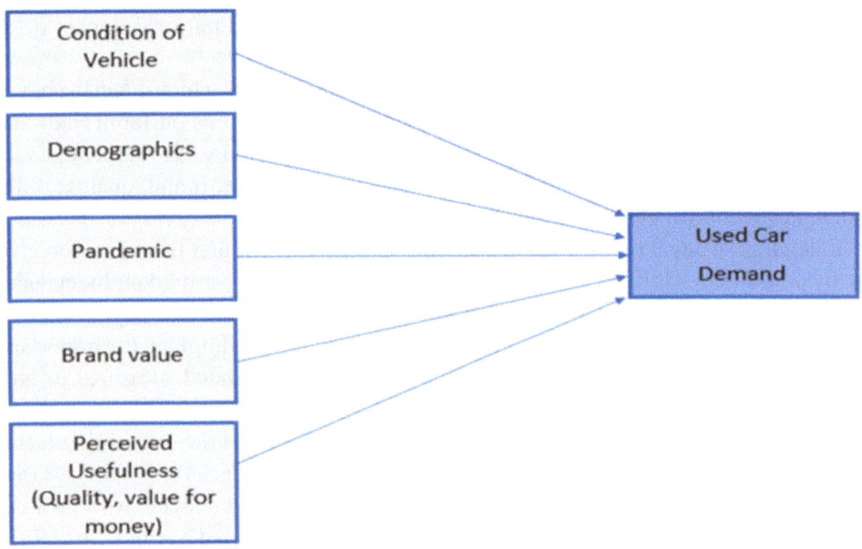

Fig. 1. Conceptual Model

The following hypotheses are formed for the study:

H1: The condition of the vehicle, and its remaining life, have a relationship with the demand for that vehicle.
H2: A change in demographics such as income, beliefs, and culture influences the demand for the vehicle.
H3: The pandemic and the lockdown have caused it to have a connection with the buying intention of the used car.
H4: The value of the vehicular brand has an impact on the demand for the used car.
H5: Perceived usefulness such as quality, value for money, return on investment, etc. has a relationship with the demand for used cars.

4 Research Methodology

The research methodology of the paper that focuses on the factors affecting the willingness to purchase used cars in the Indian automobile industry is designed as comprehensive and systematic. Employing descriptive research design, this study has targeted 150 participants from the used car market through convenience and judgment sampling techniques, and comprised of buyers, sellers, and industry experts. Data is gathered from well-designed questionnaires and structured surveys (Sher et al., 2021). The questionnaires are divided into three sections: Section A highlights the socio-demographic details of the respondent. Section B includes important predictor variables, demographic details, brand value, perceived usefulness, and the impacting effect of the pandemic on the demand factor for used cars. Section C is centered on the dependent variables i.e., the demand for the used cars.

The Likert Scale was used to identify the degree of the questions of Sections B and C with the responses ranked on a scale of 1 to 5, with 1 being Strong disapproval and 5 being strong approval. In parallel, a reliability test was conducted on the questionnaire, which also acted as the pilot test. Furthermore, interviews with industry experts help to obtain qualitative insights, complementing quantitative results obtained during research. The collected data is analyzed quantitatively as well as qualitatively, employing statistical tools such as regression analysis for the quantitative data.

Issues of ethics such as informed consent and confidentiality are accorded importance in the process of the research. During pilot testing, the survey instrument is refined to check for clarity and efficiency during data collection. The study acknowledges potential limitations while emphasizing validity and reliability. Hence, conventional quantitative survey methods and qualitative interviews with experts help in identifying and understanding the various factors that may have an impact on the purchase decisions for used cars in India.

5 Results and Discussion

5.1 Demographic Profile

The demographic profile of the respondents has been depicted in the pie chart in Fig. 2. Amidst the respondents, the maximum representation has been had in the form of the group of age covering respondents of age 55 and above, indicating a higher representation of older individuals in the survey. The data also revealed an even distribution among the middle age group (26–35 years) but it slightly increases with age whereas there are only 16 respondents that lie within the age group of '18–25 years'.

Figure 3 shows a significant gender imbalance in the data with males making up the vast majority of respondents (89%) compared to females (11%), indicating a predominant male participation in the survey.

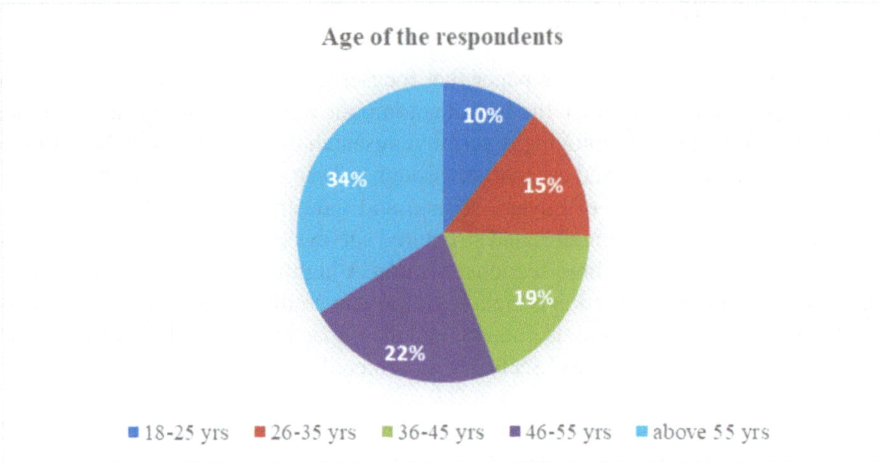

Fig. 2. Age of the respondents

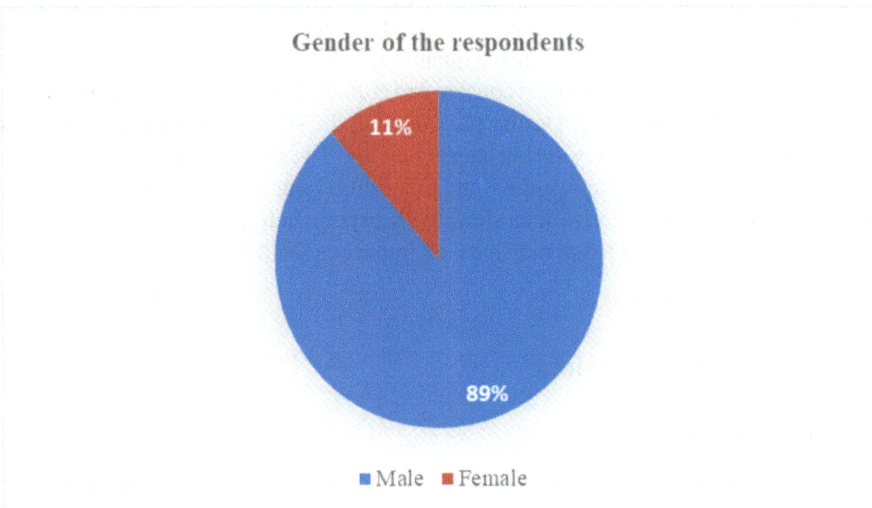

Fig. 3. Gender of the respondents

5.2 Reliability Analysis

The results of the Reliability test are indicated in Table 1, with Cronbach's Alpha test coefficient of 0.837, indicating a strong internal consistency amidst the items.

Table 1. Reliability Statistics

Cronbach's Alpha	N of items
.837	8

5.3 Regression Analysis

According to this regression model, the independent variable(s) accounts for about 41.3% of the variability factors observed in the dependent variable. The adjusted R square value depicts a numerical value of 0.375 along with a standard error estimate of 0.92910. A moderately positive association between the variables is indicated by the correlation coefficient (R) of 0.643a (Table 2). The F-Statistic value 10.788 ($p < .001$) indicates that the variables of an independent nature can explain the significant amount of variance observed in the variables of a dependent nature depicted in Table 3.

Table 2. Model Summary

Model	R	R Square	Adjusted R Square	Std. The error of the estimates
1	.643[a]	.413	.375	.92910

a. Predictors: (Constant), Conditions of Vehicles, Demographics, Pandemic, Brand Value, and Perceived Usefulness.

b. Dependent Variables: Purchase Intentions.

Table 3. ANOVA[a]

Model		Sum of Squares	df	Mean Square	F	Sig
1	Regression	55.876	6	9.313	10.788	.000[a]
	Residual	79.417	92	.863		
	Total	135.293	98			

a. Predictors: (Constant), Conditions of Vehicles, Demographics, Pandemic, Brand Value, and Perceived Usefulness.

b. Dependent Variables: Purchase Intentions.

According to Table 4, the dependent variable is positively correlated with demographic, brand value, and perceived usefulness variables ($p = 0.008$, .045, and .028 respectively). Nonetheless, there is no discernible effect of vehicle conditions and pandemic impacts on the dependent variable. These results imply that while some factors like brand reputation, demographic profile, and perceived usefulness have a significant impact on purchase intentions, other elements looked at in the model might not be reliable indicators in this particular situation.

Table 4. Coefficients

Model		Unstandardized B	Coefficients Std. Error	Standardized Coefficients Beta	t	Sig
1	(Constant)	.059	.209		.284	.777
	Economic Affairs	.105	.058	.124	1.806	.073
	Social Acceptance Factors	.490	.049	.487	9.898	**< .001**
	Environmental Factors	.162	.071	.183	2.289	**.023**
	Geographical Factors	.155	.063	.183	2.471	**.014**

Table 5 shows that the independent variable demographics such as income, beliefs, and culture influence the demand for the vehicle, and Brand value also has an impact on the demand for the used car. Similarly, perceived usefulness such as quality, value for money, return on investment, etc. has a relationship with the demand for used cars. However, the condition of the vehicle, its remaining life, and the Pandemic lockdown do not have a relationship with the demand for used cars.

Table 5. Summary of Hypotheses Testing

Hypothesis	Statement	p-value	Decision
H1	The condition of the vehicle, and its remaining life, have a relationship with the demand for that vehicle	.599	H_1 Rejected
H2	A change in demographics Such as income, beliefs, and culture influences the demand for the vehicle	**.008**	H_1 Accepted
H3	The pandemic and the lockdown have caused it to have a connection with the buying intention of the used car	.671	H_1 Rejected
H4	The value of the vehicular brand has an impact on the demand for the used car	**.045**	H_1 Accepted
H5	Perceived usefulness such as quality, value for money, return on investment, etc. has a relationship with the demand for used cars	**.028**	H_1 Accepted

6 Conclusion

Of late, a significant change has been witnessed in the Indian automobile market, marked by a growing interest in used cars. This paper extensively explores the determinants influencing the inclination to purchase used cars in the Indian automobile market. Several key factors were found based on the analysis, including economic considerations, perception of value, demographics, condition of vehicles, and the factors due to the COVID-19 pandemic. The study solidified the idea that as the socio-economic conditions of India change, so do the decision-making factors of the automobile customers. Perceived value, especially stood out as the primary reason provided by the current situation of the market and the purchasing power of the majority of the consumers. Also, as good quality cars become available, the choice of used cars as a cost-friendly substitute compared to new cars becomes more attractive to the average Indian consumer. Lastly, the paper brought about the importance of trust and credibility in the used car market, indicating a growing need for standardized, trustworthy online/offline platforms and dealers.

7 Limitations and Future Scope of the Study

To adjust marketing methods, future studies in India should examine geographical differences in used automobile buying intentions. Comparing preferences between urban and rural areas and examining the role of online platforms are crucial. Further significant insights can be obtained by investigating the effects of government laws, following changes in the industry over time, and taking psychological aspects like social status into account.

Although this study offers insightful information about the factors influencing Indians' propensity to buy old cars, its applicability is limited by several issues. Selection bias is introduced by the use of judgment and convenience sample techniques, and temporal fluctuations may not be captured by the cross-sectional design. Furthermore, response bias could result from depending too much on self-reported survey results. The study's focus on automobiles restricts its applicability to other automotive markets, and its conclusions may be impacted by assumptions about the stability of the economy and regulations. Moreover, it's possible that the study's examination of the COVID-19 pandemic underestimated its long-term impacts. Notwithstanding these drawbacks, the study establishes a framework for further research that will tackle these issues and broaden our comprehension of consumer behavior in the Indian car industry.

References

Agarwal, P., Shashikiran, L.: Influence of occupation on purchase intention of cars. Emperor Int. J. Financ. Manage. Res. 188–194 (2018)

Agarwal, R., Mehrotra, A.: Attributes influencing customers' car purchase decision–a demographical perspective. J. Appl. Manage.-Jidnyasa **10**(2), 48–61 (2018)

Bhandari, J., Phadtare, M.: Study of factors influencing pre-owned car purchase. PARIDNYA-The MIBM Res. J. **1**(1), (2016)

Chaudhuri, R., Chatterjee, S., Ghosh, A., Vrontis, D., Thrassou, A.: Sustainable innovation for shared mobility: contextual and consumer factors of an Indian car subscription business model. Int. J. Entrepreneurial Behav. Res. (2022)

David, A., Banumathi, M.: Factors influencing the purchase decision of passenger cars in Puducherry. Int. J. Exclusive Manage. Res. (IJEMR) (2014)

Dhanabalan, T., Subha, K., Shanthi, R., Sathish, A.: Factors influencing consumers' car purchasing decision in the Indian automobile industry. Int. J. Mech. Eng. Technol. **9**(10), 53–63 (2018)

Government of India. Ministry of Road Transport & Highways. Vehicle Scrappage Policy (2021). https://morth.nic.in/Circulars-Notifications-related-to-Vehicle-Scrapping-Policy. Last accessed 29 July 2024

Grover, D.P., Mandan, R.K.: Analyzing the role of social media in consumer decision-making for purchase of auto brands in India. Int. J. Manage. **8**(1), (2017)

Kaur, M.P., Bhandari, M.A.: Sustainability Development in India: The User Prospect. GURUKUL **80** (2014)

Kaushik, R., Soni, S., Swami, A., Arora, C., Kumari, N., Prajapati, R.: Sustainability of electric vehicles in India. In: 2022 International Conference on Inventive Computation Technologies (ICICT), pp. 664–667. IEEE (2022)

Kumar, G.R.: Purchase decision of Indian consumers: the factors of attraction while purchasing a car. Stud. Bus. Econ. **9**(3), 29–42 (2014)

Kumar, P., Manish, Kaushal, V.: Is India's used car market larger than previously thought? J. Technol. **11**(9), 87–91 (2023)

Maji, S., Bandyopadhyay, G.: The effect of risk perception on pre-owned car purchase decision: a logistic regression approach. J. Manag. **5**(4), 305–320 (2018)

McKinsey & Company: The future of domestic tourism in India post-COVID-19 (2020)

Ong, A.K.S., et al.: Antecedents of behavioral intentions for purchasing hybrid cars using sustainability theory of planned behavior integrated with UTAUT2. Sustainability **15**(9), 7657 (2023)

Pawar, G.B.: An analysis of new normal consumer behaviour for buying cars in Indian automotive industries in Mumbai. J. Positive Sch. Psychol. **6**(3), 6258–6270 (2022)

Rajendran Kanagavel, K., Jayakrishnan, J.: Consumer perceived risk in car purchase. ICTACT J. Manage. Stud. **4**(2), 736–741 (2019)

Saraf, N., Shastri, Y.: Development of system dynamic model for sustainability driven technology adoption in indian transport sector. In: 31st European Symposium on Computer Aided Process Engineering, pp. 939–945. Elsevier (2021). https://doi.org/10.1016/B978-0-323-88506-5.50145-5

Shende, V.: Analysis of research in consumer behavior of automobile passenger car customer. Int. J. Sci. Res. Publ. **4**(2), 1–8 (2014)

Sher, F., Curnick, O., Azizan, M.T.: Sustainable conversion of renewable energy sources. Sustainability (Switzerland) **13**(5), 1–4 (2021)

Sunitha, T., Justus, T.F.S., Ramesh, M.: Determinants of perceived risk in purchase of car. Pac. Bus. Rev. Int. **5**(2), 35–43 (2012)

Role of Corporate Social Responsibility of Banks in Emerging Markets in India

T. Srinivas[1](✉), K. Umadevi[2], V. Anitha[1], and L. Muneendra Babu[3]

[1] Priyadarshini College of Engineering and Technology, SPSR Nellore, Kanuparthipadu, Andhra Pradesh, India
sri.du.1980@gmail.com

[2] Visvesvaraya College of Engineering and Technology, Ibrahimpatnam, Telangana, India

[3] Geetanjali College of Engineering and Technology, SPSR Nellore, Nellore, Andhra Pradesh, India

Abstract. Banks have significant economic and social impact, as they play a crucial role in financing businesses, individuals, and governments. Therefore, banks are expected to be accountable for the social and environmental impacts of their lending and investment activities. This study aims to measure the effect CSRE of on Financial Performance the of Indian private banks. The study investigates how —Corporate Social Responsibility Expenditure (CSRE)|| and —Financial Performance|| (FP) are related by analyzing the Return on Assets (ROA) and Return on Equity (ROE) of five private banks over a five-year period from 2016 to 2021.The findings revealed that CSRE has a substantial beneficial effect on both ROA and ROE, which suggests that private banks in India that participate in CSR initiatives typically have stronger FP. Moreover, the study found a significant difference in the CSR expenditure, ROA, ROE, and size of selected Indian private banks. These results have significant implications for private banks in India and other emerging markets, as they suggest that CSR can be an effective strategy for improving financial performance while contributing to the societal well-being.

Keywords: Banking Sector · Corporate Social Responsibility · Expenditure · Financial Performance · Private Banks

1 Introduction

The changing environmental situation has made people recognize the relevance of CSR. As a result, they impose pressure on businesses while dealing with natural resources, conducting corporate operations, and implementing CSR policies (Stekelorum, 2020). Customers expect ethical and social behavior from companies as they become more conscious of social responsibilities and seek environmentally friendly products (Kamal, 2021). Consumers awareness of environmental issues has increased in recent years, and businesses must address this because consumers can damage a brand. To regulate the social responsibility of companies in India, GOI made CSR mandatory for businesses by amending the Companies Act. After the implementation of Amendments in the Companies Act of 2013, there has been a rise in business participation in CRS projects and CSR

B. K. Mishra et al. (Eds.): ICT4GS 2024, ISEM 56, pp. 73–80, 2025.
https://doi.org/10.1007/978-3-031-94997-5_8

governance. Establishing a secure and reliable banking system is crucial for advancing the economy, as it offers a range of benefits to society (Beck et al., 2010). Typically, banks act as financial intermediaries that promote economic growth by facilitating the movement of funds between borrowers and lenders (Aramburu & Pescador, 2017). Financial institutions also promote technological advancement by providing entrepreneurs opportunities to create and develop new products and manufacturing methods (Schumpeter, 1912). Therefore, a secure banking system is necessary to keep the economy growing steadily.

The banking sector is under enormous pressure from investors, shareholders, non-governmental organizations, the media, and its customers worldwide to do business responsibly and ethically. Various initiatives have been implemented in recent years to boost CSR activities worldwide. Indian banks have focused their CSR efforts on education, balanced growth, environmental marketing, and customer happiness (Agrawal & Sinha, 2020; Paluri & Mishra, 2018). They have taken a holistic strategy, combining CSR and customer satisfaction (McDonald & Thiele et al., 2008; Narwal, 2007). They are making efforts in the field of CSR; however, there is still a need to prioritise CSR.

1.1 Literature Review

Researchers have identified an effect of CSR on the overall FP of banks (Abilasha & Tyagi, 2019). A business's commitment to enhancing the well-being of individuals by partnering with the various sections of society to promote long-term economic growth is known as corporate social responsibility (WBCSD, 2000). It is because a business's goal is to generate profits and benefit society (Dobson, 1999; Przychodzen & Przychodzen, 2013; Smith & Rönnegard, 2016). The primary purpose of CSR is to keep business operations running efficiently while generating mutual benefits for both the company and society (Latapí Agudelo et al., 2019; Porter & Kramer, 2006). It is the approach that organisations adopt to ensure stakeholders' satisfaction (Kamatra & Kartikaningdyah, 2015; McDonald & Thiele, 2008). If CSR is appropriately done, the company's performance will improve. Stakeholders believe that companies involved in CSR demonstrate more significant concern for social and environmental challenges. They believe the company will operate without violating laws (Du et al., 2010; Lim & Greenwood, 2017). Employing CSR as a corporate strategy facilitates sustainable development (Erinle, 2019).

1.2 Financial Performance of Banks

Financial Performance (FP) refers to how well a bank does financially over a particular period and how effectively it generates resources and profits. Financial statements and annual reports are crucial to estimate a bank's FP. ROA and ROE are vital indicators for assessing a bank's financial health (Ashraf et al., 2017). In summary, FP evaluates a bank's financial position and ability to generate profits through various indicators and financial statements.

ROA is a metric to assess a company's profitability by measuring its effectiveness in using its assets (Love & Klapper, 2002). The FP is better when the ROA is higher. Higher ROA reflects a higher rate of return on investment (Riyanto, 2001). Investors prefer a company that has a high ROA. Any organization with a higher ROA is likelier to produce high corporate profits than companies with a low ROA (Ang, 2001).

The ROE measures how effectively banks manage their net assets and assesses the profitability of company shareholders' investments (Heikal et al., 2014). Ang (2001) states that a higher ROE is associated with higher profitability. Khan and Tariq (2017) found that ROA and ROE correlate positively. Nguyen et al. (2021) discovered that FP (ROA and ROE) is significantly related to corporate social responsibilities. Other studies have found similar results (Cho et al., 2019; Kaur & Singh, 2021; Mujahid & Akhtar, 2014; Nguyen, 2018; Abilasha & Tyagi, 2019; Waddock & Graves, 1997). Several studies have taken ROA and ROE as a proxy to measure FP (Kang et al., 2010; Kaur & Singh, 2021; Mishra & Saur, 2010; Mujahid & Akhtar, 2014; Nuryaman, 2013; Waddock & Graves, 1997). Based on the benefits and broad applicability, accounting-based performance metrics, such as ROA and ROE, have been employed in the study to gauge FP.

Another variable, such as —firm size||, has been chosen as control variable based on previous research (Hassouna & Salem, 2021; Hirsch et al., 2022; Kapoor & Sandhu, 2010; Lee & Hu, 2018; Manzoor et al., 2019). The studies suggest that a firm's FP variances may be due to size. Therefore, based on the analysis of Arora & Sharma (2022), Hasudungan & Bhinekawati (2022), Kaur and Singh (2021), and Prado-Lorenzo et al. (2008), total assets have been taken as a control variable.

2 CSP and FP

From the References [1–3], the connection between FP and CSR has been the subject of a great number of research (Abilasha & Tyagi, 2019). For instance, Kapoor and Sandhu (2010) conducted research on 93 companies finding that CSR had a positive influence on profitability. It revealed that prioritizing CSR could boost a company's ability to compete by raising its profitability. Similarly, Resmi et al. (2018) found that firms prioritizing CSR perform better financially, with higher equity and net income returns. A favorable correlation between CSR and FP has been shown in a number of researches, including those carried out by Abilasha and Tyagi (2019), Cho et al. (2019), Kaur and Singh (2021), Laskar and Maji (2017), Mujahid and Akhtar (2014), Nguyen (2018), and Nuryaman (2013), among others.

In contrast, Friedman (1970) discovered that being socially responsible has no economic benefits for the corporation but somewhat diminishes company stock returns. Studies by Vance (1975), Bani-Khaled (2021), Lee and Yang (2021), and Nguyen (2018) discovered that CSR has a negative impact on FP. Several other researchers further extend the contradiction in the result (e.g., Kesto, 2017; Oyewumi et al., 2018).

Although numerous researchers have reported a positive or negative association between CSR and FP, the outcomes are not yet definitive (Story & Neves, 2015). This lack of clarity implies that further investigation is needed.

3 Research Methodology

This study is based on data from five private sector banks, IndusInd, Axis, Kotak Mahindra, ICICI, and HDFC, for the financial years between 2016–2017 and 2020–2021. The data was collected from the bank's annual reports and includes information on four variables: —TA, ROA, ROE, and CSRE‖. Bank Size is a controlled variable represented by the total assets' natural logarithm. The study uses CSR expenditure as a proxy for CSR and ROA and ROE as a proxy for the banks' FP. The study uses one- way ANOVA to test the hypothesis on the significant difference in bank size, FP, and CSR expenditure among the selected banks. Post hoc analysis is performed to identify the differences in CSRE among the chosen banks. In order to evaluate the influence that CSRE has on FP, regression analysis is performed. The data for the FY 2022 was not available at the time of completing the study. Results and Interpretation

Table 1 illustrates the descriptive analysis, indicating that the mean (M) of ROA is 1.19, with a standard deviation (SD) of 0.55. The mean value for ROE is 10.29, with an SD of 4.55. Bank size has a M of 13.27 with an SD of 0.72, and CSR expenditure has a M of 4.76 with an SD of 0.97. Additionally, the data's skewness and kurtosis values indicate normal distribution, as per Klein's (1988) criteria.

Table 1. Descriptive Statistics

	Mean	SD	Skew	Kurtosis
CSRE	4.7563	.96920	-.279	-.299
ROA	1.1876	.55544	-.706	-.886
ROE	10.2960	4.54910	-.578	-.524
TA	13.2714	.71592	-.280	-1.393

4 Hypothesis Testing

Bank Size

H01: There is no significant difference in the bank size of selected banks.

In this case, the assumption of homogeneity of variance is not violated ($p = 0.790$). Further, the result significantly differs when ANOVA is applied to the bank size (TA). The analysis indicates that F-value is 46.49, with the significance value at a 99% level. This shows a significant difference between the selected banks' sizes (total assets). Thus, the research rejects the null hypotheses (Table 2).

.

H02: There is no significant difference in the CSRE of the selected banks.

When analysing the data on CSR spending using ANOVA, significant differences were observed among banks. However, —Levene's test of homogeneity of variances‖

Table 2. One way ANOVA

		Sum of Squares	Degree of freedom	Mean square	F-test	Sig
TA	B/w Groups	2.095	4	.524	46.496	.000
	Within Groups	.225	20	.011		
	Total	2.320	24			

Table 3. ANOVA

	Sum of Squares	Degrees of freedom	Mean Square	F-test	Sig
B/w the groups	3.105	4	.776	13.540	.000
Within Groups	1.147	20	.057		
Total	4.252	24			

indicated a violation of its assumption ($p < 0.001$). To prevent potential errors, the Welch test was used as a robust measure of equality of means (Table 3).

The results of the Welch test revealed a significant difference in the CSRE of the selected banks ($p < 0.001$). A —post hoc test|| is performed to locate the differences in the CSR Expenditure of the select banks. As the equality of variance is not assumed, the multiple comparisons of the CSR Expenditure of the chosen banks have been tested on Games-Howell Method (Table 4). A significant difference was observed between groups, confirmed by a one-way ANOVA test. According to the results of a Games Howell post hoc test, the CSR Expenditure of HDFC Bank is statistically and substantially different than those of IndusInd Bank ($p = .011$), Axis Bank ($p = 0.001$), Kotak Mahindra Bank ($p = 0.14$), and ICICI Bank ($p = 0.003$). According to the results of a Games Howell post hoc test, the CSR Expenditure of HDFC Bank is statistically and substantially different than those of IndusInd Bank ($p = .011$), Axis Bank ($p = 0.001$), Kotak Mahindra Bank ($p = 0.14$), and ICICI Bank ($p = 0.003$).

4.1 FP

H03: There is no significant difference in the FP of the selected banks.

The homogeneity of variances tests shows that the sig. Value is less than 0.05 for ROA and ROE, which indicates a violation of the assumption of homogeneity of variance. In addition, the Welch test (Table 6) shows a statistically significant difference in the ROA and ROE of the selected banks (Table 5).

4.2 CSRE and FP

H04: There is no significant effect of CSRE on FP (Table 7).

Table 4. Multiple Comparisons among CSRE of Banks

(I) Bank	(J) Bank	Mean Difference (I-J)	Std. Error	Sig
1	2	−.36082	.14983	.268
	3	.07855	.22306	.996
	4	−.43714	.15907	.169
	5	−.90702*	.15741	.011
2	1	.36082	.14983	.268
	3	.43937	.17002	.228
	4	−.07632	.06673	.781
	5	−.54620*	.06268	.001
3	1	−.07855	.22306	.996
	2	−.43937	.17002	.228
	4	−.51569	.17821	.151
	5	−.98557*	.17674	.014
4	1	.43714	.15907	.169
	2	.07632	.06673	.781
	3	.51569	.17821	.151
	5	−.46988*	.08236	.003
5	1	.90702*	.15741	.011
	2	.54620*	.06268	.001
	3	.98557*	.17674	.014
	4	.46988*	.08236	.003

Table 5. ANOVA

		SS	df	MS	F	Sig
ROA	Between Groups	5.860	4	1.465	18.978	.000
	Within Groups	1.544	20	.077		
	Total	7.404	24			
ROE	Between Groups	367.654	4	91.913	14.249	.000
	Within Groups	129.009	20	6.450		
	Total	496.663	24			

Table 6. Robust Tests of Equality of Means

	Statistic	df1	df2	Sig
ROA	24.297	4	8.800	.000
ROE	21.852	4	9.241	.000

Table 7. Impact of CSRE on FP.

Model			Varia ble	Coeff	S.Er	t- Stat	Prob
1 (DV = RO A)	**R**	0.550	**C**	9.927	2.887	3.439	.002
	R Sq	0.303	**CSRE**	0.562	.485	−3.250	.012
	Adj. R Sq	.240	**TA**	−0.860	0.278	−1.108	.005
	F	4.780					
	Sig	0.019					
	R	.492	**C**	71.374	24.674	2.896	.008
2 (DV = RO E)	**R Sq**	0.243	**CSRE**	4.415	.020	2.548	.019
	Adj. R Sq	0.174	**TA**	−6.185	2.376	−2.603	.016
	F	3.522					
	Sig	0.047					

Firstly, CSRE is used as the independent variable and ROA and ROE as the dependent variable in different models (Table 7). The bank size measured by total assets is considered a control variable. The linear correlation coefficient (R) is 0.550 in model 1 and 0.492 in model 2, showing a moderate correlation between the variables. The R^2 value indicates that the CSRE can explain the total variation in ROA and ROE. Model 1 explains about 30.3% ($R^2 = 0.303$) of the variation by CSRE. Model 2 explains 24.3% ($R^2 = 0.243$) of the variation by CSRE. The Sig. value in both models is indicates that the models are significant ($p < 0.001$). The constant term in both models is statistically significant and positive. The model 1 result also reveals that the coefficient of CSRE is statistically significant and positive ($\beta_1 = 0.562$) at a 5% significance level ($p = 0.012$). Similarly, in model 2, the constant coefficient of CSRE is statistically significant and positive ($\beta_1 = 71.374$) at a 5% significance level ($p=0.008$). The control variable of bank size shows a negative beta coefficient in both models. The control variable does have a significant effect in model 1 ($p = 0.05$) and model 2 ($p = 0.016$). Thus, the null hypothesis is rejected.

5 Discussion and Conclusion

This study examined how selected private banks' spending on CSR was linked to their ROA and ROE. The researchers created a model to analyse the influence of CSRE on FP, and the results supported the hypothesis (H04), indicating that CSR spending had a substantial impact on FP. Regression models showed a significant and positive correlation between CSR spending and ROA and ROE, aligning with the findings of prior research studies (Bastič et al., 2020; Cho et al., 2019; Kapoor & Sandhu, 2010; Nollet et al., 2016; Reverte et al., 2016). Also, the study examined the differences in the CSRE level, financial performance, and bank size. H1, H02, and H03 indicated a significant difference in the bank size, CSR, and financial performance among the selected banks. Post hoc analysis revealed a statistically significant difference in the

CSR expenditure of the selected banks. The HDFC bank is doing CSR Expenditure more than the other banks.

The results suggest that CSR can offer a competitive edge to banks by enhancing their profitability, and investment decisions should reflect this. The banks should engage in CSR activities that can positively affect FP. Banks that adopt CSR practices have experienced increased profits, improved goodwill, and stronger customer loyalty. Therefore, managers should focus on incorporating profitable CSR practices into their model to ensure the long-term sustainability of their bank. In addition to the legal obligation to engage in CSR activities and disclosure, banks have a moral and ethical obligation to execute logical CSR operations and report them reasonably. CSR objectives should be explicitly outlined and integrated into banking operations.

The current study could be expanded by introducing additional variables, such as net profit margin, and controlling for other variables, such as bank age, employee count, and debt-equity ratio, to gain a more thorough understanding of the effect of CSR on FP. To achieve a better understanding, the study could also be extended to include more banks over a longer period. A comparative analysis could be performed to examine the impact of CSR on FP across various industries.

As a recommendation to banks, there is significant scope for enhancing CSR practices, particularly in reporting, management system implementation, adherence to codes, environmental preservation, and the creation of sustainable financial products. To support CSR initiatives, banks should invest in projects run by organisations that prioritise CSR. The government should facilitate public-private partnerships to encourage private-sector banks to adopt CSR practices. Additionally, government policy should encourage banks to report on their sustainability efforts and outcomes and engage in international agreements.

References

1. Abilasha, N., Tyagi, P.M.: Impact of CSR on financial performance of top 10 performing CSR companies in India. IOSR J. Econ. Financ. **10**(2), 49–55 (2019). https://doi.org/10.9790/5933-1002024955
2. Ang, R.: BukuPintar Pasar Modal Indonesia. Jakarta:Mediasoft Indonesia (2001)
3. Arora, A., Sharma, D.: Do environmental, social and governance (ESG) Performance Scores Reduce the Cost of Debt? Evidence from Indian firms. Australasian Bus. Account. Financ. J. **16**(5), 4–18 (2022). https://doi.org/10.14453/aabfj.v16i5.02

Gig Economy Potential for Generating Employment in India

Sanjay Kumar Mangla(✉), Arpit Pandey, and Sakshi Rai

Symbiosis Institute of Management Studies, Symbiosis International (Deemed University), Pune, India
sam211285@gmail.com

Abstract. This paper outlines the projected growth of the gig economy in India and its potential to create lakhs of jobs, highlighting its effect across sectors like retail, transportation, and education. By mapping the current trajectory and future prospects, the paper suggests constructive policy interventions to harness this enormous dividend potential in India. The text scrutinizes potential roadblocks such as lack of job security, irregularity of incomes, and restricted access to benefits and proposes an action plan to overcome these hurdles for a resilient job market in India. This paper investigates the gig economy's potential for unemployment reduction and emphasizes how it can offer abundant job opportunities, eventually creating a better job market in India.

Keywords: Gig Economy · Policy · Platform Economy · E-Commerce · and India

1 Introduction

India, the largest populated country in the world with over 1428.62 million people, has the world's largest and youngest force. On the one side, the young force has the potential to generate a demographic dividend for the county. However, on the other side, it poses a significant challenge in creating employment opportunities. According to RBI (2019–20), unskilled labor constitutes 37.7%, whereas semi-skilled and skilled are 62.3%. Agriculture is the largest employment provider, employing 147.92 million people. However, this sector has a limitation in creating jobs, which is the availability of land for agricultural practices. The second most employment is in the service sector, constituting around 30.7% of the total employment. Employment for many people in India can be generated in a sector capable of absorbing all types of people, skilled, semi-skilled, and unskilled, which can happen in the Gig economy. However, capitalizing on the potential of the gig economic sector needs specific policy interventions to secure the interest of employees working in this sector along with the sector's growth.

In simple words, an economy in which businesses and organizations directly contract workers (independent) for short-term projects/assignments that are temporary is called a Gig economy. It can also be said that in this type of economy, workers sell their skills to companies not limited to only one; based on the skills and tasks required by the

B. K. Mishra et al. (Eds.): ICT4GS 2024, ISEM 56, pp. 81–89, 2025.
https://doi.org/10.1007/978-3-031-94997-5_9

company, people get hired on temporary contracts or fixed contract options. The gig economy can be broadly classified into two, based on the involvement of technology: the Digital economy/ Platform economy and the Physical economy. The digital economy involves online workers who work on demand via apps such as Uber, Ola, Swiggy, and Zomato.

As per Niti Aayog's India's Booming Gig and Platform Economy, India employed 6.8 million gig workers in 2020–2021, contributing to 1.3% of the total workforce in India. They also estimated that this workforce will be the size of 23.5 million workers by 2029–30, which will be 4.1% of the total workforce force and 6.7% of the non-agricultural workforce. Not only is this gig economy now being called the 4th industrial revolution, but it is also reshaping the Indian job market, leveraging technological advancement.

The freelancers were contributing nearly a third of the total workforce worldwide. The top 5 largest countries for gig work are the US, China, Brazil, Japan, and India. India is the 5th largest economy, with jobs from low-skill to highly skilled. The increase in usage of online delivery platforms and e-commerce has made India one of the largest gig employment markets, according to the Economic Survey 2020–21. (A Review Production Savings, n.d.)

Major reforms taking place in India and increasing FDI into India have made business boom, resulting in job creation and a startup culture/entrepreneurship mindset being developed in people, due to which every person wants to be their own boss, which has increased flexi work participation in India. The gig economy is becoming the most looked-into market for people with an entrepreneurship mindset, and to cater to their work, the gig workforce needs much development.

Not only this, but the gig economy also has the potential to close employment gaps in India, particularly for marginalized groups and young people looking for other ways to generate income. The gig economy simplifies access to work opportunities by lowering the entry barrier to the job market by adopting digital platforms and technology. The government's efforts to promote innovation, entrepreneurship, and job creation follow the country's economic growth goals. To address the evolving needs of gig workers, policymakers and businesses must understand its implications.

Presents challenges such as job security concerns among gig workers, exploitation of workers by businesses and organizations, and lack of strict regulatory frameworks for these work environments and workforce are hampering the potential of the gig economy. A close examination of the legal and ethical framework is required After the COVID-19 pandemic, as there has been a paradoxical shift in the workforce.

In conclusion, the gig economy in India is a dynamic and evolving sector with the huge potential to reshape traditional employment and drive economic and sustainable growth in India. By exploring all the details of the gig economy, its impact on the job market, and strategies to tackle the challenges developed, this paper aims to offer an understanding of the gig economy's role in generating job opportunities in India.

2 Literature Review

The gig economy is emerging as the new force for generating employment in India. For countries with a lower or middle income, policymakers and institutions have viewed this kind of development very positively, as it means that the country is economically developing (Wood et al., 2019). As the gig economy and technology for digit labor are increasing, this is seen as a new way of employment generation and connecting needy people to their destination, which highlights the importance of the effect that it is putting on these types of workers and their livelihood (Graham et al., 2017).

The gig economy will disrupt traditional industries and businesses, due to which the role of these businesses in the gig economy is expected to play a very crucial role in the country, as this will displace the well-established players in the market (Burtch et al., 2016).

Gig economic importance came into more highlights after the COVID-19 pandemic in India. The online retail sector is viewed as having significant turnover in the growth of the gig economy as the significant and growth is happening in gig (Joo and Shawl, 2021). When the COVID-19 pandemic was there in India, Odisha government increased employment generation through the Mahatma Gandhi National Rural Employment Guarantee Act (MGNREGA) by providing employment to many people during their time of need.

Additionally, India's micro, small, and medium enterprises (MSME) sector contribute to the GDP significantly, creating employment, exports, and sustainable development, acting as a solid backbone of the Indian economy (Shelly et al., 2020). The MSMEs play a vital role in industrial sector growth within India (Sinha and Nayak, 2025). Various case studies have highlighted the popularity of the gig economy in India and its impacts on the country (Pal, 2021). The studies have highlighted a need for support in this sector, as the growth of the MSMEs to create employment and economic development in India is linked together (Sarmah et al., 2021).

Reports suggest that the gig economy could potentially cater to up to 90 million jobs in India particularly in non-agricultural sectors. It represents a huge portion of the workforce and could contribute considerably to GDP of the country. Sectors such as transportation, delivery services and home services, the gig economy is very relevant as there is a high demand for temporary and flexible workers in these sectors.

The widespread technical advancements, especially the usage of smartphones and high-speed internet, are the significant factors for the gig economy. These advancements have enabled the generation of online platforms connecting gig workers to their potential employers in no time. Also, the day-by-day increase in preferences for flexible working schedules has escalated the growth of the gig economy.

The gig economy is expected to directly impact employment by employers and employees, extending from digital to location-based platforms, with implications for various markets (Perspectives and Recommendations on the Future of Work India's Booming Gig and Platform Economy POLICY BRIEF, 2022).

The gig economy has led to the rise of businesses like Ola, Uber, Zomato, and Swiggy, proliferating across numerous Indian towns (Gupta and Gupta, 2023). The global expansion of the gig economy, where employees take short-term, project-based jobs, is transforming work dynamics (Bihary et al., 2019).

The regulatory framework of India is becoming a hurdle; the complete operation of the gig economy is in a legal grey area, where many gig workers are classified as independent contractors instead of employees. This type of classification limits the rights and protection available to gig workers when compared to a regular employee in a business or organization. Therefore, there is an urgent need for government regulations and policies to ensure fair treatment of gig workers and prevent them from being exploitation.

3 Research Objective

The objective of this paper is to assess the role and potential of gig economy in generating employment in India along with identifying challenges in this sector and gig economy workers.

4 Paper Flow with Key Insights

See Fig. 1.

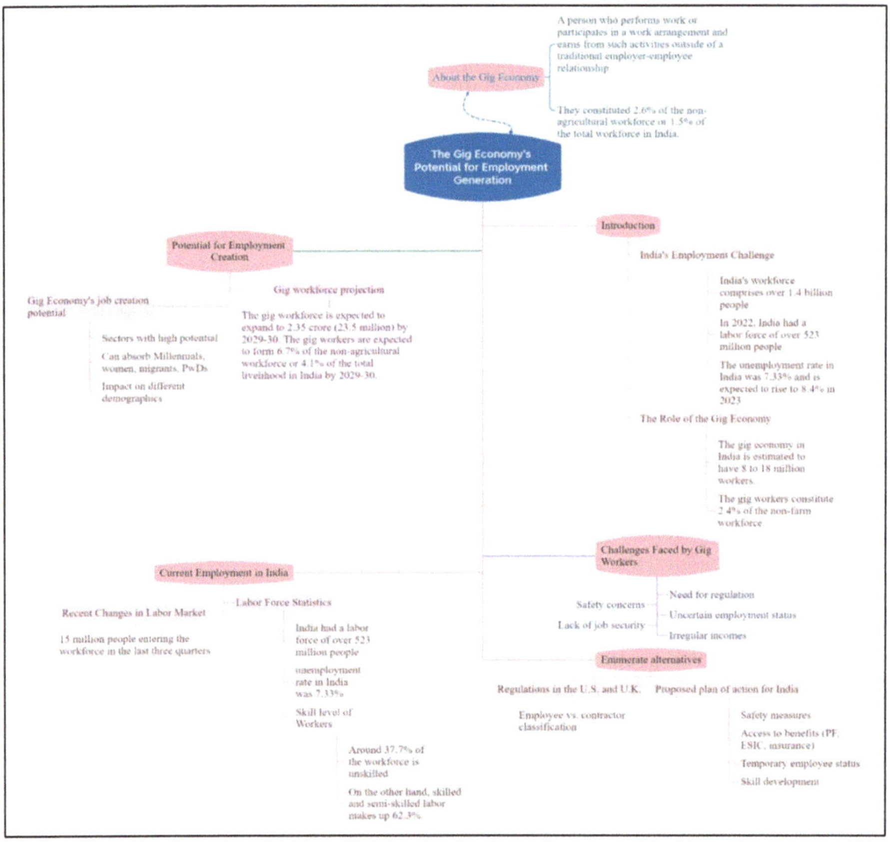

Fig. 1. Gig economy and employment generation in India

5 Present State of Employment in India

The total labour force in India was reported at 523.83 million in 2022. The division of the labour force was into agricultural, employing 147.92 million, non-financial services 140.6, 70 million people were employed in real estate and construction, and 52.76 million people were in manufacturing, public administrative services, financial services, and mining for the fiscal year 2023. The unemployment rate in India was 7.33% and, according to CMIE, will go up to 8.4% in the year 2023 (August). The demography has a young population with a median age of 28.4 years. In India, approximately 26% of the population is below 14 years and approximately 67% of the population is between the ages of 15–64 years, only 7% population is above the age of 64 years when compared to other nations, the US has around 17% of people in the age group above 65; likewise, Europe has over 21%.

Of India's total workforce of 524 million, most workers are from the "unorganized sector," forming a chunk of 85%. As this sector is unorganized, there are no formal work contracts, no social security, or any health benefits that an organized sector organization typically provides to their employees. In the last three quarters, the Indian labour market saw an increase of 15 million people, which is an indication that the Indian job market is witnessing an improvement, as per the data shared by the Centre for Monitoring Indian Economy.

The country's total unemployment saw an increase of 5.6 million due to insufficient labour demand in India, increasing the total unemployment in India from 30.5 million to 36.1 million people in the last quarter, September 2022. However, there was a significant fall in unemployment afterward; the number from 36.1 million fell to 34.2 million in the first quarter of 2023, i.e., in March 2023, it was a drop of approximately 2 million, which is a considerable number but not as compared to the total unemployment number.

The big question that arises is what should be done to create opportunities for the increasing youth population and which would be the key sectors that can drive job creation in India in the coming future.

As per the World Economic Forms, the significant job creation will be in the energy and materials section, IT and digital communications, and care, personal services and well-being, Energy and Material section because most of the economies have committed themselves to sustainable development to lower the carbon emission and together combat climate change crisis. As energy is a necessity, the transition from traditional sources of energy generation to more eco-friendly means will lead to a significant increase in job growth as the domestic renewable production capacity grows. IT and digital communications will not directly contribute to the job market but will be among the most significant driver of the job market not only in India but globally. European employment growth study has provided empirical evidence to support this. "Using a standard economic growth model with employment multipliers, the authors estimate that each additional tech job leads to the generation of around five additional jobs in other sectors in Europe," says the paper. Coming to the care, personal services, and well-being – according to ILO, this section will be adding almost 300 million jobs globally by 2035, which there will be direct job creation of 96 million in the childcare section and 136 million jobs in the long-term care section, not only this there will be indirect job creation also which will be around 67 million in other non-care sectors.

Currently, in India, the cities and sectors that are generating the greatest number of employments are the IT sectors, it has emerged as the topmost employment-generating sector in India as par a survey by Avsar, it has grown rapidly and has risen 96% in the last two quarters. Other sectors that contributed the highest to the job creation in 2022 were E-commerce, FMCG, and BFSI. The top Indian cities that created the greatest number of jobs were Bangalore, Delhi-NCR, Mumbai, Pune, and Hyderabad, these cities offered wide range of jobs to people.

6 Status of Gig Economy in India

According to the Code on Social Security (COSS) 2020, a gig worker is "a person who performs works or participates in a work arrangement and earns from such activities outside of a traditional employer-employee relationship." A platform worker has a similar line definition but differs in the fundamental way of working, it is defined as "a person engaged in or undertaking platform work, while platform work is a work arrangement outside of a traditional employer-employee relationship in which organizations or individuals use an online platform to access other organizations or individuals to solve specific problems or to provide specific services or any such other activities which may be notified by the Central Government, in exchange for payment." Furthermore, the reliance on temporarily filled independent contracts and freelancers is greater in a gig economy, and a similar trend is true for India.

Around the world, a new economic revolution is being brought in by the quickly expanding gig workforce. With a nearly one billion-strong workforce, the most youthful population in the world, accelerating urbanization, and widespread adoption of smartphones and related technologies, India is emerging as a leader. The gig economy has demonstrated its resilience and promise by creating millions of jobs and building relationships within communities during the COVID-19 pandemic. India's workforce is evolving due to the growth of the gig economy. There are currently between 8 and 18 million gig economy jobs estimated in India; in the next eight to ten years, this number is expected to rise more tthan 90 million jobs in the non-agricultural sector. (Tiwari et al., 2019) (BCG and Michael and Susan Dell Foundation, 2021).

As per the latest Niti Aayog reports, there were 6.8 million people in India who were working as gig workers as of 2019–20, by both principal and subsidiary status means, they combined formed 2.4% of the non-agricultural workforce or 1.3% of India's total workforce. The report also mentions that there will be 23.5 million new jobs in the gig economy by 2029–30. With this, the total contribution of the gig workforce towards India will be 6.7% of the non-agricultural workforce. Currently, in India, most gig work requires a medium skill level, with 47% of the jobs demanding a medium skill set. Of the rest, 22% require high skills, and 31% low skills. As per the current trend in the job market, the concentration is declining continuously for medium skills, and demand for the other two is increasing.

Employment Potential in Gig Economy in India.

In a gig/platform economy, any individual who has an access to high speed internet and smartphones can use them together to become a gig economy worker and generate money in order to live a good life. Hence gig economy is more than just providing a wide range of services to consumers but also providing a healthy livelihood opportunity to every section of workers, which includes Persons with Disabilities (PwDs), women, migrants and the youth.

Sectors such as travel and delivery, retail, personal care and home care are offering innovative solutions, which are again creating a source of income for those who are willing to work. The gig economy has no restrictions upon designation, as anybody here is welcomed and can work according to their time schedules. A classic example of this would be students, store owners and homemakers, who are either pre occupied in their work or engaged in informal sector but still the flexibility has given them an opportunity to work as per their efficiency and timelines. And the fact that smartphones and internet are the basic amenities that are required, which is now available to almost everyone in this country, makes it even easier for any individual can work in this industry. These kind of low entry barriers gives the gig economy the potential to hold and utilize the Indian demography of over 400 million people in millennials (Morgan Stanley, 2017).

However, it has been seen that these platforms generate self-employment and opportunity to increase self-income. But the flexibility and choice of work/working makes it more attractive for most of the people. As we know India is the world's top two-wheeler market because of the boom in hyperlocal economy in past few years. The share rides in form of Bike taxies such as Rapido provides the last mile connectivity to the public where a four-wheeler may not be able to enter, also it is comparatively cheaper for the consumers and an easy way out to skip long traffic. A study which was done in 2020 estimated that the bike taxies in India have the power to offer over 2 million jobs for the riders (Raman, 2020).

One of the fundamental aspects of the gig economy is its "rapid expansion" across various sectors, which includes restaurants and food delivery chains. Platform like Zomato has a network of a total of 161.63 thousand active employees (Alawadhi & Abrar, 2021). Whereas on the other hand Swiggy, which has also diversified its services and has included delivery of grocery items, packages and food delivery, engaging a workforce of 150 thousand delivery partners as of 2021. Recently Zomato acquiring Blinkit is also a positive addition for its delivery partners. Along with this, the micro, small, and medium-sized platforms are also offering travel and delivery services for their customers across the country, which is providing jobs to 3.32 million people in India. (Raman et al., 2021).

In around 300 cities in India, which are diverse in sizes, the gig economy provides services and earning opportunities to the customers and workers. As these platforms cover a wide range of urban services like e-commerce, logistic fulfilment, medicines groceries and food-based delivery, they cover a wide range of Urban services, facilitating the smooth functioning of workers and customers across the markets.

After the COVID-19 pandemic the MNCs started to opt for flexible hiring strategies, particularly for "specialized project" or on contract basis for some specific projects as a means to cut down their operational expenses. This major shift is playing a vital role in cultivating the gig economy in India.

7 Challenges Faced by Gig Workers

The gig economy also poses several challenges. Concerns include the absence of job security and benefits for the workers working for these businesses in the gig economy. Unlike traditional employees, gig workers often lack access to benefits such as health insurance, paid leaves, and retirement plans—lacking these essential amenities can deter some workers from entering the gig economy.

Despite the attractiveness of the gig economy workers face a lot of challenges on daily basis. The benefits that are mentioned can only be fully utilized if there is a proper regulation in the gig economy sector.

- Internet services and digital technology access can pose certain limitations for workers who are interested in doing jobs in the gig economy.
- The biggest hurdles identified in the gig economy sector are the absence of job security, inconsistent wages, and non-characterized employment. The irregular work and income may lead to extreme stress and pressure for gig workers.
- The relationship between the platform owner and the workers is defined as something other than employment. In such cases, gig workers are referred to as "independent contractors"; they cannot get any protections or designations associated with their workplace.
- Around 83% of workers in the app-based taxi sector and 89% in the app-based delivery sector have reported concerns about safety at work. Incidences like theft, physical assault, or accidents are some frequent issues for them.

These challenges clearly point out the importance of regulation in this sector. If these issues are addressed, a huge number of individuals may move from their traditional work platforms to the gig economy. The gig economy has the potential to create and absorb a diverse workforce, including skilled, semiskilled, and unskilled laborers. With complete regulation, the workers' irrespective of their skill sets can enter into the gig economy and earn a livelihood from themselves.

8 Conclusion

This article emphasizes the gig economy's transformative potential, highlighting its ability to unlock a demographic dividend and significantly reduce unemployment in India. However, regulatory measures and policy interventions are crucial to secure the gig economy's growth and ensure worker welfare while creating millions of jobs. In conclusion, India's gig economy holds immense promise as a pivotal force in addressing the country's burgeoning employment challenges and unlocking its demographic dividend. With the potential to generate 5 million jobs in the next two years, the gig economy's influence transcends retail, transportation, and education sectors. To harness this transformative potential, it is imperative for policymakers to enact comprehensive regulations that afford gig workers essential benefits and protections, including temporary employee status, access to PF and ESIC, safety measures, and opportunities for skill development. Such measures would not only fortify the gig economy but also enable the country to tap into its abundant and youthful workforce, ultimately fostering a resilient job market and bolstering economic growth.

References

Alawadhi, N., Abrar, P.: Zomato's revenue from operations doubles to Rs 2,604 cr amid pandemic. Business Standard. April 28. 2021 (2021). https://www.business-standard.com/article/companies/failure-of-govt-stepsvaccinations-may-cause-economicdisruption-zomato-121042800994_1.html

Bihary, Z., Csóka, P., Kerényi, P., Szimayer, A.: Self-respecting worker in the precarious gig economy: A dynamic principal-agent model. (2019). https://doi.org/10.2139/ssrn.3866721

Burtch, G., Carnahan, S., Ross, S.M., Greenwood, B.N.: Working Paper Can You Gig it? An Empirical Examination of the Gig-Economy and Entrepreneurial Activity Can You Gig It? An Empirical Examination of the Gig-Economy and Entrepreneurial Activity * (2016). http://ssrn.com/abstract=2744352

Graham, M., Hjorth, I., Lehdonvirta, V.: Digital labour and development: impacts of global digital labour platforms and the gig economy on worker livelihoods. Transfer **23**(2), 135–162 (2017). https://doi.org/10.1177/1024258916687250

Gupta, R., Gupta, R.: Lost in the perilous boulevards of gig economy: making of human drones. South Asian J. f Human Resour. Manage. **10**(1), 85–106 (2023). https://doi.org/10.1177/23220937221101259

Joo, B.A., Shawl, S.: COVID-19 Pandemic and the Rising Gig Economy: An Emerging Perspective. Global Economics Science, 16–23 (2021). https://doi.org/10.37256/ges.232021917

Stanley, M.: India's Millennials to Drive Growth in Four Key Sectors. Morgan Stanley. April 2017 (2017). https://www.morganstanley.com/ideas/Indiamillennials-growth-sectors

Pal, B.: Rising Popularity in Gig Economy: A Case Study from India. Int. J. Relig. Cultur. Stud. **3**(2) (2021). https://doi.org/10.34199/ijracs.2021.09.08

Perspectives and Recommendations on the Future of Work India's Booming Gig and Platform Economy POLICY BRIEF (2022)

Raman, A., Ramachandran, S.: Safety net: Social security for Indian workers. Fortune India. November 13, 2020 (2020a). https://www.fortuneindia.com/opinion/safety-net-social-security-for-indianworkers/104836

Raman, A., Ramachandran, S., Sindhu, R.: Women in the Platform Economy: Breaking Barriers and Driving Inclusive Growth. Ola Mobility Institute & Children's Investment Fund Foundation. September 2021 (2021). https://olawebcdn.com/ola-institute/women-in-platform-economy.pdf

Sarmah, A., Saikia, B., Tripathi, D.: Can unemployment be answered by micro small and medium enterprises? evidences from Assam. Indian Growth and Development Review **14**(2), 199–222 (2021). https://doi.org/10.1108/IGDR-09-2020-0140

Shelly, R., Sharma, T., Bawa, S.S.: Role of micro, small and medium enterprises in Indian Economy. Int. J. Econo. Fin. Iss. **10**(5), 84–91 (2020). https://doi.org/10.32479/ijefi.10459

Sinha, G., Nayak, B.S.: Rise of Hindutva Politics, Demonetisation, and Its Impact on Micro, Small and Medium Enterprises in India (n.d.). www.intechopen.com

Tiwari, Ram, Roy: Three Gigs. Three Men. Three Stories. What it is like to work in a gig economy job. Times of India. May 29. 2019 (2019). https://timesofindia.indiatimes.com/india/what-it-is-like-to-work-in-agig-economy-job/articleshow/69371217.cms

Wood, A.J., Graham, M., Lehdonvirta, V., Hjorth, I.: Good gig, bad gig: autonomy and algorithmic control in the global gig economy. Work Employ Soc. **33**(1), 56–75 (2019). https://doi.org/10.1177/0950017018785616

Robust Identification and Meta Agnostic Visualization of Pest, Weed, and Disease in Tomato Plant Using Deep Convolutional Neural Network

N. Sasikaladevi(✉) and A. Santhosh Kumar

Department of CSE, School of Computing, SASTRA Deemed University, Thanjavur, TN, India
sasikalade@gmail.com

Abstract. India, a prominent global producer of tomatoes, plays a vital role in meeting the worldwide demand and supporting the livelihoods of farmers. Tomatoes are rich in essential vitamins and antioxidants, contributing to food security. Tomato production has been negatively affected by climate change, poor water management, and excessive pesticide use, resulting in reduced yields. To tackle these issues, we suggest the implementation of an advanced deep convolutional neural network with tailored machine learning techniques to identify and categorize weeds, pests, and diseases that impact tomato plants. Through the training and assessment of this model using holdout validation on an extensive dataset of tomato leaves, our goal is to achieve higher validation accuracy and enhance current methodologies in the analysis of tomato plants.

Keywords: pest · deep convolutional neural networks · tomato plant disease

1 Introduction

Tomato plants are essential in global agriculture and nutrition, providing vital nutrients like vitamins A and C, potassium, and antioxidants, and serving as a significant income source for farmers. Effective disease management is crucial to sustain tomato crops, as they thrive in various climates but require well-drained, fertile soil, sufficient sunlight, and moisture for optimal growth. Implementing targeted cultural practices and chemical treatments is necessary to protect yields and prevent economic losses. Neglecting disease control can jeopardize individual harvests and disrupt the global supply chain, underscoring the importance of proactive management. Early detection of diseases is critical for efficient disease management, making it vital to understand the origins and impacts of these diseases to ensure sustainable tomato cultivation and global nutritional security.

1.1 Bacterial Spot

Bacterial spot in tomato plants manifests as dark lesions on leaves and fruit, caused by the Xanthomonas campestris pv. Vesicatoria bacteria in warm and humid conditions. Early Blight.

B. K. Mishra et al. (Eds.): ICT4GS 2024, ISEM 56, pp. 90–103, 2025.
https://doi.org/10.1007/978-3-031-94997-5_10

Early blight, commonly observed in tomato plants, appears as dark lesions on lower leaves, caused by the fungus Alternaria solani in warm and moist conditions. It results in leaf discoloration, premature loss, and reduced fruit yield, posing significant challenges to cultivation.

1.2 Late Blight

Late blight, caused by Phytophthora infestans, causes water-soaked lesions on tomato leaves, spreading rapidly in humid and cool conditions. It damages foliage, reduces fruit yield and quality, and poses challenges to cultivation.

1.3 Leaf Mold

Leaf mold, caused by the fungus Fulvia fulva, manifests as yellowish or brownish patches on the underside of tomato leaves. Flourishing in warm and humid environments, it spreads rapidly and hampers fruit production.

1.4 Septoria Leaf Spot

Septoria leaf spot, triggered by the fungus Septoria lycopersici, causes small dark spots on tomato leaves, swiftly spreading in warm, humid conditions. This leads to leaf loss and diminished fruit output, necessitating the implementation of efficient management techniques.

2 Related Work

In recent years, a variety of methodologies have been investigated in academic literature for detecting diseases in tomato plants using machine learning, computer vision, and deep learning techniques.

Islam et al. (2024) evaluated VGG, ResNet, and DenseNet for tomato plant disease identification. Their custom deep neural network achieved over 99% accuracy on datasets with 10 classes, showing promise for efficient agricultural disease detection. Chen et al. (2024) published thier research on GAN-based Transformer model for augmenting tomato leaf images, achieving 99.45% accuracy on PlantVillage data and outperforming previous models, showcasing its robustness for crop disease control. Roy et al. (2023) used PCA DeepNet combined with F-RCNN, achieving a classification accuracy of 99.60% and an average precision of 98.55%. The dataset utilized contains a mixture of healthy and diseased tomato leaf images, and the hybrid framework incorporates PCA and GAN techniques for data enhancement.

Anim-Ayeko et al. (2023) utilized a ResNet-9 model for the classification of early and late blight diseases in potato and tomato plant leaves. The accuracy achieved was 99.25%, with overall precision of 99.67%, recall of 99.33%, and F1-score of 99.33%. Saliency maps were employed to explain the model's predictions, revealing consideration of leaf shape, diseased areas, and healthy regions. Kumar et al. (2023) utilised DenseNet201 to classify plant diseases including tomato spider mites, pepper bell healthy, tomato late blight, and others. It achieved a validation accuracy of 98.67% and a loss of 0.04. Precision, recall, and F1 score were 0.98, 0.99, and 0.98, respectively.

Tarek et al. (2022) used MobileNetV3 Small and MobileNetV3 Large for detecting tomato leaf diseases. MobileNetV3 Small attained an accuracy rate of 98.99%, whereas MobileNetV3 Large achieved a higher accuracy of 99.81% in detecting tomato leaf diseases. Kaur et al. (2022) used is a modified Mask R-CNN for autonomous segmentation and detection of tomato plant leaf diseases. It achieved an accuracy of 0.98, with F1-score of 0.912 and Mean Average Precision (mAP) of 0.88.

Sakkarvarthi et al. (2022) The models used Convolutional Neural Network (CNN), InceptionV3, ResNet152, and VGG19. The classification is done on tomato crops, and the achieved accuracy is 98% for training and 88.17% for testing. Özbılge et al. (2022) utilized a compact CNN with six layers for tomato disease identification, outperforming pre-trained deep networks on the PlantVillage dataset. It achieved high accuracy, including 99.70% for F1 score, 98.49% for Matthews correlation coefficient, 98.31% for true positive rate, and 99.81% for true negative rate on unseen test images.

Altalak et al. (2022) proposed model is a hybrid deep learning system integrating a Convolutional Neural Network (CNN), Convolutional Attention Module (CBAM), and Support Vector Machines (SVM) for early detection and classification of tomato plant leaf diseases. Abbas, 2021 The proposed model is a combination of Conditional Generative Adversarial Network (C-GAN) for synthetic image generation and DenseNet121 for classification. It achieved accuracies of 99.51%, 98.65%, and 97.11% for classifying tomato leaf images into 5, 7, and 10 disease categories, respectively (Table 1).

Table 1. Deep learning and machine learning methods for tomato disease detection.

Author	Method Used	Accuracy
(Islam, 2024)	Visual Geometry Group, Residual Net, and DenseNet	99%
(Chen, 2024)	CNN	99.45%
(Ullah, 2023)	EfficientNet and MobileNet	99.92%
(Roy, 2023)	Faster R-CNN	99.60%
(Anim-Ayeko, 2023)	ResNet-9	99.25%
(Chong, 2023)	ResNet-50	93.33%
(Kumar, 2023)	DenseNet201	98.67%
(Baser, 2023)	CNN	98.19%
(Tarek, 2022)	MobileNetV3	98.99% for MobileNetV3 Small and 99.81% for MobileNetV3 Large
(Kaur, 2022)	Mask R-CNN	98%
(Islam, 2022)	attention-based dilated CNN logistic regression (ADCLR)	testing accuracy of 100%, and validation accuracy of 96.6%

(continued)

Table 1. (*continued*)

Author	Method Used	Accuracy
(Sakkarvarthi, 2022)	CNN	88.17%
(Anandhakrishnan, 2022)	DCNN	98.4%
(Özbılge, 2022)	CNN	99.70%
(Altalak, 2022)	CNN,CBAM,and SVM	97.2%
(Tan, 2022)	ResNet34	99.7%
(Abbas, 2021)	C-GAN + DenseNet121	99.51% for 5 classes, 98.65% for 7 classes, and 97.11% for 10 classes
(Zhao, 2021)	DCNN	96.81% on the tomato leaf diseases dataset and 99.24% on the grape leaf diseases dataset
(Prabhakar, 2020)	ResNet101	94.6%
(Agarwal, 2020)	CNN	98.4%

3 Proposed Framework

The proposed deep learning model includes fine-tuned feature extraction, machine learning-based classification, as well as training, validation, and meta-visualization phases. The images of tomato leaf diseases were obtained from the PlantVillage dataset, and they were divided into 70% for training and 30% for validation purposes. The framework diagram provided below depicts the proposed approach (Fig. 1).

3.1 Feature Extraction from Tomato Plant Leaves Dataset

EfficientNet B0 demonstrates exceptional performance in accurately categorizing tomato plant diseases into ten distinct classes through its refined feature extraction process. This is achieved by employing a unique compound scaling technique that effectively balances model depth, width, and resolution. Compared to alternative architectures, this method allows for high performance outcomes with a reduced number of parameters, rendering it particularly suitable for deployment in resource-limited environments such as edge devices and mobile platforms. The capability of EfficientNet B0 to capture intricate image details enables precise discrimination between different disease types, thereby supporting early detection and intervention efforts. Its simplified structure also leads to quicker inference times, enabling real-time or near-real-time disease identification in agricultural scenarios (Fig. 2).

3.2 Machine Learning Based Classification

Support Vector Machines (SVM) handle high-dimensional data and maximize the margin between categories to efficiently classify tomato leaves into nine disease categories and

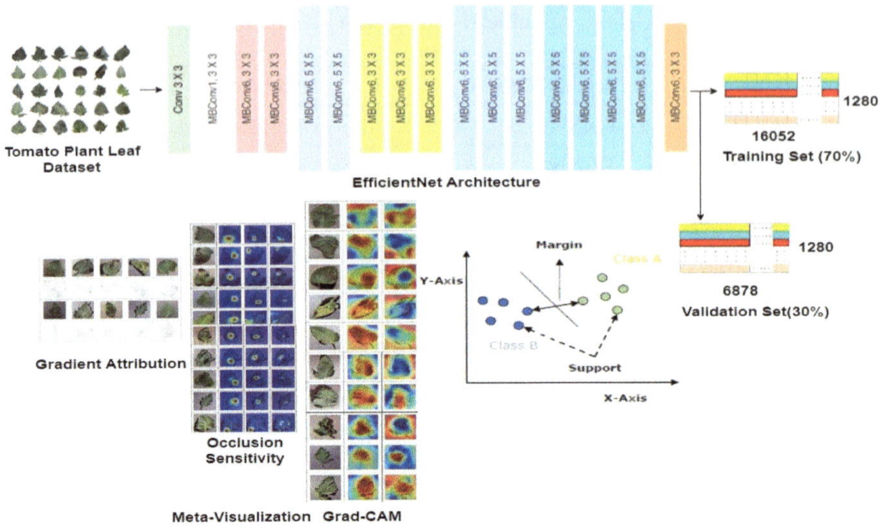

Fig. 1. Proposed framework for tomato plant disease detection.

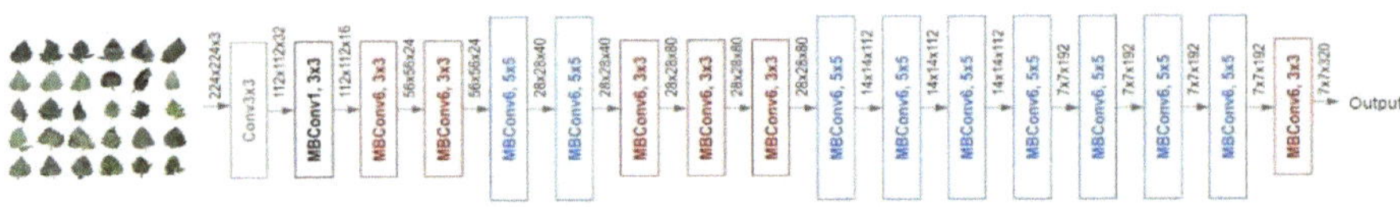

Fig. 2. Architecture of EfficientNet B0.

one healthy class. SVM is accurate in differentiating between diseased and healthy leaves, facilitating early detection and increasing crop yield. This is due to its flexibility to different features and robustness against noise.

Objective Function: The objective function of the quadratic SVM is a trade-off between maximizing the margin and minimizing the classification error. It's defined as follows:

$$\text{Minimize } \frac{1}{2}w^T w + C\sum\nolimits_{i=1}^{N} \xi_i^2 \tag{1}$$

where:

- w is the weight vector.
- C is the regularization parameter.
- ξ_i are slack variables, representing the degree of misclassification for each sample

Constraints: The constraints ensure that each sample is correctly classified within the margin:

$$y_i\left(w^T x_i + b\right) \geq 1 - \xi_i \, for \; i = 1, 2, \ldots, N \tag{2}$$

where:

- yi is the class label of sample xi.
- b is the bias term.

Quadratic Programming: The optimization problem in the quadratic SVM formulation is a quadratic programming problem, which can be solved using various optimization techniques. The goal is to find the optimal values of w and b that minimize the objective function while satisfying the constraints.

Decision Function: Once the optimization problem is solved, the decision function for predicting the class label of a new sample x is given by:

$$y(x) = sign\left(w^T x + b\right) \tag{3}$$

where the sign function assigns the class label based on the sign of $w^T x + b$

This detailed formulation allows for a flexible decision boundary, incorporating both linear and non-linear aspects through the kernel trick.

3.3 Meta-visualization

When using the Plant Village dataset for tomato disease detection, meta-visualization is essential for assessing the interpretability and efficacy of methods such as gradient attribution, Grad-CAM, and occlusion sensitivity.

Occlusion Sensitivity. Occlusion sensitivity identifies crucial regions in tomato leaf images by systematically blocking parts of the input and observing changes in prediction confidence. This technique reveals important features for model decisions and highlights areas where the model may be uncertain or less accurate. Given a neural network $f(x)$ that maps an input x to an output, and assuming the input x is an image represented by a matrix, we can denote the output of the network as $y = f(x)$.

Now, let's consider occluding a small region of the input image x with a patch P centered at position (i, j). We denote the occluded input as $x_{i,j}$. The output of the network with the occluded input will be,

$$y_{i,j} = f\left(x_{i,j}\right) \tag{4}$$

Occlusion sensitivity $s(i, j)$ at position (i, j)measures the difference in output between the original input and the occluded input:

$$s(i, j) = y - y_{i,j} \tag{5}$$

Now, let's derive $y_{i,j}$ mathematically:

1. Extract the patch P from the input x centered at position (i, j), resulting in the occluded input $x_{i,j}$.
2. Compute the output of the network with the occluded input: $y_{i,j} = f(x_{i,j})$

Finally, plug $y_{i,j}$ into the occlusion sensitivity equation:

$$s(i,j) = y - f\left(x_{i,j}\right) \tag{6}$$

This equation quantifies how much the output of the network changes when a specific region of the input is occluded. Higher values of s(i, j) indicate that occluding the region centered at position. (i, j) has a greater impact on the network's output, suggesting that this region is more important for the network's decision-making process.

Grad-CAM (Gradient-weighted Class Activation Mapping). Grad-CAM (Gradient-weighted Class Activation Mapping) creates a heatmap to show significant image regions for CNN predictions. In order to determine the significance of each feature map for the target class, it computes the gradient of the target class score y c y c with respect to the feature maps A c A k of the final convolutional layer.

The gradient of y^c with respect to A^k is computed as follows:

$$\frac{\partial y^C}{\partial A^k} \tag{7}$$

Then, these gradients are weighted by the global average pooling of the gradients across spatial locations to obtain the importance weights:

$$\alpha^k = \frac{1}{\sum_i \sum_j z} \frac{\partial y^C}{\partial A^k} \tag{8}$$

where z is the number of spatial locations in the feature map.

The final heatmap L^c is obtained by taking the element-wise multiplication of the importance weights α^k and the corresponding feature maps A^k, followed by a ReLU activation to discard negative values:

$$L^C = ReLU\left(\sum_k \alpha^k A^k\right) \tag{9}$$

This heatmap L^c highlights the regions of the input image that are most relevant for predicting the target class c. It can be overlaid on the input image to visualize the important regions.

Gradient Attribution. Gradient attribution methods aim to quantify the contribution of each input feature to the output of a model. One common method is the Gradient * Input technique, which multiplies the gradient of the output with respect to the input by the input itself.

Mathematically, for a model f with parameters θ and input features x, and assuming a scalar output, the Gradient * Input attribution for a specific input feature xi can be expressed as:

$$Atrribution(x) = \frac{\partial f}{\partial x} x_i \tag{10}$$

where:

- $\frac{\partial f}{\partial x}$ represents the gradient of the output with respect to the input feature x_i.
- xi is the value of the input feature.

This formula provides a quantitative measure of how much each input feature contributes to the model's output. By analysing these attributions, we can gain insights into which features are most influential in driving the model's predictions.

4 Results

The dataset utilized for the model implementation is the PlantVillage dataset, which consists of nine folders dedicated to different tomato diseases and one folder for healthy leaves (Table 2).

Table 2. Distribution of samples in each class of tomato plant village dataset.

S. No	class name	Total samples	Training samples	Validation samples
1	Bacterial spot	2552	2127	425
2	Early Blight	2880	2400	480
3	Late blight	2777	2314	463
4	Leaf Mold	2822	2352	470
5	Septoria leaf spot	2617	2181	436
6	Two spotted spider mites	2611	2176	435
7	Target Spot	2741	2284	457
8	Tomato Yellow Leaf Curl Virus	2941	2451	490
9	Tomato Mosaic Virus	2686	2238	448
10	Healthy	2888	2407	481

4.1 Performance Analysis of Feature Extraction Phase

In deep neural networks, pooling layers play a crucial role in improving feature extraction by reducing dimensionality and computational burden, as well as consolidating features from preceding layers.

4.2 Performance Analysis of Classification Phase

Support Vector Machines (SVM) are highly proficient in classification tasks because of their capacity to efficiently manage high-dimensional data and determine the best decision boundary with maximum margin, thereby improving generalization performance (Figs. 3 and 4).

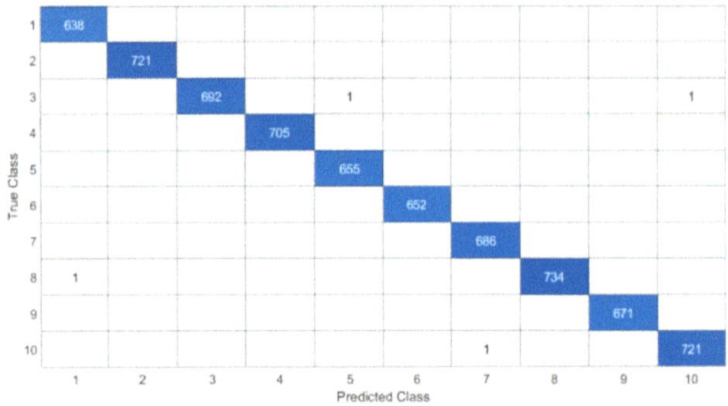

Fig. 3. Confusion matrix for Quadratic SVM.

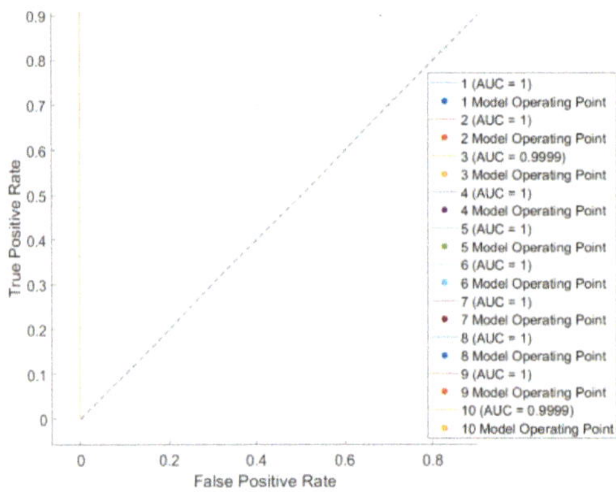

Fig. 4. ROC Curve for Quadratic SVM.

4.3 Deep Learning Metrics for Tomato Plant Disease Detection

The metrics utilized in our deep learning evaluation encompass precision, sensitivity, specificity, accuracy, and F1 score (Table 3).

$$Precision = \frac{True\,Positives}{True\,Positives + False\,Positives}$$

$$Sensitivity = \frac{True\,Positives}{True\,Positives + False\,Negatives}$$

$$Specificity = \frac{True\,Negatives}{True\,Negatives + False\,Positives}$$

$$Accuracy = \frac{True\,Positives + True\,Negatives}{Total\,Population}$$

$$F1 = 2 * \frac{Precision * Sensitivity}{Precision + Sensitivity}$$

Table 3. Deep Learning Metrics for Tomato Disease Detection.

Name	Bacterial spot	Early blight	Late blight	Leaf Mold	Septoria leaf spot	Two spotted spider mite	Target Spot	Tomato Yellow Leaf Curl Virus	Tomato Mosaic Virus	healthy	macro AVG	micro-AVG
true_ positive	637	714	690	706	651	652	681	731	671	720	685.3	685.3
false_ positive	8	2	4	0	0	3	6	0	0	2	2.5	2.5
false_ negative	1	6	4	0	3	1	4	4	0	2	2.5	2.5
true_ negative	6232	6156	6180	6172	6224	6222	6187	6143	6207	6154	6187.7	6187.7
precision	0.987596899	0.997206704	0.994236311	1	1	0.995419847	0.991266376	1	1	0.997229917	0.996295605	0.996365222
sensz zzitivi ty	0.998432602	0.991666667	0.994236311	1	0.995412844	0.998468606	0.994160584	0.994557823	1	0.997229917	0.996416535	0.996365222
specificity	0.998717949	0.999675219	0.999353169	1	1	0.999518072	0.999031164	1	1	0.999675114	0.999597069	0.999596136
accuracy	0.996365222	0.996365222	0.996365222	0.996365222	0.996365222	0.996365222	0.996365222	0.996365222	0.996365222	0.996365222	0.996365222	0.996365222
F-measure	0.992985191	0.994428969	0.994236311	1	0.997701149	0.996941896	0.99271137	0.997271487	1	0.997229917	0.996350629	0.996365222

To determine the overall metrics (precision, sensitivity, specificity, accuracy, and F1 score) for the complete dataset, we can consolidate the values from all categories. Consequently, the precision, sensitivity, specificity, accuracy, and F1 score for the entire dataset are approximately 0.996287, 0.997736, 0.999571, 0.995929, and 0.996367 respectively.

4.4 Performance Analysis of Meta-Visualization Phase

Occlusion Sensitivity. Occlusion sensitivity involves systematically occluding parts of the input image to observe how it affects the model's confidence, providing insights into which regions are crucial for the model's decision-making process (Table 4).

Grad-CAM Visualization. Grad-CAM, also known as Gradient-weighted Class Activation Mapping, employs the gradients of a specific class in relation to the convolutional feature maps of the model. This technique generates a heatmap that emphasizes the significant areas in the input image, which play a crucial role in the model's prediction for that particular class (Table 5).

Gradient Attribution. Gradient attribution quantifies the contribution of each input feature to the model's output by multiplying the gradient of the output with respect to the input by the input itself, aiding in understanding feature importance in model predictions (Table 6).

Table 4. Occlusion sensitivity for tomato plant disease detection.

S. No	Name	Original Image	Low Occlusion	Medium Occlusion	High Occlusion
1	Bacterial spot		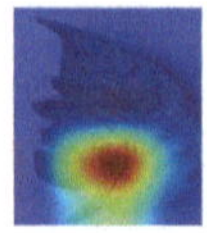	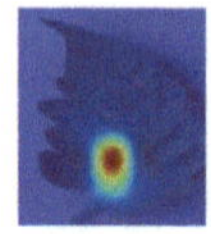	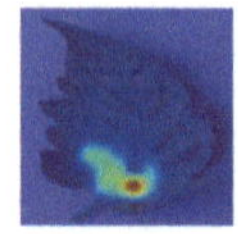
2	Early blight		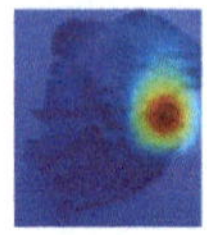	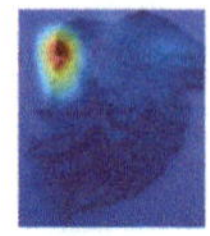	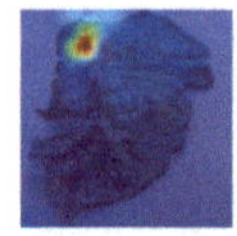
3	Late blight		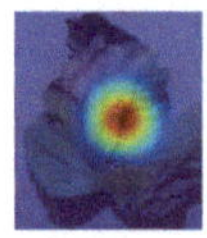	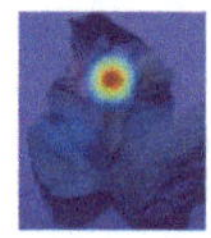	
4	Leaf Mold		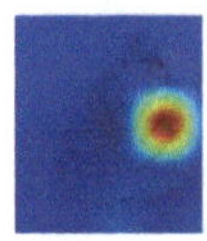	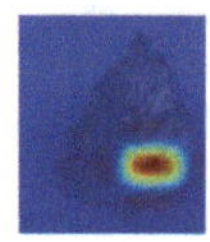	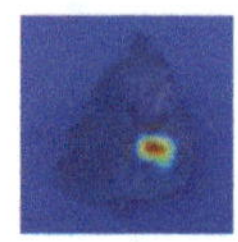
5	Septoria leaf spot		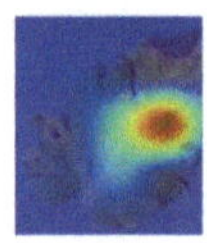	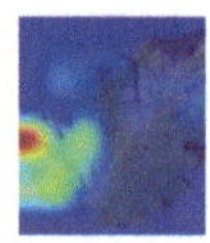	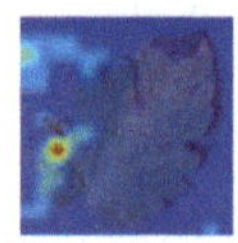
6	Spider mites Two spotted spider mite		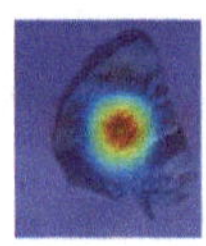	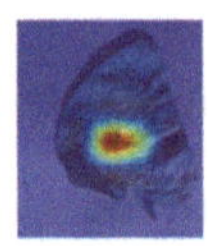	
7	Target Spot		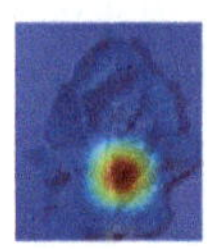	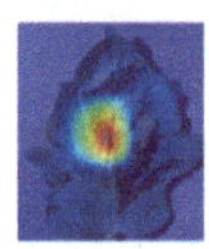	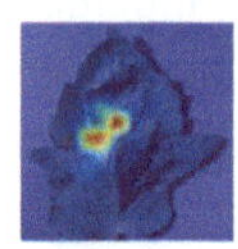
8	Tomato Yellow Leaf Curl Virus		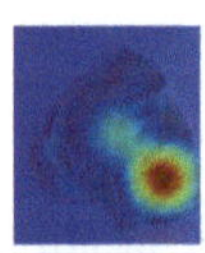	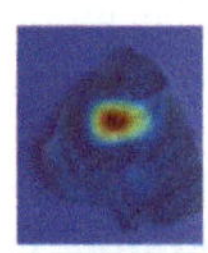	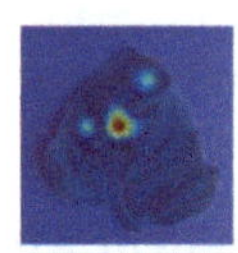
9	Tomato Mosaic Virus		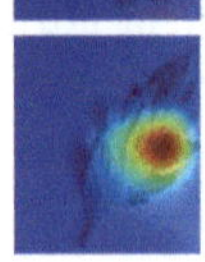	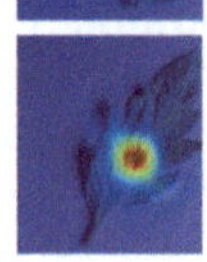	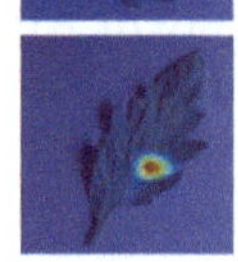
10	healthy		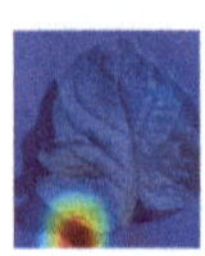	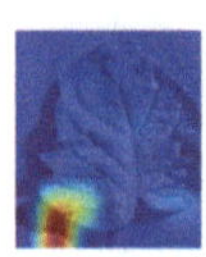	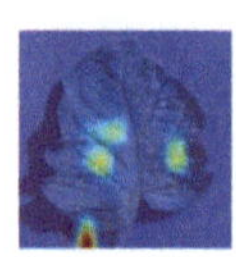

Table 5. Grad-CAM Visualization for tomato plant disease detection.

S. No	Name	Original Image	Convolu-tional 2D	Sigmoid Layer
1	Bacterial spot		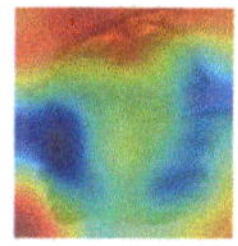	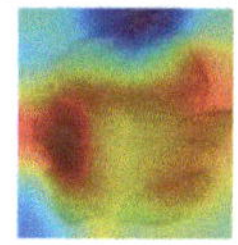
2	Early blight		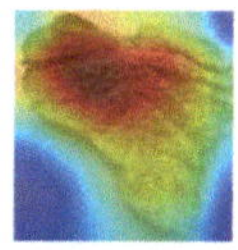	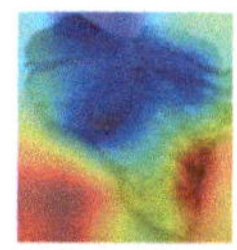
3	Late blight		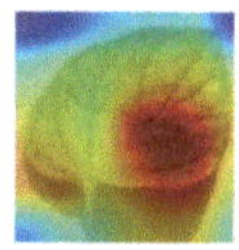	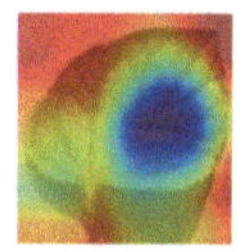
4	Leaf Mold		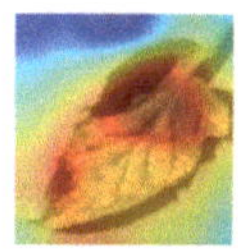	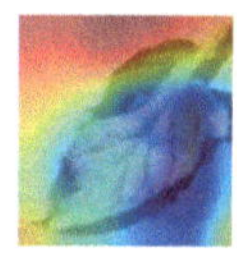
5	Septoria leaf spot		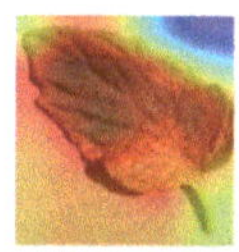	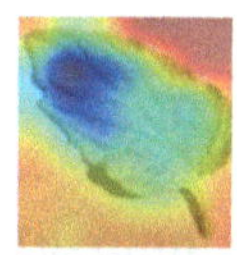
6	Spider mites Two spotted spider mite		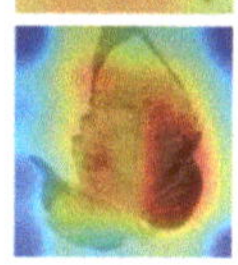	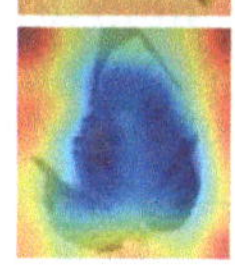
7	Target Spot		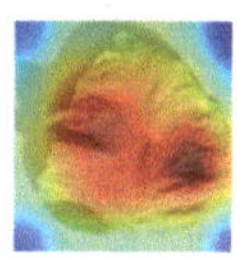	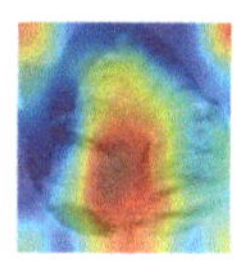
8	Tomato Yellow Leaf Curl Virus		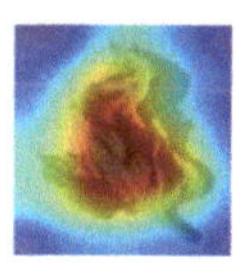	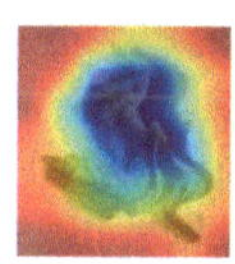
9	Tomato Mosaic Virus		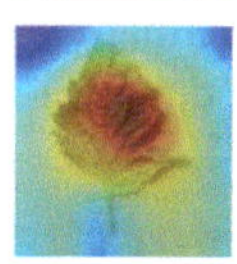	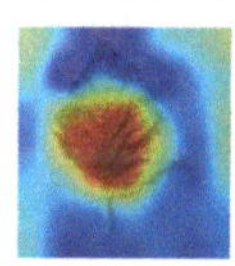
10	healthy			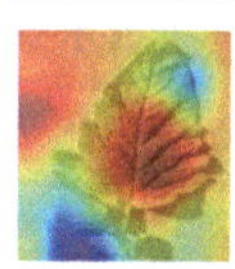

Table 6. Gradient attribution for tomato plant disease detection.

5 Conclusion

Tomato plant health and productivity can be greatly affected by diseases. Our proposed approach involves using EfficientNetB0 to extract features and SVM for classification to identify nine different types of diseases in tomato plant leaf images. These diseases include Bacterial spot, early blight, late blight, leaf mold, septoria leaf spot, two-spotted spider mite, target spot, yellow leaf curl virus, and tomato mosaic virus, as well as the absence of these diseases. Our model demonstrates strong performance based on precision, sensitivity, specificity, accuracy, and F1 score metrics. Expanding this method to cover diseases affecting other plants like corn, potato, and pepper, such as common scab, bacterial wilt, leaf curl, and bacterial spot, could improve its effectiveness. Additionally, using deep learning models on edge devices like smartphones enables farmers and stakeholders to access and utilize them efficiently. Integrating this model with a user-friendly interface that allows farmers to capture real-time images of their affected crops for immediate diagnosis and intervention holds promise for future research and development.

References

Islam, S.U., Zaib, S., Ferraioli, G., Pascazio, V., Schirinzi, G., Husnain, G.: Enhanced deep learning architecture for rapid and accurate tomato plant disease diagnosis. AgriEngineering **6**(1), 375–395 (2024)

Chen, Z., Wang, G., Lv, T., Zhang, X.: Using a hybrid convolutional neural network with a transformer model for tomato leaf disease detection. Agronomy **14**(4), 673 (2024)
Ullah, Z., Alsubaie, N., Jamjoom, M., Alajmani, S.H., Saleem, F.: EffiMob-Net: A deep learning-based hybrid model for detection and identification of tomato diseases using leaf images. Agriculture **13**(3), 737 (2023)
Roy, K., et al.: Detection of tomato leaf diseases for agro-based industries using novel PCA Deep-Net. IEEE Access **11**, 14983–15001 (2023)
Anim-Ayeko, A.O., Schillaci, C., Lipani, A.: Automatic blight disease detection in potato (Solanum tuberosum L.) and tomato (Solanum lycopersicum, L. 1753) plants using deep learning. Smart Agricultural Technology **4**, 100178 (2023)
Chong, H.M., Yap, X.Y., Chia, K.S.: Effects of different pretrained deep learning algorithms as feature extractor in tomato plant health classification. Pattern Recognit Image Anal. **33**(1), 39–46 (2023)
Kumar, Y., Singh, R., Moudgil, M.R., Kamini: A systematic review of different categories of plant disease detection using deep learning-based approaches. Archives of Computational Methods in Engineering **30**(8), 4757–4779 (2023)
Baser, P., Saini, J.R., Kotecha, K.: TomConv: An improved CNN model for diagnosis of diseases in tomato plant leaves. Procedia Computer Science **218**, 1825–1833 (2023)
Tarek, H., Aly, H., Eisa, S., Abul-Soud, M.: Optimized deep learning algorithms for tomato leaf disease detection with hardware deployment. Electronics **11**(1), 140 (2022)
Kaur, P., Harnal, S., Gautam, V., Singh, M.P., Singh, S.P.: An approach for characterization of infected area in tomato leaf disease based on deep learning and object detection technique. Eng. Appl. Artif. Intell. **115**, 105210 (2022)
Islam, M.S., et al.: Multimodal hybrid deep learning approach to detect tomato leaf disease using attention based dilated convolution feature extractor with logistic regression classification. Sensors **22**(16), 6079 (2022)
Sakkarvarthi, G., Sathianesan, G.W., Murugan, V.S., Reddy, A.J., Jayagopal, P., Elsisi, M.: Detection and classification of tomato crop disease using convolutional neural network. Electronics **11**(21), 3618 (2022)
Anandhakrishnan, T., Jaisakthi, S.M.: Deep Convolutional Neural Networks for image based tomato leaf disease detection. Sustainable Chemistry and Pharmacy **30**, 100793 (2022)
Özbılge, E., Ulukök, M.K., Toygar, Ö., Ozbılge, E.: Tomato disease recognition using a compact convolutional neural network. IEEE Access **10**, 77213–77224 (2022)
Altalak, M., Uddin, M.A., Alajmi, A., Rizg, A.: A hybrid approach for the detection and classification of tomato leaf diseases. Appl. Sci. **12**(16), 8182 (2022)
Tan, L., Lu, J., Jiang, H.: Tomato leaf diseases classification based on leaf images: a comparison between classical machine learning and deep learning methods. AgriEngineering **3**(3), 542–558 (2021)
Abbas, A., Jain, S., Gour, M., Vankudothu, S.: Tomato plant disease detection using transfer learning with C-GAN synthetic images. Comput. Electron. Agric. **187**, 106279 (2021)
Zhao, S., Peng, Y., Liu, J., Wu, S.: Tomato leaf disease diagnosis based on improved convolution neural network by attention module. Agriculture **11**(7), 651 (2021)
Prabhakar, M., Purushothaman, R., Awasthi, D.P.: Deep learning based assessment of disease severity for early blight in tomato crop. Multimedia Tools and Applications **79**, 28773–28784 (2020)
Agarwal, M., Gupta, S.K., Biswas, K.K.: Development of Efficient CNN model for Tomato crop disease identification. Sustainable Computing: Informatics and Systems **28**, 100407 (2020)

Unregulated Intensification and Key Challenges in Sustainable Aquaculture Production in Some Selected Regions of Ganjam District of Odisha

Bisoyi Rama Chandra[1(✉)], Biswal Nirmal Chandra[1], and Samant Devi Prasad[2]

[1] Department of Life Science (Zoology), GIET University, Gunupur, Rayagada, Odisha, India
ramchandrabisoyi862@gmail.com

[2] Aska Science College, Aska, Ganjam, Odisha, India

Abstract. Aquaculture sector is growing at amazing pace to provide nutrient rich food to the fast growing population of the developing world. Our present study encompasses aquaculture enterprises developed by female aquaculture entrepreneurs are often found in the Ganjam districts of Odisha. Out of 22 revenue blocks investigated initially with respect to large scale fish production six blocks are found to be maximally involved in aquaculture enterprises. Out of this two pond types as semi-intensive and intensive culture taken in Aska and Surada each. Four women SHGs with 12 members each involved in aquaculture enterprise. It has been observed that, women aquaculture entrepreneurs now trying to increase their aquaculture production which made a tragic shift of aquaculture from extensive to semi-intensive and extensive system. Semi-intensive culture ponds A and B and Intensive culture ponds P and Qtaken for observation for fish production over three years 2021–2023. The mean value of the production as per ANOVA was at par. But, Initial productivity in high input intensive culture pond was high and found to be 1.915 tonnes and 1.898 tonnes /acre reduced to 1.461 tonnes and 1.458 tonnes/acre between 2021 to 2023. Reverse was observed in Semi-intensive ponds as found to be increased from 1.58 tonnes to 1.64 tonnes/acre and 1.59 tonnes to 1.625 tonnes/acre respectively in the same time period. Similarly net profit (mean) per annum was found to be high of INR 2,25,171.3 and INR 2,25,438 in semi-intensive system over three years as compared to intensive system which is found to be low as INR 2,13.591.3 and INR 2,01,441.3. Water quality parameter such as DO,pH, dissolved carbondioxide is found to beyond safe limit in intensive culture ponds as compared to semi-intensive ponds. Due to lack of proper knowledge about intensive aquaculture female stake holders facing problem for optimum output from intensive aquaculture which can be optimized with proper regulation of feeding and maintenance of water quality parameters by acquisition of proper technical knowledge.

Keywords: Unregulated · intensification · Key challenges · Sustainable aquaculture. Ganjam. Odisha First Section

B. K. Mishra et al. (Eds.): ICT4GS 2024, ISEM 56, pp. 104–114, 2025.
https://doi.org/10.1007/978-3-031-94997-5_11

1 Introduction

Explosive growth of human population and considerable increase in the affluence in most developing countries is executing an immense pressure of global food production sector. UN's Zero hunger targets is facing greater challenge due to drastic climate change, biodiversity losses threatening all aspect of human existence by threatening jobs, livelihood options and food security as a whole. The food crisis is so grim today that, nearly 811 million people are suffering from hunger and nearly 3 billion cannot afford to healthy diet. For survival of the large section of population and ensure food and nutritional security and provision of healthy affordable diet, agricultural system needs significant transformations. This will also safeguard natural resources. Among all agricultural sectors, aquaculture comes in the forefront as a food production sector as it plays major role in supply the high quality animal protein source as fish and additional high value nutritional components like Omega-3 and Omega-6 fatty acids and many micronutrients [1]. The demand for food is increasing day by day. Among all foods animal food is in the first priority. Food production sector is a major driver of global environmental change and the system of production becomes unsustainable [2].

Over last thirty years aquaculture developed from a state of obscurity to a major enterprise and a significant component of the global food production system. The aquaculture today, supplying more than 50% of the total fish supplied for direct human consumption in global scale [3]. In the same perspective the capture fisheries became stagnant from mid- 1990s and with maximum sustainable effort for maximizing production in the natural environment, the worldwide supply from natural fish stock the contribution cannot go beyond 10% of the total fish supply [4]. In this light aquaculture seems to be the only option which can substantiate and fulfill the future demand for fish [3, 4]. Fisheries and aquaculture increasingly givers a quantitative contribution towards food and nutritional security and employment generation. Fisheries and aquaculture production reached a record high of 214 million tons of worth of USD 424 billion. This is about 60% higher as compared to the 1990s average value and it is mainly due to higher growth of aquaculture production.

Aquatic foods are in greater demand as we are eating double of the same of 20.2 kg per capita as compared to that of 50 years ago. Aquatic food nearly provides 17% of the animal protein and in Asia and Africa reaching 50 percent. It employs 58.5 million people in primary sector including 21% women [1].

Although intensification of food production through aquaculture is most common in almost all regions but mostly it is not coming under the radar. Farmers of most developing nations are not aware about it. In consequence it reduces the aquaculture production and induce drastic change in hydrological parameters which impairs ecosystem functioning. Aquaculture feeds and feeding pattern resulted in the increase of nutrient and organic loads may lead to differential plankton [5]. There is a possibility of environmental damage due to increased volume of fish excretory materials, chemical used for disease prevention and therapy and also the surplus nutrients. Waste food accumulated in the in the aquatic environment altering the redox potential with high organic material and also N and P and estimated that about 50% of nitrogen supplied and 28% P supplied wasted and dissolved in water [6]. Aquaculture is one of the most profitable enterprises as it can meet the hidden hunger of the world [7]. It provides the food to a sizeable section of global populace and contributes to food and nutritional security by establishing a stronger value chain [8, 9]. Aquaculture brings severe enviormental repurcussions and adversely alters the ecological footprints mostly for intensive production system [10]. It is imperative to design and implement innovative and new productive aquaculture interventions strategies having very insignificant negative environmental impact [11, 12]. Keeping the environmental sustainability in mind all the aquaculture interventions must be formulated towards improvement of system designs with reducing cost of investments [13, 14].

Rural women fish farmers in many regions were at the beginning phase of their aquaculture enterprises and mostly prefer to adapt to traditional methods of aquaculture using traditional knowledge. They have basic skills and rely on locally available fish feeds to boost their aquaculture production.. But, due to introduction of intensive culture in the wake of blue revolution for enhancing aquaculture production.

2 Materials and Methods

2.1 Research Sites and Aquaculture Interventions

Ganjam is among the top in fisheries and aquaculture production among 30 districts of Odisha. Six different blocks of the Ganjam district including Aska, Bellguntha, Bhanjanagar, Polasara, Rangeilunda and Surada were included for the overall study of aquaculture intervention in terms of feed use and fish production. A baseline study was conducted from June 2021 to May 2023 and a total of four ponds selected two in each site of Aska and surada with one pond under semi-intensive and one pond under intensive system in each site. Moderate sized ponds of about one acre each were selected as fish culture ponds for investigation of fish production under different pond management condition. Type 1 pond managed with indigenous and farm made feed application with a limited amount of commercial fish feeds. Type 2 ponds managed with scientifically formulated farm made feed components with locally available components with a larger quantity of commercial fish feeds. The investment incurred per acre per annum in semi-intensive fishery intervention is about INR 42,502 and in intensive intervention is about INR 59752. The trial was run with 1500 fingerlings per acre in semi-intensive culture and 3000 fingerlings per acre in intensive culture (Table 1).

Table 1. Total investment in pond aquaculture per year in INR

Sl. No.	Component	Total cost in INR
Type A and Type B ponds*	Fingerling of total 1500@ INR 3.5	5250
	Fish feed	15552
	Lime + Urea + Biofit	21700

(*continued*)

Table 1. (*continued*)

Sl. No.	Component	Total cost in INR
Total		**42502**
Type P and TypeQ ponds**	Fingerling of total 3000@ INR 3.5	10500
	Fish feed	27552
	Lime + Urea + Biofit	21700
Total		**59752**

* Type A and B ponds –Semi intensive culture ponds.
** Type P and Q ponds – Intensive culture with use of more commercial fish feed

2.2 Methods of Data Collection

Four independent women SHGs with 48 stake holders those were directly or indirectly involved in fishery enterprise were interviewed with pre-structured questionnaire using PRA tools. This is highly useful for generating a comprehensive data for the present study.

2.3 Statistical Analysis

Data collected from the two aquaculture ponds i.e. two semi-intensive ponds and two intensive ponds were separately arranged in excel sheets. Two separate data were compared by using the ANOVA test to know the significance of difference between two types of aquaculture interventions if any. The results for investments, production and cost benefit calculation are made for finding statistical significance if any.

3 Results and Discussion

During investigation in two blocks Aska and Surada we found that female stake holders as a part of SHGs are largely involved in fishery enterprises and activities like cultivation and sales (Figs. 1, 2 and 3) and male members of the family are involved in pond preparation and management. Investments costs per annum in semi-intensive culture system is INR 45202 and that in the intensive culture system is INR 59752 and the total fish production at site A and B (semi-intensive culture) is found to be increased from 1.58 tonnes to 1.64 tonnes and 1.59 tonnes to 1.625 tonnes respectively between year 2021 to 2023. Further, in intensive culture system the production in site P and Q decreased from 1.915 tonnes to 1.461 tonnes and 1.898 tonnes to 1.458 tonnes between the years 2021 to 2023. Data pertaining to the semi intensive and intensive aquaculture system was collected and statistically analysed which have been presented in Tables 3, 4, 5, 6, 7 and 8 and Fig. 6, respectively. The mean of production and standard deviation in both semi-intensive and intensive culture system is reflected in descriptive statistics Table 3 and ANOVA Table 4. It indicates that the intensive culture without proper regulation leads to decrease in production over the years. The major cause of this is found to be alteration of water quality parameters in intensive culture ponds P and Q. In the semi-intensive system the gross revenue generated in type A aquaculture system increased from INR 263340 to INR 273070 and that in type B aquaculture ponds increased from 264870 to 270909 Gross revenue generated in semi-intensive culture system is found to be between INR 267673.3 to INR 267940 and that of intensive culture system is between INR 271193.3 to INR 273343.3(Table 2). The mean of gross revenue generated and standard deviation is reflected in Table 5 and ANOVA Table 6.

Table 2. Comparision of output of intensive aquaculture system and semi-intensive aquaculture system

Pond type	Pond number	Investment costs	Year	R + C + M weight in tonnes/Acre	Gross Revenue in INR	Average of Gross Revenue	Profit per annum in INR
Semi-intensive	A	1,27506@42502 per year	2021	0.537 + 0.526 + 0.517 = 1.58	179180 + 84160 = 263340	267673.3	2,25,171.3
			2022	0.545 + 0.539 + 0.516 = 1.60	180370 + 86240 = 266610		
			2023	0.562 + 0.556 + 0.521 = 1.64	184110 + 88960 = 273070		
	B	1,27506@42502 per year	2021	0.546 + 0.526 + 0.517 = 1.59	180710 + 84160 = 264870	267940	2,25,438
			2022	0.565 + 0.531 + 0.512 = 1.608	183090 + 84960 = 268050		
			2023	0.571 + 0.535 + 0.519 = 1.625	185300 + 85600 = 270900		
Intensive	P	179256@59752 per year	2021	0.695 + 0.623 + 0.597 = 1.915	219640 + 99680 = 319320	273343.3	2,13,591.3
			2022	0.529 + 0.512 + 0.502 = 1.543	175270 + 81920 = 257190		
			2023	0.497 + 0.485 + 0.479 = 1.461	165920 + 77600 = 243520		
	Q	179256@59752 per year	2021	0.686 + 0.623 + 0.589 = 1.898	216750 + 99680 = 316430	271193.3	2,01,441.3
			2022	0.522 + 0.505 + 0.498 = 1.525	173400 + 80800 = 254200		
			2023	0.501 + 0.491 + 0.466 = 1,458	164390 + 78560 = 242950		

R- Rohu
C –Catla
M – Mrigala

This is further clarified by DATA collected

In intensive fish culture ponds fish mortality is observed in some cases as seen in intensive culture pond Q (Fig. 5) that is due alteration of water quality parameters. This is mainly seen is hot summer months due to low water depth, decrease in DO(mean value of 3.6 mg/l observed in pond Q and > 4.2 mg/l observed in pond P in seven days observed period) and accumulation of suspended solid, increase in CO_2 concentration (mean value of 7.2 mg/l observed in pond Q and > 5.5 mg/l observed in pond P in seven days observed period).Fish growth also retarded in third year in both ponds P and Q (Fig. 4).However, The calculated value of water quality parameter in semi-intensive ponds are found within the safe limit.

Fig. 1. Women stake holders applying fish feeds

Fig. 2. Women stake holders monitoring the growth of fishes

It is evident from the table the production of fish was more in intensive aquaculture system than semi intensive aquaculture system. But as evident from the ANOVA table, both the means of fish production were found to statistically at par and no significant difference was observed between the two means of production. Likewise, same trend was observed for revenue generation of both intensive and semi intensive aquaculture system and both the means were found to be statistically at par. However, as seen in the Fig. 6, investment cost was more in intensive system than semi intensive system and the cumulative profit over the consecutive three year period was more in semi-intensive system than intensive system.

Fig. 3. Female stake holders involved in value chain of fishery enterprise (sales)

Fig. 4. Low growth rate in intensive ponds (comparative)

Fig. 5. Fish mortality seen in intensive ponds

Table 3. Descriptive Statistics of production of fish in two different aquaculture systems

Aqua.Systems	N	Mean	Std. Deviation	Std. Error	95% Confidence Interval for Mean		Minimum	Maximum
					Lower Bound	Upper Bound		
1-Semi intensive	6	1.6068	.02214	.00904	1.5836	1.6301	1.58	1.64
2-Intensive	6	1.6333	.21435	.08751	1.4084	1.8583	1.46	1.92
Total	12	1.6201	.14594	.04213	1.5274	1.7128	1.46	1.92

Table 4. ANOVA for Production of Fishes

	Sum of Squares	df	Mean Square	F	Sig
Between Groups	.002	1	.002	.091	.769^{NS}
Within Groups	.232	10	.023		
Total	.234	11			

Table 5. Descriptive Statistics of Gross Revenue generated in two different aquaculture systems

	N	Mean	Std. Deviation	Std. Error	95% Confidence Interval for Mean		Minimum	Maximum
Aqua. Systems					Lower Bound	Upper Bound		
1-Semi intensive	6	267806.66	3669.81	1498.19	263955.4	271657.9	263340	273070
2-Intensive	6	272268.33	35788.16	14610.4	234711.0	309825.7	242950	319320
Total	12	270037.5	24366.55	7034.02	254555.7	285519.3	242950	319320

Table 6. ANOVA for revenue generated in different aquaculture system

VAR00004					
	Sum of Squares	df	Mean Square	F	Sig
Between Groups	59720000.00	1	59720000.00	.092	.768NS
Within Groups	6471000000.00	10	647100000.00		
Total	6531000000.00	11			

Table 7. Descriptive Statistics of Gross Profit generated in two different aquaculture systems

	N	Mean	Std. Deviation	Std. Error	95% Confidence Interval for Mean		Minimum	Maximum
Aqua. Systems					Lower Bound	Upper Bound		
1-Semi intensive	6	225304.67	3669.82	1498.20	221453.43	229155.90	220838.00	230568.00
2-Intensive	6	212516.33	35788.17	14610.46	174958.96	250073.71	183198.00	259568.00
Total	12	218910.50	25157.55	7262.36	202926.15	234894.85	183198.00	259568.00

Table 8. ANOVA for gross profit generated in two different aquaculture system

VAR00005					
	Sum of Squares	df	Mean Square	F	Sig
Between Groups	490624408.33	1	490624408.33	0.76	.404NS
Within Groups	6471302016.67	10	647130201.67		
Total	6961926425.00	11			

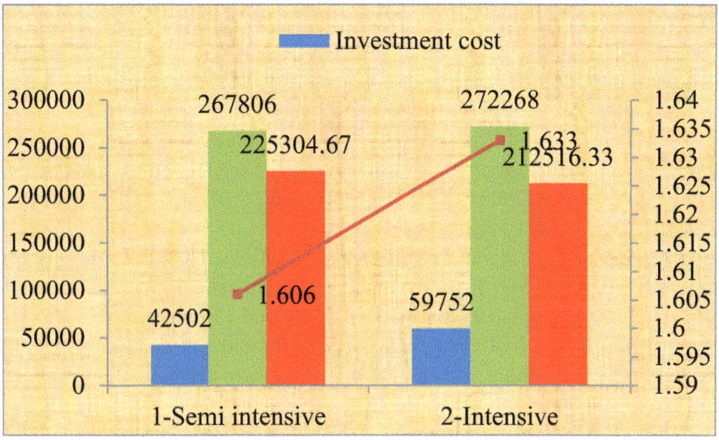

Fig. 6. Comparison of investment, production, revenue and profit in semi-intensive and intensive system.

4 Conclusion

Fishery enterprise is the second most prioritized option in Odisha after agriculture for boosting the rural economy especially female stake holders operating through SHGs. Female stake holders of different SHGs are involved in aquaculture interventions and enterprises with the introduction of Indian Major Carps (IMCs) i.e. polyculture with inclusion of three species including Rohu, Catla and Mrigal. They are mainly accustomed to old/traditional methods of aquaculture which is low input and low productive. With emergence of blue revolution female SHGs are actively participating in the movement. Some of them adapt to semi-intensive system with moderate input and moderate productive system with the use of locally available farm made fish feeds with small amount of commercial fish feeds. But taking it as a more lucrative enterprise some of them opt for very high input and high productive intensive system with larger quantities of commercial fish feeds. But due to lack of proper technical expertise excess feed accumulation sometimes alter the water quality parameters such as seen in intensive culture ponds reducing fish production over the years. It is imperative for rural female stake holders to opt for intensive culture after gaining deeper knowledge about the challenges of intensive culture and heading for sustainability of intensive culture system. Training, technical knowhow of the can largely enhance the aquaculture production over the years and can make a smooth transition from semi intensive to sustainable intensive system.

Acknowledgement. I am very thankful to all the members of SHGs of Aska and Surada Block for providing information about their aquaculture enterprises. Further the help of the SPCB for providing laboratory facility for calculation water quality parameters helped us to study the challenges in aquaculture system and draw a conclusion about decrease in productivity in the intensive culture ponds.

References

1. FAO: The State of World Fisheries and Aquaculture 2022. Towards Blue Transformation. FAO, Rome (2022). https://doi.org/10.4060/cc0461en
2. Henriksson, P.J.G., Belton, B., Jahan, K.M., Rico, A.: Measuring the potential for sustainable intensification of aquaculture in Bangladesh using life cycle assessment. Proc. Natl. Acad. Sci. U S A. **115**(12), 2958–2963 (2018). https://doi.org/10.1073/pnas.1716530115. Epub 2018 Mar 5. PMID: 29507224; PMCID: PMC5866565
3. Food and Agriculture Organization of the United Nations, FAO Fisheries and Aquaculture Department (2016) Fish Stat J–software for fishery statistical time series. Version 2.11.4. Available at www.fao.org/fishery/statistics/en. Accessed 22 January 2017
4. The World Bank: Fish to 2030: Prospects for fisheries and aquaculture (World Bank Group, Washington, DC), Agriculture and environmental services discussion paper no. 3, report no. 83177-GLB (2013)
5. ICES: Report of the working group on environmental interactions of mariculture, ICES headquarters, Copenhagen (2002)
6. Mente, E., Pierce, G.J., Santos, M.B., Neofitou, C.: Effect of feed and feeding in culture of salmonids on the marine aquatic environment: A synthesis for European aquaculture. Aquaculture International **14**, 499–522 (2006). https://doi.org/10.1007/s10499-006-9051-4
7. FAO: The state of world fisheries and aquaculture 2014. FAO, Rome, Italy (2014)
8. Ogello, E.O., Munguti, J.M.: Aquaculture: a promising solution for food insecurity, poverty and malnutrition in Kenya. African Journal of Food Agriculture, Nutrition and Development **16**(4), 11332–11351 (2016). https://doi.org/10.18697/ajfand.76.15900
9. Obiero, K., et al.: The contribution of fish to food and nutrition security in Eastern Africa: Emerging trends and future outlooks. Sustainability **1636**(11), 1–15 (2019). https://doi.org/10.3390/su11061636
10. Alleway, H.K., Gillies, C.L., Bishop, M.J., Gentry, R.R., Theuerkauf, S.J., Jones, R.: The ecosystem services of marine aquaculture: valuing benefits to people and nature. Bioscience **69**(1), 1 (2019). https://doi.org/10.1093/biosci/biy137
11. Hasan, M.R.: Nutrition and feeding for sustainable aquaculture development in the third millennium. In: Subasinghe, R.P., Bueno, P., Phillips, M.J., Hough, C., McGladdery, S.E., Arthur, J.E. (eds.) Aquaculture in the Third Millennium, pp. 193–219. FAO, Rome, Italy (2001). https://doi.org/10.1057/9780333985557_13
12. Naylor, R.L., et al.: Effect of aquaculture on world fish supplies. Nature **405**, 1017–1024 (2000). https://doi.org/10.1038/35016500
13. Musyoka, S.N., Liti, D.M., Ogello, E., Waidbacher, H.: Utilization of the earthworm, Eisenia fetida (Savigny, 1826) as an alternative protein source in fish feeds processing: A review. Aquac. Res. **50**(9), 2301–2315 (2019). https://doi.org/10.1111/are.1409
14. Waite, R., et al.: Increasing productivity of aquaculture. Working paper, installment 6 of creating a sustainable food future. World Resources Institute, Washington, D.C. (2014)

Data-Driven Solutions for Sustainable Urban Development: The Role of Smart Cities

Tejinder Kaur[1(✉)], Anita Venugopal[2], Sukhvir Kaur[3], Kirti Vasdev[4], Kiran Rani Panwar[5], and Kiran Rani Panwar[6]

[1] Department of MMICTBM, Maharishi Markandeshwar (Deemed To Be University), Mullana, Ambala-Haryana 133207, India
teji7890@gmail.com

[2] Department of IT, Dhofar University, Sultanate, Oman
anita@du.edu.om

[3] Department of Computer Science and Engineering, Chandigarh University, Gharuan, Punjab 140413, India
sukhvirb0@gmail.com

[4] Distinguished Engineer, Network Engineering, Verizon Data Services, Tampa, FL 33637, USA
kirtivasdev12@gmail.com

[5] Subharti Department of Liberal Arts and Humanities (FASS), Swami Vivekanand Subharti University, Meerut, U.P., India
kirantomer79@gmail.com

[6] Division of Research and Development, Lovely Professional University, Phagwara, India
mukesh.research24@gmail.com

Abstract. Urbanization is accelerating at an unprecedented rate, posing significant challenges for sustainable development. It examines key components, technologies, and strategies that drive smart city initiatives, focusing on their potential to enhance environmental, economic, and social sustainability. The study also addresses the challenges and future directions in the implementation of smart cities, providing a holistic view of their impact on urban living. Cities around the world can learn from each other's experiences and best practices, fostering innovation and improving urban sustainability on a global scale. These technologies enable the real-time collection, analysis, and dissemination of data, facilitating efficient urban management and service delivery.

Keywords: Smart cities · urban sustainability · technology · data-driven approaches

1 Introduction

The rapid urbanization witnessed globally presents both opportunities and challenges for sustainable development. With over half of the world's population residing in urban areas, cities are at the forefront of addressing critical issues such as resource management, pollution, and social equity. Smart cities, defined as urban areas that use various types of electronic data collection sensors to supply information used to manage

B. K. Mishra et al. (Eds.): ICT4GS 2024, ISEM 56, pp. 115–123, 2025.
https://doi.org/10.1007/978-3-031-94997-5_12

assets and resources efficiently, are emerging as a strategic response to these challenges (Abdelsabour et al., 2019). Smart cities emerge as a pivotal solution to these urban challenges. Block chain, smart cities aim to optimize urban operations, enhance the quality of life for residents, and promote sustainability. The concept of smart cities revolves around using digital technologies to improve urban infrastructure and services, thereby making cities more efficient, sustainable, and livable (Aboalsamh et al., 2023). This paper provides a comprehensive analysis of smart cities and their role in promoting urban sustainability. It examines the key components that constitute a smart city, including information and communication technology (ICT) infrastructure, sustainable energy systems, smart mobility solutions, and efficient waste and water management systems. Furthermore, the paper delves into technological innovations that underpin smart city initiatives and explores strategic approaches for achieving urban sustainability. By addressing the benefits and challenges associated with smart cities, this study aims to offer insights into how urban areas can transition towards more sustainable and resilient futures. The subsequent sections will explore these themes in detail, starting with an overview of the essential components of smart cities and followed by an analysis of technological innovations. Strategies for promoting urban sustainability through smart city initiatives will be discussed, along with the challenges that cities may face in this transition. Finally, the paper will consider future directions for the development and implementation of smart cities, highlighting the potential for continued innovation and global collaboration in creating sustainable urban environments (Adel et al., 2023).

2 Key Components of Smart Cities (Arner et al., 2022)

Smart cities are complex ecosystems that integrate a variety of technologies and strategies to enhance urban living and sustainability. The key components of smart cities include robust ICT infrastructure, advanced data analytics and AI, sustainable energy systems, smart mobility solutions, and efficient waste and water management systems. Each of these components plays a crucial role in optimizing urban operations, improving quality of life, and promoting sustainability (Alexander et al., 1997).

2.1 Information and Communication Technology (ICT) Infrastructure

A strong ICT infrastructure is the backbone of any smart city. It includes high-speed internet connectivity, widespread deployment of sensors and IOT (Internet of Things) devices, and advanced data processing and storage capabilities (Allam et al., 2021).

2.2 Data Analytics and Artificial Intelligence

(Benton-Short et al., 2017). Data analytics and artificial intelligence (AI) are essential for making sense of the vast amounts of data generated by smart city technologies. These tools help predict trends, optimize resource use, and enhance decision-making processes. For instance, AI can improve traffic management by analyzing patterns and suggesting optimal routes, thereby reducing congestion and emissions. Big data analytics can also provide insights into energy consumption patterns, helping cities to implement more efficient and sustainable energy policies (Berrou et al., 2020).

2.3 Sustainable Energy Systems

Sustainable energy systems are a core component of smart cities. Smart grids enhance the efficiency and resilience of urban energy supplies by using advanced sensors and meters to monitor and manage energy flows in real time. Energy storage solutions, such as batteries, are also crucial for balancing supply and demand and ensuring a stable energy supply (Berrou et al., 2020).

2.4 Smart Mobility Solutions (Calthorpe et al., 1993)

Transportation is a major focus area for smart cities, with initiatives aimed at promoting sustainable mobility. Smart mobility solutions include the development of electric vehicle (EV) infrastructure, intelligent public transportation systems, and shared mobility services. These solutions leverage technology to reduce traffic congestion, lower emissions, and improve accessibility (Charitonidou et al., 2021).

2.5 Waste and Water Management (Charitonidou et al., 2021)

Efficient waste and water management systems are essential for urban sustainability. Smart cities employ technologies such as smart meters, automated waste collection, and water recycling systems to optimize the use and disposal of resources. Smart meters can monitor water usage and detect leaks, while automated waste collection systems can optimize routes and schedules for garbage trucks. Water recycling systems can treat and reuse wastewater, reducing the strain on freshwater resources (Chen et al., 2020).

2.6 Public Safety and Security (Chen et al., 2020)

Public safety and security are paramount in smart cities. Technologies such as surveillance cameras, emergency response systems, and predictive policing tools help enhance the safety of urban environments (Chen et al., 2020). These systems can provide real-time data to law enforcement and emergency services, enabling quicker response times and more effective interventions. Additionally, AI and data analytics can be used to predict and prevent crime, ensuring a safer living environment for residents (Chen et al., 2020).

2.7 E-Governance and Citizen Engagement (Clarinval et al., 2020)

E-governance and citizen engagement are critical for the success of smart cities. E-governance platforms enable efficient, transparent, and accessible government services, making it easier for residents to interact with public authorities. Citizen engagement initiatives encourage community participation in decision-making processes, ensuring that the needs and preferences of residents are taken into account. Digital platforms, such as mobile apps and online portals, can facilitate these interactions and provide a channel for feedback and collaboration (Duranton et al., 2020).

3 Technological Innovations in Smart Cities

Technological innovations are at the heart of smart cities, driving advancements that enhance urban efficiency, sustainability, and quality of life. Key innovations include the Internet of Things (IOT), block chain technology, autonomous systems, and big data analytics. These technologies enable cities to collect and analyze data, automate processes, and provide intelligent services to residents (Duranton et al., 2020).

3.1 Internet of Things (IOT) ((Duranton et al., 2020)

The Internet of Things (IOT) refers to a network of interconnected devices that communicate and exchange data. In smart cities, IOT devices are used for various applications, including environmental monitoring, energy management, public safety, and transportation. These devices provide real-time data that helps in making informed decisions and improving urban services.

3.1.1 Environmental Monitoring

IOT sensors can monitor air and water quality, noise levels, and weather conditions. This data is crucial for addressing environmental issues and promoting public health. For instance, air quality sensors can detect pollution hotspots, allowing city authorities to implement targeted measures to reduce emissions (Gold et al., 2019).

3.1.2 Smart Lighting (Gold et al., 2019)

Smart lighting systems use IOT sensors to adjust street lighting based on real-time conditions. These systems can dim or brighten lights depending on the time of day, weather conditions, and pedestrian or vehicle presence. This not only saves energy but also enhances public safety (Gold et al., 2019).

3.2 Block chain Technology (Gold et al., 2019)

Block chain technology offers a secure and transparent way of managing data and transactions in smart cities. It can be used for various purposes, such as securing smart contracts, ensuring data privacy, and enhancing the transparency of municipal operations (Gold et al., 2019).

3.2.1 Smart Contracts (Gold et al., 2019)

In smart cities, they can be used for various applications, such as managing utilities, processing payments, and ensuring compliance with regulations. Block chain ensures that these contracts are executed securely and transparently (Kaur et al., 2019).

3.2.2 Data Privacy (Kaur et al., 2019)

Block chain technology can enhance data privacy by providing a decentralized and secure way of storing data. It ensures that data is tamper-proof and only accessible to authorized parties. This is particularly important for protecting sensitive information in smart city applications (Kaur et al., 2019).

3.3 Autonomous Systems (Kaur et al., 2019)

Autonomous systems, including self-driving cars and drones, are transforming urban mobility and logistics. These technologies promise to enhance the efficiency and safety of transportation systems while reducing environmental impact.

3.3.1 Self-driving Cars (Lehmann et al., 2016)

Self-driving cars use sensors, cameras, and AI to navigate and operate without human intervention. These vehicles can optimize routes; reduce traffic congestion, and lower emissions. They also enhance road safety by reducing human error (Lehmann et al., 2016).

3.3.2 Drones (Lehmann et al., 2016)

Drones can be used for various urban applications, such as delivering goods, monitoring infrastructure, and conducting surveillance. They offer a flexible and efficient way to perform tasks that would otherwise be time-consuming or dangerous for humans (Lehmann et al., 2016).

3.4 Big Data Analytics (Panlasigui et al., 2021)

Big data analytics involves processing large volumes of data to uncover patterns and insights. In smart cities, this technology is used to analyze data from various sources, such as social media, sensors, and public records, to inform policy-making and improve urban services (Panlasigui et al., 2021).

3.4.1 Predictive Analytics (Panlasigui et al., 2021)

In smart cities, it can be used to predict traffic patterns, energy consumption, and public health trends. This helps city authorities to plan and allocate resources more effectively.

3.4.2 Real-Time Data Analysis (Panlasigui et al., 2021)

Real-time data analysis allows cities to respond quickly to emerging situations. For instance, real-time traffic data can be used to adjust traffic signals and reduce congestion. Similarly, real-time energy data can help manage supply and demand in smart grids (Panlasigui et al., 2021).

3.5 Artificial Intelligence (AI) (van Hoof et al., 2021)

Artificial Intelligence (AI) is crucial for interpreting data and making autonomous decisions in smart cities. AI algorithms can optimize various urban processes, from traffic management to energy distribution (van Hoof et al., 2021).

3.5.1 Traffic Management (van Hoof et al., 2021)

AI can analyze traffic data to optimize signal timings, predict congestion, and suggest alternative routes.

3.5.2 Energy Management

This enhances the efficiency and sustainability of urban energy systems.

3.6 Cloud computing (van Hoof et al., 2021)

It enables real-time data processing, collaboration, and remote access to urban management tools (van Hoof et al., 2021).

3.6.1 Data Storage and Processing (van Hoof et al., 2021)

Cloud platforms offer virtually unlimited storage and processing capabilities.

3.6.2 Collaboration and Accessibility (van Hoof et al., 2021)

Cloud-based solutions enable collaboration among various stakeholders, including city authorities, businesses, and residents. They also provide remote access to urban management tools, facilitating efficient and effective decision-making.

4 Strategies for Urban Sustainability (van Hoof et al., 2021)

4.1 Integrated urban planning (van Hoof et al., 2021)

Integrated urban planning involves the coordinated development of urban infrastructure, land use, and services. This approach ensures that various aspects of urban living, such as housing, transportation, and public spaces, are planned in a cohesive manner. Integrated planning helps in creating sustainable, livable cities that can adapt to future challenges (van Hoof et al., 2021).

4.2 Public-Private Partnerships (PPPs) (Zheng et al., 2020)

These collaborations leverage the strengths of both sectors, combining public oversight with private sector innovation and investment. PPPs can help in financing smart city initiatives, ensuring efficient project execution, and fostering innovation (Zheng et al., 2020).

4.3 Community Engagement (Zheng et al., 2020)

Community engagement ensures that the needs and preferences of residents are considered, fostering a sense of ownership and participation. Methods such as public consultations, participatory planning, and digital platforms for feedback can enhance community involvement.

4.4 Regulatory Frameworks

Effective regulatory frameworks are necessary to guide the development and operation of smart cities. These frameworks should address issues such as data privacy, cyber security, and the ethical use of AI. Clear regulations ensure that smart city initiatives are implemented responsibly and sustainably ((Zheng et al., 2020).

5 Challenges in Implementing Smart Cities (Zheng et al., 2020)

5.1 High Initial Costs (Zheng et al., 2020)

The development of smart cities requires significant initial investment in technology and infrastructure. These costs can be a barrier, especially for cities with limited financial resources. Innovative financing models and international cooperation can help address this challenge.

5.2 Data Privacy and Security

Ensuring that data is collected, stored, and used in a secure and transparent manner is critical. Robust cyber security measures and transparent data governance policies are necessary to protect residents' privacy.

5.3 Technological Integration

Integrating various technologies and systems in a seamless manner can be challenging. Interoperability issues, legacy systems, and the complexity of urban environments can hinder the effective implementation of smart city solutions. Standardization and the use of open platforms can help overcome these challenges.

5.4 Social Equity

There is a risk that technology-driven solutions may exacerbate existing inequalities if not implemented inclusively. Policies that prioritize social equity and inclusion are essential for the success of smart cities.

6 Future Directions

6.1 Sustainable Urban Development (Zetzsche et al., 2019)

Future smart cities will place an even greater emphasis on sustainability. This includes the widespread adoption of renewable energy sources, circular economy practices, and nature-based solutions. Sustainable urban development will be key to addressing the environmental challenges of the 21st century.

6.2 Global Collaboration (Zetzsche et al., 2019)

International collaboration and knowledge-sharing will be crucial for the advancement of smart city initiatives.(Zetzsche et al., 2019).

6.3 Citizens-Centric Approaches (Zetzsche et al., 2019)

This includes the development of personalized services, improved public participation, and the use of technology to address specific community needs (Yan et al., 2019).

7 Conclusion

Smart cities represent a transformative approach to urban sustainability, leveraging technology and data to address the complex challenges of modern urbanization. By focusing on key components such as ICT infrastructure, data analytics, sustainable energy systems, and smart mobility, smart cities can enhance environmental, economic, and social sustainability. However, the successful implementation of smart city initiatives requires overcoming challenges related to cost, data privacy, technological integration, and social equity. Future directions in smart city development will be shaped by advancements in AI, a stronger emphasis on sustainability, global collaboration, and citizen-centric approaches.

References

Abdelsabour, I.: Investigating bio-morphism approach to enhance structure's creativity and efficiency. J. Eng. Appl. Sci. **66**(5), 491–513 (2019). https://www.researchgate.net/publication/339043001

Aboalsamh, H.M., Khrais, L.T., Albahussain, S.A.: Pioneering perception of green fintech in promoting sustainable digital services application within smart cities. Sustainability **15**(14), 11440 (2023)

Adel, A.: Unlocking the future: fostering human–machine collaboration and driving intelligent automation through industry 5.0 in smart cities. Smart Cities **6**(5), 2742–2782 (2023)

Arner, D.W., Ahmed, S.M., Gazi, S.: Building regulatory and supervisory technology ecosystems: for Asia's financial stability and sustainable development. University of Hong Kong Faculty of Law Research Paper, (2022/51) (2022)

Alexander, C.: A pattern language. Oxford University Press, Oxford (1977)

Allam, Z., Jones, D.S.: Future (post-COVID) digital, smart and sustainable cities in the wake of 6G: Digital twins, immersive realities and new urban economies. Land Use Policy **101**, 105201 (2021). https://doi.org/10.1016/j.landusepol.2020.105201

Benton-Short, L., Keeley, M., Rowland, J.: Green infrastructure, green space, and sustainable urbanism: Geography's important role. Urban Geogr. **40**(3), 330–351 (2017). https://doi.org/10.1080/02723638.2017.1360105

Berrou, Y., et al.: Smart city development strategy profile: Use case modeling based on simplicial complexes. ISPRS Annals of the Photogrammetry, Remote Sensing and Spatial Information Sciences **6**(4), 25–32 (2020). https://doi.org/10.5194/isprs-annals-VI4-W2-2020-25-2020

Calthorpe, P.: The next American metropolis: Ecology, community, and the American dream. Princeton Architectural Press, New York (1993)

Charitonidou, M.: Revisiting Giancarlo De Carlo's participatory design approach: From the representation of designers to the representation of users. Heritage **4**(2), 985–1004 (2021). https://doi.org/10.3390/heritage4020054

Chen, Y., Zhu, M., Lu, J., Zhou, Q., Ma, W.: Evaluation of ecological city and analysis of obstacle factors under the background of high-quality development: Taking cities in the Yellow River Basin as examples. Ecol. Ind. **118**, 106771 (2020). https://doi.org/10.1016/j.ecolind.2020.106771

Clarinval, A., Simonofski, A., Vanderose, B., Dumas, B.: Public displays and citizen participation: a systematic literature review and research agenda. Transforming Government: People, Process and Policy **15**(1), 1–35 (2020). https://doi.org/10.1108/TG-12-2019-0127

Duranton, G., Puga, D.: The economics of urban density. J. Econ. Perspect. **34**(3), 3–26 (2020). https://doi.org/10.1257/jep.34.3.3

Gold, J.R.: Athens Charter (CIAM), 1933. In: Orum, A.M. (ed.) The Wiley Blackwell encyclopedia of urban and regional studies, pp. 1–3. Chichester: Wiley-Blackwell (2019). Retrieved from https://www.researchgate.net/profile/JohnGold-3/publication/332425567_Athens_Charter_CIAM_1933/links/5d4f5715a6fdcc370a8c2d53/Athens-CharterCIAM-1933.pdf

Kaur, H., Garg, P.: Urban sustainability assessment tools: a review. J. Clean. Prod. **210**, 146–158 (2019). https://doi.org/10.1016/j.jclepro.2018.11.009

Lehmann, S.: Sustainable urbanism: towards a framework for quality and optimal density? Future Cities and Environment **2**, 8 (2016). https://doi.org/10.1186/s40984-016-0021-3

Panlasigui, S., Spotswood, E., Beller, E., Grossinger, R.: Biophilia beyond the building: Applying the tools of urban biodiversity planning to create biophilic cities. Sustainability **13**(5), 2450 (2021). https://doi.org/10.3390/su13052450

The New Leipzig Charter. The transformative power of cities for the common good (2020). Retrieved from https://futurium.ec.europa.eu/sites/default/files/2021-03/new_leipzig_charter_en.pdf

van Hoof, J., Marston, H.R.: Age-friendly cities and communities: State of the art and future perspectives. Int. J. Environ. Res. Public Health **18**(4), 1644 (2021). https://doi.org/10.3390/ijerph18041644

Zheng, C., Yuan, J., Zhu, L., Zhang, Y., Shao, Q.: From digital to sustainable: a scientometric review of smart city literature between 1990 and 2019. J. Clean. Prod. **258**, 120689 (2020)

Zetzsche, D.A., Buckley, R.P., Arner, D.W.: FinTech for financial inclusion: driving sustainable growth. Sustainable Development Goals: Harnessing Business to Achieve the SDGs through Finance. Technology, and Law Reform, 177–203 (2019)

Yan, D., Kong, Y., Ren, X., Shi, Y., Chiang, S.: The determinants of urban sustainability in Chinese resource-based cities: a panel quantile regression approach. Sci. Total Environ. **686**, 1210–1219 (2019)

Selection of Electrical Vehicle: An m-Polar Fuzzy TOPSIS Approach

Madan Jagtap(✉) and Seema Ghangale

Symbiosis Institute of Operations Management, Constituent of Symbiosis International (Deemed University), Nashik Campus, Pune 412115, Maharashtra, India
madan.jagtap@siom.in

Abstract. The intensity of global warming is increasing at an alarming rate, posing an increasingly serious threat to the world. Life on Earth will be jeopardized due to the effects of global warming unless swift and coordinated action is taken to reduce greenhouse gas emanations. The launch of electric vehicles (EVs) is significant development popular reducing the impact of global warming. Many nations are moving toward the use of electric vehicles, which encourages sustainable development. Even though EVs are becoming more and more popular globally, choosing the best option among the many available models can be challenging. In order to opt and grade the finest substitute for an electric vehicle, this study presents an integrated strategy that combines the m-polar fuzzy set (mFs) and Technique for Order of Preference by Similarity to Ideal Solution (TOPSIS) technique as a multi-criterion decision-making tool (MCDM). The data of the decision matrix is used as m-polar fuzzy input for the decision making. The weight coefficients of the criterion are obtained exercising the AHP methodology, and the TOPSIS method is used to assess the choice of EVs alternative. A sample of six workable options has been taken into consideration for the investigation. The combined m-polar fuzzy TOPSIS strategy, which hasn't been used in this field before, is what makes this work original. This study contributes by evaluating the best option among the electric vehicles which is KONA ELECTRIC vehicle helps the customer to purchase the electric vehicle. This study also explains the application of the m-polar fuzzy TOPSIS technique to resolve the MCDM glitch.

1 Introduction

The upsurge in global temperatures is bringing raised sea levels, melted ice caps and glaciers, more frequent and stronger heatwaves, and altered weather patterns. These changes pose serious hazards to human society, jeopardizing food and water security, intensifying natural disasters, and uprooting millions of people from their homes, in addition to endangering ecosystems and biodiversity. Greenhouse gas releases are one of the key grounds of climate change (Contini et al., 2023), and we can drastically reduce them by switching from conventional internal combustion engine automobiles to electric vehicles.

Since EVs have no tailpipe emissions, they reduce pollution, which worsens global warming and adds to the greenhouse effect (Chandran et al., 2021). Moreover, as EVs

B. K. Mishra et al. (Eds.): ICT4GS 2024, ISEM 56, pp. 124–133, 2025.
https://doi.org/10.1007/978-3-031-94997-5_13

can be charged with clean energy, the environmental advantages of EVs are increased as sources of renewable energy. Adopting electric vehicles is a practical response to the urgent environmental issues of our day. It not only lessens the severity of global warming but also encourages sustainability and resilience in the face of it. One inevitable mitigating measure to address global warming is the usage of renewable energy (Bernai, 2013). The transportation industry has contributed the maximum share of greenhouse gas discharges to environment alteration (I Vassileva, 2017). As stated by the International Energy Agency, the transportation industry contributes approximately 24.6% of the carbon dioxide linked to energy in the world (Ren et al., 2021). The introduction of alternative fuels is thought to be a replacement for fossil fuels like gasoline. These fuels include natural gas, ethanol, methanol, propane, electricity, biodiesel, and hydrogen. The global shift to hybrid vehicles is being driven by a declining supply of liquid fuel and growing environmental consciousness (Panday et al., 2016). Consequently, encouraging the use of electric vehicles can help achieve sustainable development goals and help prevent or lessen global warming (S Deuten, 2020). Electric vehicles (EVs) are a key component of the decarbonization process and are acknowledged as the most practical technology for reducing CO2 emissions through the use of alternative energy sources (Chandran et al., 2021). Electric vehicles are becoming more and more popular these days for a number of reasons, such as their effective battery capacity, decreased emissions of harmful gases, government subsidies and other incentives for buying them, enhanced vehicle performance, and other environmental benefits (N Lefèvre, 2020). Electric vehicles (EVs) are a key component of the decarbonization process and are acknowledged as the most practical technology for reducing CO2 emissions through the use of alternative energy sources (Friedlingstein et al., 2020).

Furthermore, projections indicate that the overall bazaar segment of electric vehicles may expand by up to 75% by 2050 as a result of both technological breakthroughs and rising fossil fuel costs. These days, a lot of government agencies are encouraging people to buy electric cars by putting in place different laws, grants, and other financial aid to lower CO2 emissions (John Brady, 2011). As a result, there may be a decrease in the environmental release of gases such as carbon dioxide and carbon monoxide, which would lessen the issue of global warming. Public and policy makers alike are adopting electric vehicles at a rapid pace. From this vantage point, several nations think of creative strategies that promote the spread of relevant technologies (Guler and Yomralioglu, 2018). The majority of the researchers used experimental analysis to draw their conclusions about how Electric cars can contribute to a decrease in greenhouse gas emissions. Electric vehicles can help reduce greenhouse gas emissions in various geographic zones (B Yagcitekin, 2015). With this eco awareness people start demanding the electric vehicles and to suffice this demand there came up many global automakers, such as General Motors, Ford, Hyundai, Fiat, BMW, Tata Motors, Mitsubishi, and others, who have already begun producing electric cars with standout features for their commercial vehicles. Due to this, the buyer must select the best option from a variety of EV vehicles that have these unique attributes. There available variability of customer preferences to choose an electric vehicle and in a similar vein, different performance assessment criteria were offered for different brands of electric vehicles. Customers thus find it challenging to select the best option and the relevant criteria. Consequently, in order to close this research gap, a decision-making tool must be introduced in order to choose performance measures—which often have contradicting nature—and the best options among the various electric car variations that are now on the market. Our research is based on multi-criteria decision-making tools that provide consumers with a quick overview of their preferences before they buy an electric car. In order to choose and prioritize the optimum option for an electric car, a unified approach of the AHP-MABAC technique has been employed as a multi-criteria decision-making instrument. Fuzzy AHP was created for this topic since decision-making problems generally have fuzzy properties (Guler and Yomralioglu, 2018). It offers mathematical improbability and ambiguity as well as solutions for numerous issues involving intrinsically inaccurate measurements (Kahraman, 2004).

Despite the large body of research on the adoption of electric vehicles has evident, major studies mainly addressed the issue of climate change or government subsidies or incentive programs (R Vidhi, 2018). Even if EVs are becoming more and more popular globally, it might be challenging to choose the best option from the range of EVs that are readily available. There isn't a thorough analysis in the literature to choose the best EV options. Annual sales of EVs in the global market surpass one million units. According to the material now in publication, they have a major impact on the environment (Loganathan, 2022). However, doing an assortment of the best EV and suitable decision making there found hardly any exploration. The intent of this analysis is to help decision-makers choose an appropriate electric car by applying an integrated AHPMABAC technique. India is our unit of analysis for this study since several Indian cities have already seen a rise in the sales of electric vehicles.

When the ratios of the selectivity criteria are absent, evaluation and weight determination become challenging. One of the MCDM approaches, the AHP method, can be used in these circumstances. AHP resolves complicated issues with the aid of the hierarchical organizational structure, which is produced by substitutes. The comparisons between pairs produced in the hierarchy is primarily determined by assessments, as opposed to as a comprehensive collection of all judgments and standards (Saaty, 1980). This study considers a number of factors, including torque, driving array, pricing, battery volume, charging period, and seating space. Furthermore, six distinct electric automobiles that are commonly accessible in India have been chosen by way of substitutes for consideration and order inside the passenger car category. This work is stimulating since it uses the combined AHP-MABAC technique, which hasn't been used in this field before. This study has three impressions to offer. In order to give a true preference for an extensive set of selection criteria for electric vehicles, it first uses a decision-making tool. Next the study also evaluates which of the electric automobiles on the Indian market is the best option. Thirdly, the low-performing EVs can evaluate their standards with the use of this analysis. This study might also aid manufacturers who are substantially investing in electric vehicles by examining potential customer preferences before developing electric automobiles.

Results and evidence regarding EV selection can be found through systematic review analysis and analysis with minimum variables, but a methodical and scientific approach is available to delve deeply into the issue in order to comprehend the interrelation, complexity, and polarity of the various variables and factors taken together.

Though EVs are popular nowadays, it is challenging and sometime frenzied to go for best option. In this scenario previous studies have looked at a single component and hence the need of the present research has evidential reasoning where the multiple factors are considered with diverse group, with different extent of features. In light of the many variables at play and the difficult terrain to navigate, it became necessary to use a multilevel evaluative technique while analyzing the data. When it comes to tackling real-world problems that call for particular complications, the m-Polar Fuzzy ELECTRE-I practice has various novelties that make it very beneficial. It has improved flexibility and versatility compared to traditional MCDM (Multi-Criteria Decision Making) procedures (Ecer, 2021). The bipolar fuzzy set, created by is expanded upon by the m-polar fuzzy set. They resolved real-life hitches with multiple negotiators, multiple matters, multiple traits, multipolar information, also improbability (Chen et al., 2014). The approaches cannot handle real-world problems needing several features and statistics with multipolar information (Akram et al., 2019). The fuzzy m-polar the ELECTRE-I approach stood established to solve a variety of features and data, including multipolar information. They identified an airport, located a diesel plant, and evaluated the efficacy of a physical science instructor, among other practical problems. It was investigated by Jagtap (Jagtap et al., 2021) how to use an m-polar fuzzy set to choose non-conventional milling techniques. Jagtap et al. (Jagtap and Karande, 2022) applied an m-polar fuzzy ELECTRE-I algorithm to industrial robot assortment difficulties. Vector normalization is a suitable method for decision matrix normalization, per Jagtap & Karande's (Jagtap and Karande, 2022) investigation of the effect of normalizing on the m-polar fuzzy ELECTRE-I algorithm. The objective of present project is to incorporate a appropriate weight adder method into the ELECTRE-I m-polar fuzzy set procedure. Scientific computation of parameter weights can be achieved by the m-polar fuzzy set ELECTRE-I integrated AHP. The m-polar fuzzy set ELECTRE-I integrated AHP technique can be applied to any concept pertaining to parameters or pole size. Because of its ability to handle multipolar data, tolerate ambiguity, and offer useful and flexible solutions for difficult real-world decision-making scenarios, the m-Polar Fuzzy ELECTRE-I is unique. In this paper, the m-polar fuzzy TOPSIS method is consumed to unravel difficulties of customers in the selection of electric vehicles.

2 Research Methodology

The m-polar fuzzy TOPSIS method is useful for the addressing the MCDM problems. The following is the stepwise methodology explained for the m-polar fuzzy TOPSIS.

Let, O be the set of alternatives and P be the set of variables,

First step is to write the set of alternatives available as shown in Eq. (1).

$$\text{Set of alternatives}\, B = \left(b_1, b_2, b_3, \ldots\ldots\ldots.b_p\right) \tag{1}$$

Second step is to write the set of variables considered for the decision as shown in Eq. (2).

$$\text{Set of variables}\, C = \left(c_1, c_2, c_3, \ldots\ldots\ldots\ldots c_q\right) \tag{2}$$

Third step is to write the decision matrix considering alternatives and variables as shown in Eq. (3).

Let, D be set of alternatives and variables, Decision Matrix

$$D = \left\{d_{jk}^{1}, d_{jk}^{2}, d_{jk}^{3}, \ldots\ldots\ldots\ldots d_{jk}^{m}\right\} \text{ where } d_{jk}^{i} = d_{jk}\left(b_j, c_k\right) \tag{3}$$

$$j = 1, 2, 3, \ldots\ldots.p, k = 1, 2, 3, \ldots\ldots q \text{ and } i = 1, 2, 3, \ldots\ldots\ldots\ldots m$$

Fourth step is to write the criteria weights for the available criteria as shown in Eq. (4).

W is the set of all the criteria weights obtained by either the AHP method or Shannon's entropy weight calculation method, Weights of criteria

$$W = (w_1, w_2, w_3, \ldots\ldots.w_k) \text{ where } k = 1, 2, 3, \ldots\ldots.q, \tag{4}$$

After having decision matrix and criteria weight, the weight multiplied decision matrix is obtained as shown in Eq. (5).

F, is the weight multiplied decision matrix, Weight multiplied decision matrix

$$F = \left[\left(f_{jk}^{1}, f_{jk}^{2}, f_{jk}^{3}, \ldots\ldots\ldots\ldots f_{jk}^{m}\right)\right] \tag{5}$$

$$\text{Where } f_{jk}^{1} = w_1 d_{jk}^{1}, \ldots\ldots\ldots\ldots\ldots\ldots f_{jk}^{m} = w_k d_{jk}^{m}$$

The weight multiplied decision matrix will be processed further to obtain the positive ideal solution as shown in Eq. (6) and negative ideal solution as shown in Eq. (7).

The m-polar fuzzy positive ideal solution and the m-polar fuzzy negative ideal solution of alternatives, mP_{IS} and mN_{IS} are the positive ideal solution and negative ideal solution of alternatives

$$mP_{IS} = \left\{\left(\left(d_1^1\right)^+, \left(d_1^2\right)^+, \ldots\ldots\ldots\ldots, (d_1^m)^+\right), \left(\begin{matrix}(\left(d_2^1\right)^+, \left(d_2^2\right)^+, \ldots\ldots\ldots\ldots, (d_2^m)^+)) \ldots\ldots\ldots\ldots, \\ \left(\left(d_q^1\right)^+, \left(d_q^2\right)^+, \ldots\ldots\ldots\ldots, \left(d_q^m\right)^+\right)\end{matrix}\right)\right\} \tag{6}$$

$$mN_{IS} = \left\{\left(\left(d_1^1\right)^-, \left(d_1^2\right)^-, \ldots\ldots\ldots\ldots, (d_1^m)^-\right), \left(\begin{matrix}(\left(d_2^1\right)^-, \left(d_2^2\right)^-, \ldots\ldots\ldots\ldots, (d_2^m)^-)) \ldots\ldots\ldots\ldots, \\ \left(\left(d_q^1\right)^-, \left(d_q^2\right)^-, \ldots\ldots\ldots\ldots, \left(d_q^m\right)^-\right)\end{matrix}\right)\right\} \tag{7}$$

Where,

$$(d_{\mathrm{k}}^{1})^+ = \max\{d_{\mathrm{jk}}^{1}\}, (d_{\mathrm{k}}^{2})^+ = \max\{d_{\mathrm{jk}}^{2}\} \ldots (d_{\mathrm{k}}^{m})^+ = \max\{d_{\mathrm{jk}}^{m}\}$$

$$(d_{\mathrm{k}}^{1})^- = \max\{d_{\mathrm{jk}}^{1}\}, (d_{\mathrm{k}}^{2})^- = \max\{d_{\mathrm{jk}}^{2}\} \ldots (d_{\mathrm{k}}^{m})^- = \max\{d_{\mathrm{jk}}^{m}\}$$

After getting the positive ideal solution and negative ideal solution, the positive Euclidean distance and negative Euclidean distance for each alternative are calculated with the help of Eqs. (8) and (9).

Now, $D_e(b_j, P_{IS})$ be the positive Euclidean distances of each alternative and $D_e(b_j, N_{IS})$ be the negative Euclidean distances of each option, which can be calculated as,

$$D_e(b_j, P_{IS}) = \sqrt{\frac{1}{m}q\sum_{k=1}^{q}[(d_{jk}^1 - (d_j^1)^+)^2 + (d_{jk}^2 - (d_j^2)^+)^2 + \ldots + \left(d_{jk}^m - (d_j^m)^+)^2\right]}, \quad (8)$$

$$D_e(b_j, N_{IS}) = \sqrt{\frac{1}{m}\sum_{k=1}^{q}[(d_{jk}^1 - (d_j^1)^-)^2 + (d_{jk}^2 - (d_j^2)^-)^2 + \ldots + (d_{jk}^m - (d_j^m)^-)^2]}, \quad (9)$$

With the help of positive Euclidean distance and negative Euclidean distance, the relative closeness coefficient should be calculated as shown in Eq. (10).

The relative closeness coefficient of each alternative can be computed as R_j be the relative closeness coefficient for each option.

$$R_j = \frac{D_e(b_j, N_{IS})}{D_e(b_j, P_{IS}) + D_e(b_j, N_{IS})}, j = 1, 2, 3, \ldots.., p. \quad (10)$$

Values of relative closeness coefficient can be used to rank the alternatives.

3 Case Study Implementation

The case study mentioned below will be solved with the m-polar fuzzy TOPSIS methodology. The m-polar fuzzy TOPSIS methodology was developed to solve the MCGDM problem, in this case study, our objective is to verify whether the MCGDM algorithm can be implemented to the MCDM problems. The same case study was solved with the help of MCDM methodology. We are comparing the results obtained from the MCDM implementation with the results obtained from the MCGDM algorithm implementation. Selecting the greatest electric car alternative available on the market is a challenging task. Products are chosen in concert, considering variables like cost and charge duration. The case study utilized here is used to rate the various performance measures using AHP. The rankings for the various categories of electric vehicles are also available on the market. This work's author has considered all relevant and irrelevant factors for a difficult real-world problem. The alternatives' ranking was determined using the recently developed AHP-MABAC approach. As mentioned earlier, this study uses a numerical example to investigate the impact of normalization procedures. The decision matrix used to select an electric vehicle can be found in Sonar and Kulkarni's work. The author has chosen six distinct electric vehicle options from various businesses, including Hyundai. Based on many factors named c_1-Price, c_2-Driving Range, c_3-Charging Time, c_4-Battery capacity, c_5-Seating Capacity, and c_6-Toque. b_1 = KONA ELECTRIC, b_2 = Mahindra e Verito, b_3 = Mahindra e2o, b_4 = MG ZS, b_5 = Tata Tigor EV, and b_6 = Tata Nexon EV. Table 1, displays the matrix of decisions. The decision problem's attribute weights are W = (0.036, 0.054, 0.154, 0.134, 0.233, and 0.39).

Weights of the criteria are given as shown in Table 2. The criteria weights are taken from the same research paper of Sonar and Kulkarni.

Table 1. Decision matrix for Electrical Vehicle selection

Alternatives	Criteria					
	c_1	c_2	c_3	c_4	c_5	c_6
b_1	23.86	452	6.1	39.2	5	395
b_2	13.95	140	11.3	18.55	5	91
b_3	8.23	110	6	10.08	4	70
b_4	20.88	340	16	44.5	5	353
b_5	10.61	213	11.5	21.5	5	105
b_6	14.99	312	8.5	30.2	5	245

Table 2. Criteria weight for the Electrical Vehicle criteria

Criteria	c_1	c_2	c_3	c_4	c_5	c_6
Weights	0.036	0.054	0.154	0.134	0.233	0.39

The weighted normalized decision matrix is obtained by multiplying the criteria weight to the normalized decision matrix. The weighted normalized decision matrix is as shown in Table 3.

Table 3. Weighted Normalized Decision matrix (vector)

Alternatives	Criteria					
	c_1	c_2	c_3	c_4	c_5	c_6
b_1	0.09422	0.25072	0.11747	0.07190	0.02273	0.02354
b_2	0.15186	0.07765	0.08633	0.03402	0.02273	0.00542
b_3	0.18513	0.06101	0.11807	0.01848	0.01819	0.00417
b_4	0.11155	0.1886	0.05819	0.08162	0.02273	0.02103
b_5	0.17129	0.11815	0.08513	0.03943	0.02273	0.00625
b_6	0.14581	0.17306	0.10310	0.05539	0.02273	0.01460

The m-polar fuzzy positive model key and the m-polar fuzzy negative model key for the weighted normalised input matrix can be calculated using Eqs. (6) and (7). Positive ideal solution mHPIS for the alternatives is shown in Table 4.

Negative ideal solution mHNIS for the alternatives is shown in Table 5.

The Eqs. (8) and (9) can be utilised to find the Euclidean distance of alternatives from positive ideal solution and negative ideal solution. The values of the Euclidean distances of alternatives from positive ideal solution and negative ideal solution are shown in Table 6.

Table 4. M-polar fuzzy positive model key

Alternatives	b_1	b_2	b_3	b_4	b_5	b_6
mHPIS	0.0067	0.0090	0.0043	0.0029	0.0008	0.0008

Table 5. Negative model key mHNIS

Alternatives	b_1	b_2	b_3	b_4	b_5	b_6
mHNIS	0.0034	0.0022	0.0021	0.0007	0.0007	0.0002

Table 6. Euclidean distance of alternatives from the positive ideal solution and negative ideal solutions

	$D_e\ (b_j,P_{IS})$	$D_e\ (b_j,N_{IS})$
b_1	0.001	0.003
b_2	0.003	0.001
b_3	0.003	0.002
b_4	0.002	0.002
b_5	0.002	0.001
b_6	0.001	0.002

The Eq. (10) can be utilized to calculate the relative closeness coefficient of each alternative using the Euclidean distance of alternatives from the positive ideal solution and negative ideal solutions.

Table 7. Relative closeness coefficient for alternatives

Rank	Score	m-polar fuzzy TOPSIS	Sonar and Kulkarni
1	0.693	b_1	b_1
2	0.596	b_6	b_6
3	0.560	b_4	b_5
4	0.413	b_5	b_3
5	0.351	b_3	b_4
6	0.268	b_2	b_2

Table 7, shown above presents the rank obtained for the electrical vehicle based on the relative closeness coefficient. It shows that the KONA ELECTRIC is the best electric vehicle from the given available electric vehicle alternatives. Table 7, also compares the rank of alternatives obtained from the m-polar fuzzy TOPSIS and the rank obtained by the Sonar and Kulkarni.

Conclusions

The above study compares the electric mobility vehicles used in the Indian market. It compares six different alternatives for the six criteria and the m-polar fuzzy TOPSIS method for this case study resulted in the descending rank order of electric vehicles as KONA ELECTRIC rank one followed by Tata Nexon EV, MG ZS, Tate tigor EV, Mahindra e2o, Mahindra e Verito. It shows that the rank obtained by the m-polar fuzzy TOPSIS method is comparatively easier and accurate as compared with the rank obtained by the method used previously. When we compared KONA ELECTRIC with all other alternatives, it shows that the it is superior to other substitutes particular to the norms considered for the selection purpose. Previously, researchers have proven the use of the m-polar fuzzy TOPSIS method for the multi-criteria group decision-making (MCGDM). In this study we have proven that the m-polar fuzzy TOPSIS method can also be consumed to crack the MCDM challenges having single valued real time observations.

References

Yagcitekin, B., Uzunoglu, M.: Assessment of electrically-driven vehicles in terms of emission impacts and energy requirements: a case study for Istanbul, Turkey. J. Clean. Prod. (2015)

Bernai, R.: Renewable Energy Sources and Climate Change Mitigation, Special Report of the Intergovernmental Panel on Climate Change (IPCC). JSTOR (2013)

Vassileva, I., Campillo, J.: Technology assessment of the two most relevant aspects for improving urban energy efficiency identified in six mid-sized European cities from case studies in Sweden. Applied Energy (2017)

John Brady, M.O.: Travel to work in Dublin. The potential impacts of electric vehicles on climate change and urban air quality. Transport and Environment (2011)

Kahraman, C.C.: Multi-attribute comparison of catering service companies using fuzzy AHP. Int. J. Prod. Econ. (2004)

Loganathan, M.K.: Review and selection of recycling technology for lithium-ion batteries made e for EV application- A life cycle perspective. International Conference on Green Energy and Technology. IOP Conference Series: Earth and (2022)

Vidhi, R., Shrivastava, P.: A review of electric vehicle lifecycle emissions and policy recommendations to increase EV penetration in India. Energies MDPI (2018)

Deuten, S., Vilchez, J.: Analysis and testing of electric car incentive scenarios in the Netherlands and Norway. Technological Forecasting and Social Change (2020)

Saaty, T.: The analytic hierarchy process (AHP). The J. Operat. Res. Soc. (1980)

Akram, M., Dudek, W.A., Ilyas, F.: Group decision-making based on pythagorean fuzzy TOPSIS method. Int. J. Intell. Syst. **34**(7), 1455–1475 (2019)

Chen, J., Li, S., Ma, S., Wang, X.: Polar fuzzy sets: an extension of bipolar fuzzy sets. The Sci. World J. (2014)

Ecer, F.: A consolidated MCDM framework for performance assessment of battery electric vehicles based on ranking strategies. Renew. Sustain. Energy Rev. **143**, 110916 (2021)

Friedlingstein, P., et al.: Global carbon budget 2020. Earth Sys. Sci. Data Discuss. **2020**, 1–3 (2020)

Jagtap, M., Karande, P.: Application of m-polar fuzzy set algorithm for nontraditional machining process selection. In: Digitalization of Society, Economics and Management: A Digital Strategy Based on Post-pandemic Developments, pp. 221–233. Springer (2022)

Jagtap, M., Karande, P., Athawale, V.M.: Rank assessment of robots using m-polar fuzzy ELECTRE-I algorithm. Proceedings of the International Conference on Industrial Engineering and Operations Management, pp. 246–255 (2021)

Chandran, V., et al.: State of charge estimation of lithium-ion battery for electric vehicles using machine learning algorithms. World Electric Vehicle Journal **12**(1) (2021). https://doi.org/10.3390/wevj12010038

Contini, G., Peruzzini, M., Bulgarelli, S., Bosi, G.: Developing key performance indicators for monitoring sustainability in the ceramic industry: The role of digitalization and industry 4.0 technologies. J. Clean. Prod. **414**(June), 137664 (2023). https://doi.org/10.1016/j.jclepro.2023.137664

Guler, D., Yomralioglu, T.: Gis and fuzzy ahp based area selection for electric vehicle charging stations. Int. Archi. Photogr. Remote Sensing and Spatial Info. Sci. - ISPRS Archi. **42**(4), 315–320 (2018). https://doi.org/10.5194/isprs-archives-XLII-4-249-2018

Panday, A., Om Bansal, H., Bansal, H.O.: Regular paper multi-objective optimization in battery selection for hybrid electric vehicle applications. In: JES Journal of Electrical **12**(2) (2016). https://www.researchgate.net/publication/304942767

Ren, X., Sun, S., Yuan, R.: A study on selection strategies for battery electric vehicles based on sentiments, analysis, and the MCDM model. Math. Probl. Eng. (2021). https://doi.org/10.1155/2021/9984343

Lefèvre, J., Briand, Y., Pye, S., et al.: A pathway design framework for sectoral deep decarbonization: the case of passenger transportation. Climate Policy **21**, 93–106 (2021)

AI-Driven Ensemble Strategy for Discovering Neurotherapeutic Medicinal Plants

N. Sasikaladevi(✉) and A. Santhosh Kumar

Shanmugha Arts Science Technology and Research Academy, SASTRA Deemed University, Thanjavur, Tamil Nadu, India
sasikalade@gmail.com

Abstract. Central nervous system (CNS) diseases have witnessed an alarming rise globally, affecting millions of individuals and imposing significant healthcare expenses. Throughout history, indigenous medicinal plants have played a vital role in addressing various ailments, including CNS disorders. In this research, we propose a cutting-edge machine learning approach to accurately predict the effectiveness of medicinal plants in treating CNS diseases. Leveraging the VNPlant200 dataset comprising plant images and associated metadata, we train a convolutional neural network (CNN) to extract profound features. To amplify the discriminative power of these features, we employ matrix-based discriminant analysis, thereby augmenting our model's predictive capabilities. Furthermore, we integrate an ensemble technique that combines multiple classifiers, resulting in improved accuracy with a remarkable rate of 100%. Additionally, we have developed a user-friendly mobile application empowering individuals to identify and classify medicinal plants based on their potential for treating CNS diseases. This innovative work holds great promise in efficiently and cost-effectively identifying and harnessing the ability of conventionally used medicinal herbs to treat CNS disorders.

Keywords: Medicinal plants · central nervous system diseases · machine learning · convolutional neural networks · discriminant analysis · ensemble learning · mobile application

1 Introduction

The human nervous system, which comprises the central nervous system (CNS), plays a crucial role in processing sensory inputs, controlling movements, and regulating behaviors, with the spinal cord and brain serving as its primary components. CNS disorders such as Parkinson's disease, multiple sclerosis, epilepsy, and spinal cord damage have a significant impact on motor control, sensory perception, and overall function. Traditional medicines and medicinal plants have been historically used to address these conditions, and recent research has shed light on their potential for the development of new CNS treatments (Baell & Walters, 2014; Kennedy et al., 2018). The model's remarkable accuracy rate of over 99% in predicting medicinal plants for CNS diseases

B. K. Mishra et al. (Eds.): ICT4GS 2024, ISEM 56, pp. 134–148, 2025.
https://doi.org/10.1007/978-3-031-94997-5_14

underscores its robustness, driven by InceptionResNetV2 for feature extraction and Support Vector Machine (SVM) for classification. These findings highlight the real-world applicability of the model and the significant potential of machine learning in identifying therapeutic plants for CNS disorders.

2 Related Work

Several notable studies have explored the predictive analysis of medicinal plants for central nervous system (CNS) diseases using machine learning techniques. DeepPlant-A Deep Learning Approach for Medicinal Plant Analysis in CNS Diseases. liMedLeaf-Deep Learning-based Identification of Medicinal Plants for CNS Diseases Li et al. (2023) proposed MedLeaf and their approach combined CNN and attention mechanisms to extract discriminative features from leaf images, achieving an impressive accuracy rate of 96% on a large-scale dataset. These related works demonstrate the growing interest in employing machine learning techniques to analyze the effectiveness of medicinal plants in treating CNS diseases, showcasing significant advancements in accuracy and predictive capabilities (Table 1).

Table 1. State-of-the-Art Approaches for Classification of Medicinal Plants

Authors	Method Used	Dataset	Performance
Chen et al. (2022)	DeepPlant (CNN + RNN)	Various CNS medicinal plant images	92%
Zhang et al. (2020)	PlantNet (CNN)	CNS medicinal plant leaf images	90%
Liu et al. (2021)	NeuroHerb (Deep Learning with graphs)	Botanical and chemical data	93%
Wang et al. (2022)	HerbCNS (Graph Convolutional Networks)	Molecular structures and disease associations	94%
Li et al. (2023)	MedLeaf (CNN + Attention)	Large-scale medicinal plant leaf images	96%
Sharma et al. (2022)	LeafNet (CNN)	CNS medicinal plant leaf images	88%
Kumar et al. (2021)	NeuroPlant (Ensemble of CNN and Random Forest)	Medicinal plant images and chemical properties	91%
Park et al. (2022)	CNSPlant (Deep Learning with Graph Convolutional Networks)	Molecular structures and gene expression data	93.5%
Jiang et al. (2023)	HerbMed (CNN + LSTM)	Historical medicinal plant usage data	89.6%

Chen et al. (2022) – DeepPlant (CNN + RNN): One limitation of DeepPlant lies in its potential lack of interpretability. This complexity could hinder the model's explainability and transparency. Zhang et al. (2020) – PlantNet (CNN): PlantNet, relying heavily on visual features, faces challenges in handling variations in leaf conditions. Liu et al. (2021) – NeuroHerb (Deep Learning with Graphs): NeuroHerb's reliance on graph-based features introduces a limitation related to data quality. Wang et al. (2022) – HerbCNS (Graph Convolutional Networks): HerbCNS faces potential challenges with sparse datasets or incomplete molecular structures. Li et al. (2023) – MedLeaf (CNN + Attention): MedLeaf's limitation lies in its sensitivity to diverse plant structures and conditions. Sharma et al. (2022) – LeafNet (CNN): LeafNet's potential limitation stems from sensitivity to noisy or inconsistent data in CNS medicinal plant leaf images.

3 Proposed Framework

Three essential phases are involved in the suggested algorithm for predicting medicinal plants for CNS diseases: collecting and preprocessing data; using InceptionResNetV2 for deep feature extraction; and using matrix-based discriminant analysis for feature extraction.

3.1 Data Collection and Preprocessing

The dataset known as VNPlant-200, obtained from the National Institute of Medicinal Materials, consists of 20,000 pictures representing 200 different medicinal plant species, each containing 100 images. Its purpose is to facilitate machine learning applications in plant categorization. This dataset tackles issues such as cluttered backgrounds and overlapping foliage by integrating preprocessing methods like noise reduction, normalization, and data augmentation. The images are available in resolutions of 128 × 128, 256 × 256, and 512 × 512 pixels, with 70% allocated for training and 20% for testing (Figs. 1 and 2, Table 2).

(a) Abrus precatorius (b)Ageratum conyzoides

Fig. 1. Sample images of Central Nervous System (CNS) medicinal plants from VNPlant-200 dataset

Table 2. References of the medicinal plant available in the VNPlant-200 data set

Indian medicinal plant	Plant part	Therapeutical use identifiers	References
Abrus precatorius	leaf	MESH:D002493, UMLS:C0007682, DOID:331	ISBN:9770972795006
Artemisia vulgaris	aerial part, leaf	MESH:D009422, UMLS:C0027765, DOID:863, ICD-11:8E7Z	ISBN:9788172361266
Abutilon indicum	root	MESH:D009422, UMLS:C0027765, DOID:863, ICD-11:8E7Z	ISBN:9770972795006, ISBN:9788172360481
Ageratum conyzoides		MESH:D002493, UMLS:C0007682, DOID:331	ISBN:9770972795006
Curculigo orchioides	rhizome	MESH:D009422, UMLS:C0027765, DOID:863, ICD-11:8E7Z	ISBN:9770972795006
Cymbopogon citratus	leaf	MESH:D002493, UMLS:C0007682, DOID:331	ISBN:9770972795006
Dimocarpus longan	stem	MESH:D002583, UMLS:C0007682, DOID:385	ISBN:9770972795006
Ficus racemosa	fruit	MESH:D009422, UMLS:C0027765, DOID:863, ICD-11:8E7Z	ISBN:9770972795006
Gymnema sylvestre	leaf	MESH:D009422, UMLS:C0027765, DOID:863, ICD-11:8E7Z	ISBN:9770972795006
Hibiscus rosa-sinensis	flower	MESH:D009422, UMLS:C0027765, DOID:863, ICD-11:8E7Z	Standardization of Single Drugs of Unani Medicine Part – III

(*continued*)

Table 2. (*continued*)

Indian medicinal plant	Plant part	Therapeutical use identifiers	References
Hibiscus sabdariffa	fruit	MESH:D009422, UMLS:C0027765, DOID:863, ICD-11:8E7Z	ISBN:9770972795006
Iris domestica	whole plant	MESH:D002493, UMLS:C0007682, DOID:331	ISBN:9770972795006
Ixora coccinea	whole plant	MESH:D002493, UMLS:C0007682, DOID:331	ISBN:9770972795006
Jasminum sambac	whole plant	MESH:D002493, UMLS:C0007682, DOID:331	ISBN:9770972795006
Lawsonia inermis	leaf	MESH:D002493, UMLS:C0007682, DOID:331	ISBN:9770972795006
Litsea glutinosa	fruit, leaf	MESH:D002493, UMLS:C0007682, DOID:331	ISBN:9770972795006
Melastoma malabathricum	flower	MESH:D009422, UMLS:C0027765, DOID:863, ICD-11:8E7Z	ISBN:9770972795006
Moringa oleifera	Fruit, root	MESH:D009422, UMLS:C0027765, DOID:863, ICD-11:8E7Z	ISBN:9770972795006
Ocimum gratissimum		MESH:D002493, UMLS:C0007682, DOID:331	ISBN:9770972795006
Oenanthe javanica	whole plant	MESH:D002493, UMLS:C0007682, DOID:331	ISBN:9770972795006
Platycladus orientalis	whole plant	MESH:D002493, UMLS:C0007682, DOID:331	ISBN:9770972795006

(*continued*)

Table 2. (*continued*)

Indian medicinal plant	Plant part	Therapeutical use identifiers	References
Rauvolfia tetraphylla	fruit	MESH:D009422, UMLS:C0027765, DOID:863, ICD-11:8E7Z	Medicinal Plants Of Nagpur And Wardha Forest Divisions (Maharashtra)
Senna alata	fruit	MESH:D002493, UMLS:C0007682, DOID:331	ISBN:9770972795006

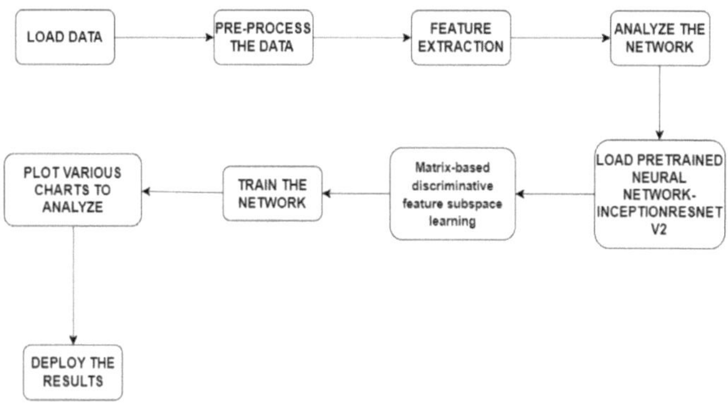

Fig. 2. Workflow diagram

3.2 Feature Extraction Using InceptionResnetV2

InceptionResNetV2 starts with convolutions represented by $Y = f\ (W * X + b)$ Y = f(W ∗ X + b), where W W are filter weights, b b are biases, and f f is the activation function. This allows the model to be trained in advance and extract abstract features from images. Feature map dimensions are reduced by pooling layers, such as max pooling, as demonstrated by Y = max(X) = max(Y). It combines the outputs of several convolutional filters applied in parallel branches of inception modules by integrating them along the channel dimension. Direct information flow is facilitated by residual connections when Y = X + F(X) and $Y = X + F\ (X)$. This method transforms unprocessed photos into feature representations that are necessary for tasks like classifying medicinal plants.

3.3 Learning Decision Trees with Random Sampling

By generating subsets D i based on random selection with replacement from the training data, Random Forest enhances model performance through random sampling. The size of the subset is determined by parameter M. Using criteria like Gini impurity or information gain and algorithms like CART, each subset is used to train a decision tree (DTTi). Averaging the predictions from all N decision trees yields the final prediction for an input x; majority voting is usually used for classification tasks. By using a variety of subsets and combining multiple tree predictions, this approach improves performance and generalizability while lessening the effect of outliers and peculiarities in individual subsets.

4 Results

In this section, we present the experimental setup and the results obtained from evaluating and validating the proposed predictive analysis model for Indigenous plants used in the treatment of CNS diseases. The performance of the model is compared to several machine-learning algorithms commonly used in medicinal plant research.

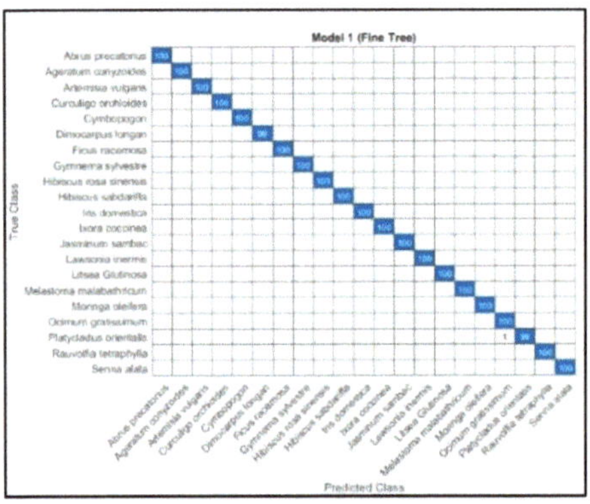

a: Fine tree - Confusion matrix

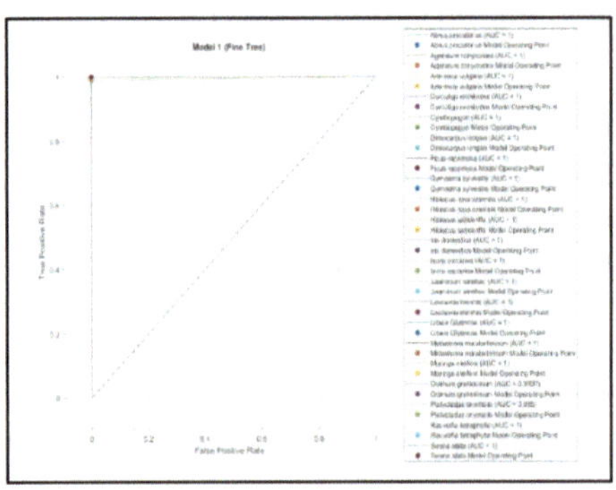

b: Fine Tree– ROC Curve

Fig. 3. Fine Tree (a) Fine tree – Confusion matrix; (b) Fine Tree– ROC Curve

4.1 Experimental Set-Up and Evaluation

System Configuration

The experiments were done on the computer system comes with a 512 GB NVMe solid-state drive, an Intel Core i5-10300H processor with a base clock speed of 2.50 GHz, 4 cores, 8 logical processors, 8 GB DDR4 RAM, and an NVIDIA GeForce GTX 1650 graphics card (Fig. 3).

The ROC (Receiver Operating Characteristic) curve is a graphical representation that illustrates the trade-off between sensitivity and specificity for different classification thresholds in Fig. 4 (Fig. 5).

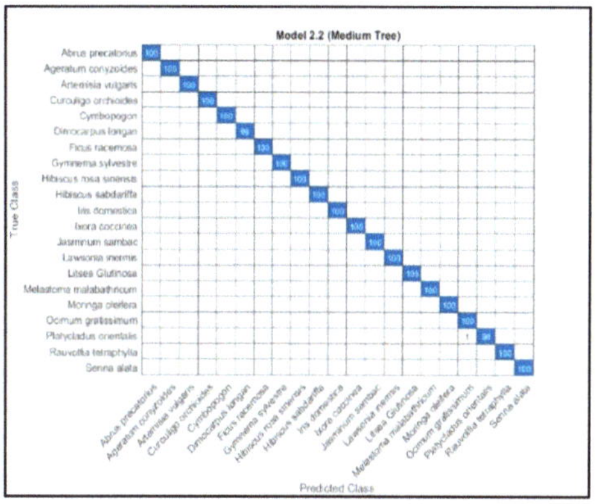

a: Medium Tree - Confusion matrix

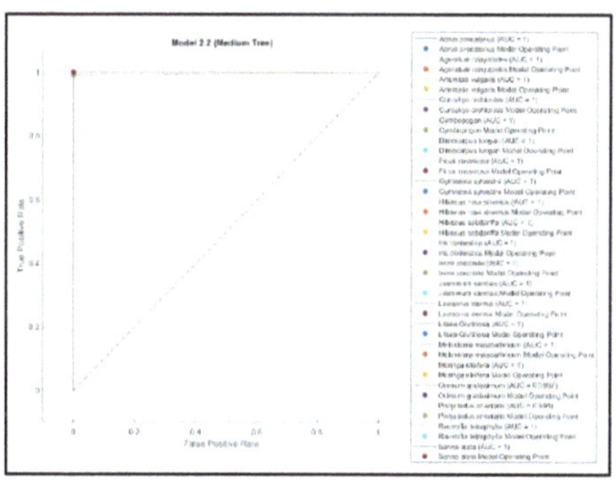

b: Medium Tree – ROC Curve

Fig. 4. Medium Tree. (a) Medium Tree – Confusion matrix; (b) Medium Tree – ROC Curve

A higher AUC (Area Under the Curve) value indicates better overall model performance. An AUC of 0.90 suggests that the model has a 90% chance of correctly distinguishing between positive and negative cases (Fig. 6).

The Subspace Discriminant model's ROC curve is a visual gauge of its discrimination performance. Customized for the SVM Kernel, the ROC curve assesses the equilibrium between sensitivity and specificity. A sharp curve suggests stronger discrimination. A

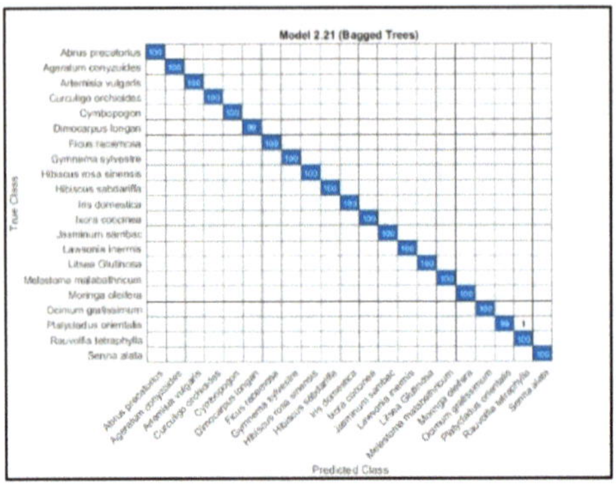

a: Bagged tree - Confusion matrix

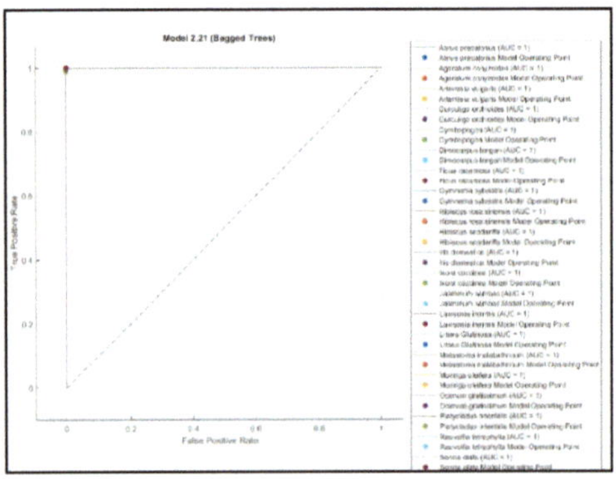

b: Bagged Tree – ROC Curve

Fig. 5. (a) Bagged tree – Confusion matrix; (b) Bagged Tree – ROC Curve

higher AUC highlights the SVM Kernel's proficiency in accurately classifying instances, making it suitable for complex datasets and various classification scenarios (Tables 3, 4 and 5).

Table 7 highlights the Ensemble Bagged Tree model's impressive 100.0% accuracy, outperforming all existing deep learning-based medical classification methods in identifying medicinal plants. As shown in Table 6, Ensemble Bagged Tree achieves the maximum training accuracy of 100.0%. The test sample findings are shown in Table 7, together with the necessary labels.

4.2 Interpretation

The model's impressive accuracy exceeding 99% in predicting medicinal plants for CNS diseases demonstrates its robustness, achieved through InceptionResNetV2 for detailed feature extraction and Support Vector Machine (SVM) for precise classification.

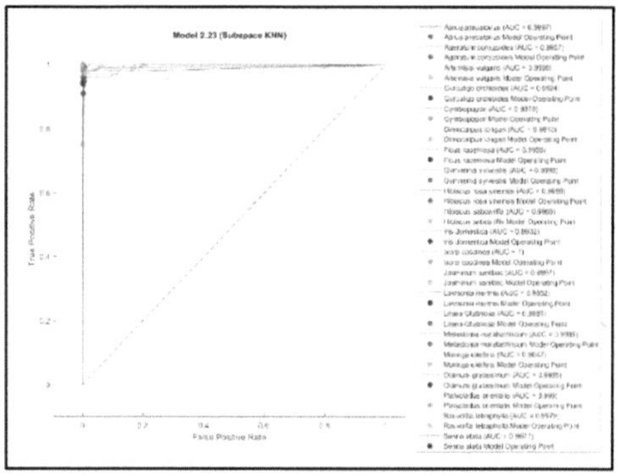

a: Subspace KNN- ROC

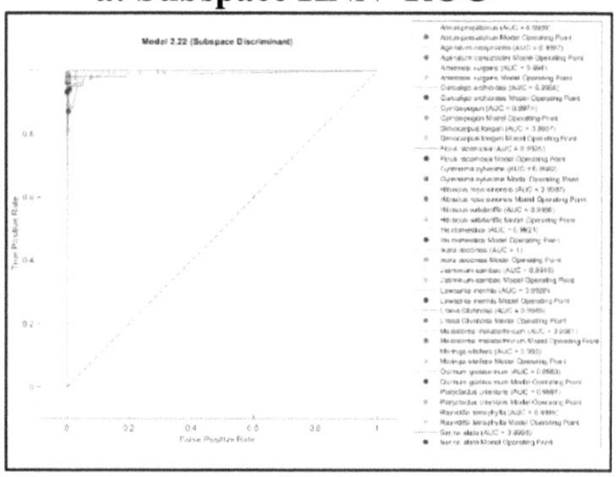

b: Subspace Discriminant-ROC

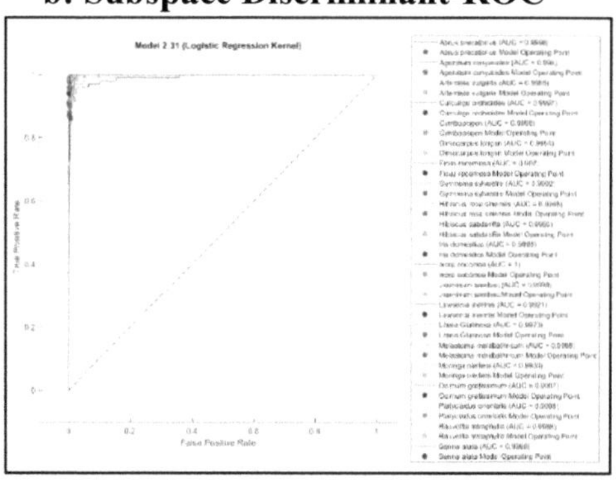

c: Logistic Regression Kernel- ROC

Fig. 6. ROC Curve. (a) Subspace KNN-ROC; (b) Subspace Discriminant-ROC; (c) Logistic Regression Kernel-ROC; (d) SVM Kernel-ROC

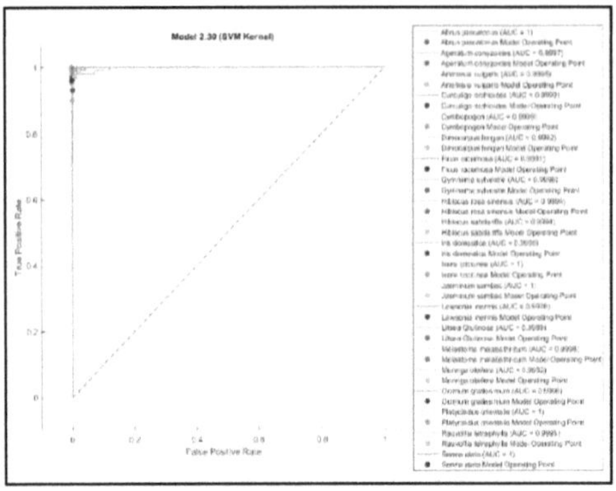

d: SVM Kernel- ROC

Fig. 6. (*continued*)

Table 3. Classification performance with machine learning models

Sr. No	Models Trained	Total Validation Cost	Training Time (in seconds)	Prediction Speed	Validation Accuracy
1	**Tree: Fine Tree**	**1**	**77.572**	**1400/sec**	**100.0%**
2	**Tree: Medium Tree**	**1**	**77.844**	**1300/sec**	**100.0%**
3	Tree: Coarse Tree	1599	39.14	1400/sec	23.8%
4	Linear Discriminant	NA	NA	NA	Failed
5	Quadratic Discriminant	NA	NA	NA	Failed
6	Naive Bayes: Gaussian	NA	NA	NA	Failed
7	Naive Bayes: Kernel	1093	10251	0.93/sec	47.9%
8	SVM: Linear	204	291.99	86/sec	90.3%
9	SVM: Quadratic	163	327.78	72/sec	92.2%
10	SVM: Cubic	169	344	67/sec	91.9%
11	SVM: Fine Gaussian	1602	521.33	71/sec	23.7%

(*continued*)

Table 3. (*continued*)

Sr. No	Models Trained	Total Validation Cost	Training Time (in seconds)	Prediction Speed	Validation Accuracy
12	SVM: Medium Gaussian	332	517.8	91/sec	84.2%
13	SVM: Coarse Gaussian	387	548.58	73/sec	81.60%
14	KNN: Fine	290	561.87	250/sec	86.20%
15	KNN: Medium	336	557.16	260/sec	84.00%
16	KNN: Coarse	624	605.92	130/sec	70.30%
17	KNN: Cosine	372	614.98	130/sec	82.30%
18	KNN: Cubic	356	795.41	37/sec	83.0%
19	KNN: Weighted	300	662.23	160/sec	85.70%
20	Ensemble: Boosted Trees	1999	695.74	1400/sec	4.80%
21	**Ensemble: Bagged Trees**	**1**	**1321.7**	**720/sec**	**100.0%**
22	Ensemble: Subspace Discriminant	105	1137.1	27/sec	95.0%
23	Ensemble: Subspace KNN	100	1133.9	27/sec	95.20%
24	Ensemble: RUSBoosted Trees	1999	1163	1400/sec	4.80%
25	Neural Network: Narrow	NA	1166.9	1400/sec	82.70%
26	Neural Network: Medium	NA	1185.9	1300/sec	88.50%
27	Neural Network: Wide	NA	1196.1	1300/sec	90.80%
28	Neural Network: Bilayered	NA	1262.5	1200/sec	81.70%

(*continued*)

Table 3. (*continued*)

Sr. No	Models Trained	Total Validation Cost	Training Time (in seconds)	Prediction Speed	Validation Accuracy
29	Neural Network: Trilayered	NA	1350.4	1200/sec	78.70%
30	Kernel: SVM	61	11092	1.9/sec	97.10%
31	Kernel: Logistic Regression	143	8960.3	1.3/sec	93.20%

Table 4. Classification performance with discriminant analysis

Sr. No	Models Trained	Total Validation Cost	Training Time (in seconds)	Prediction Speed	Validation Accuracy
1	Linear Discriminant	NA	NA	NA	Failed
2	Quadratic Discriminant	NA	NA	NA	Failed
3	**Ensemble: Subspace Discriminant**	**105**	**1137.1**	**27/sec**	**95.0%**

Table 5. Classification performance with Ensemble models

Sr. No	Models Trained	Total Validation Cost	Training Time (in seconds)	Prediction Speed	Validation Accuracy
1	Ensemble: Boosted Trees	1999	695.74	1400/sec	4.80%
2	**Ensemble: Bagged Trees**	**1**	**1321.7**	**720/sec**	**100.0%**
3	Ensemble: Subspace Discriminant	105	1137.1	27/sec	95.0%
4	Ensemble: Subspace KNN	100	1133.9	27/sec	95.20%
5	Ensemble: RUSBoosted Trees	1999	1163	1400/sec	4.80%

Table 6. Classification performance with Ensemble Subspace Discriminant

Sr. No	Models Trained	Total Validation Cost	Training Time (in sec)	Prediction Speed	Learner type	Validation Accuracy
1	Ensemble: Subspace KNN	105	1137.1	27/sec	Nearest Neighbours	95.0%
2	**Ensemble: Subspace Discriminant**	**100**	**1133.9**	**27/sec**	**Discriminant**	**95.20%**

Table 7. Classification performance with Ensemble Trees

Sr. No	Models Trained	Total Validation Cost	Training Time (in seconds)	Prediction Speed	Validation Accuracy
1	Ensemble: Boosted Trees	1999	695.74	1400/sec	4.80%
2	**Ensemble: Bagged Trees**	**1**	**1321.7**	**720/sec**	**100.0%**

5 Conclusion

A new method for predicting medicinal plants for CNS diseases using machine learning was proposed in this research. The method involves extracting features from plant images using InceptionResnetV2 and then classifying the images into different categories using SVM. Data augmentation strategies were also explored to enhance the model's performance. The proposed approach achieved an accuracy of over 99%, showing its effectiveness in predicting medicinal plants for CNS diseases. The study demonstrates the potential of machine learning methods in identifying plants with therapeutic characteristics. Future studies could consider additional deep learning models or ensemble techniques to improve the classification model's accuracy. Overall, the study presents a promising approach for predicting medicinal plants and has significant implications for the discovery of new therapeutic agents for CNS diseases.

References

Almazaydeh, L., Alsalameen, R., Elleithy, K., et al.: Herbal leaf recognition using mask-region convolutional neural network (mask r-CNN). J. Theoret. Appl. Inform. Technol. **100**(11) (2022)

Kumar, G., Kumar, V., et al.: Herbal plant leaf image classification using deep learning models based on augmentation approach. Available at SSRN 4292344 (2022)

Kumar, D., Bhat, Z.A., Shah, M.Y., et al.: Neuroprotective effect of dry fruits of *Ficus carica* L. Mice. Indian J. Pharmaceut. Sci. **75**(4), 424–429 (2013)

Parveen, A., Akhtar, M., Zainuddin, M., Khan, N.A., et al.: Neuroprotective effect of ethanol extract of gentiana kurroo royle on aluminum chloride-induced neurotoxicity in rats. BMC Complement. Altern. Med. **18**(1), 1–10 (2018)

Bora, K.S., Sharma, A., Singh, S., et al.: Evaluation of antidepressant-like activity of *Glycyrrhiza glabra* L. in experimental models of depression. Pharmacognosy Res. **3**(1), 21–25 (2011)

Sarubbo, F., Ramis, M.R., Kienzer, C., Aparicio, S., Esteban, S., Miralles, A., Moranta, D., et al.: Neuroprotective effects of flavonoids on transient cerebral ischemia in mice: structural requirements for activity. Pharmacol. Res. **100**, 363–373 (2015)

Sivarajan, I.B., et al.: Ayurvedic Drugs and Their Plant Sources. CBS Publishers & Distributors Pvt Ltd, India (2017)

Sachar, S., Kumar, A., et al.: Deep ensemble learning for automatic medicinal leaf identification. Int. J. Inf. Technol. **14**(6), 3089–3097 (2022)

Ngoc, T.M., Li, M.S., Liu, S.H., Wu, C.F., Yen, J.H., Su, Y.J., et al.: Neuroprotective effects of Alpinia oxyphylla extract against glutamate-induced oxidative stress and apoptosis in PC12 cells. Nutrients **11**(10), 2469 (2019)

Shailendra, R., et al.: An IoT and machine learning based intelligent system for the classification of therapeutic plants. Neural Process. Lett. **54**(5), 4465–4493 (2022). https://doi.org/10.1007/s11063-022-10818-5

Das, S., Mohapatra, S.K., Mishra, S.K., et al.: An in silico approach to evaluate the potential of Withania somnifera bioactives for inhibiting SARS-CoV-2 viral proteins. Informat. Med. Unlocked **24**, 100579 (2021)

Abdollahi, J., et al.: Identification of medicinal plants in Ardabil using deep learning: identification of medicinal plants using deep learning. In: 2022 27th International Computer Conference, Computer Society of Iran (CSICC), pp. 1–6. IEEE (2022)

Gao, K., Xu, A., Krulwich, T.A., et al.: Roles of resveratrol in improving the antioxidant deficiency of chronic fatigue syndrome. Nutrients **10**(10), 1395 (2018)

Ariful Hassan, M., Sydul Islam, M., Mehedi Hasan, M., Shorif, S.B., Tarek Habib, M., Uddin, M. S., et al.: Medicinal plant recognition from leaf images using deep learning. In: Computer Vision and Machine Learning in Agriculture, vol. 2, pp. 137–154. Springer, Singapore (2022)

Li, D., et al.: Deep learning and machine intelligence: New computational modeling techniques for discovery of the combination rules and pharmacodynamic characteristics of Traditional Chinese Medicine. European J. Pharmacol. **933**, 175260 (2022). https://doi.org/10.1016/j.ejphar.2022.175260

Azadnia, R., Al-Amidi, M.M., Mohammadi, H., Cifci, M.A., Daryab, A., Cavallo, E., et al.: An ai-based approach for medicinal plant identification using deep CNN based on global average pooling. Agronomy **12**(11), 2723 (2022)

Tajidini, F.: A comprehensive review of deep learning in lung cancer. arXiv preprint arXiv:2308.02528 (2023)

Tomitaka, A., Vashist, A., Kolishetti, N., Nair, M., et al.: Machine learning assisted nanomedicine using magnetic nanoparticles for central nervous system diseases. Nanoscale Adv. **5**(17), 4354–4367 (2023)

Altemimi, A., Lakhssassi, N., Baharlouei, A., Watson, D.G., Lightfoot, D.A., et al.: Phytochemicals: Extraction, isolation, and identification of bioactive compounds from plant extracts. Plants **6**(4), 42 (2017)

Tiwari, V., Joshi, R.C., Dutta, M.K., et al.: Deep neural network for multi-class classification of medicinal plant leaves. Expert. Syst. **39**(8), e13041 (2022)

Jangra, A., Dwivedi, V., Sharma, M., Joshi, R., et al.: Centella Asiatica attenuates neurobehavioral, neurochemical, and neurophysiological abnormalities in MPTP-induced parkinsonism in mice. Metabolic Brain Dis. **34**(6), 1781–1794 (2019)

A Real Time Application for Crime Trends Prediction Using ML Algorithms

S. Chaithra[1] and R. Pushpalatha[2](✉)

[1] Department of Computer Science and Engineering, VTU Centre for PG Studies Mysuru, Mysuru, India
Chaithra.ambale@gmail.com

[2] Department of Computer Science and Engineering, VTU Centre for PG Studies Mysuru, Mysuru, India

Abstract. Now day's crimes are increasing rapidly and there is no region or city without crimes. So we require an automation for crime detection and prevention. System should stop crimes before it starts. As crimes are increasing, precautionary measures to be taken to stop crimes. Finding frequent crimes and related correlations is a tedious and high effort task in the current crime sector. This issues needs to be addressed. Current real-time system isn't supports automation for crime prediction in real time. Machine learning or AI is the emerging technology to solve this issue. Efficient Unsupervised learning algorithms used to process crime training datasets and frequent crimes and their relationships are identified [5]. Proposed system build to reduce crimes and applicable to crime sector. We are building a real time application where it is useful for crime sectors to reduce the crimes. Currently no one of the crime sector applications does this and we use efficient data science algorithms to predict crimes with better results.

Keywords: Data Science · Machine Learning · Apriori Algorithm · Apriori TID Algorithm · Crime Types · Patterns

1 Introduction

In real time crimes are increasing more and there are various types of crimes such as pickpocket, murder, kidnap etc.… Several crime related works have proposed many of techniques to solve the crimes that used too many applications [3]. Crimes are increasing more in day today life. All cities and all over world facing this crime problems. Crime type depends on many factors such as location, Population, school zone, hospital zone etc… It is very significant to identify the frequent crimes and take necessary actions to reduce the crimes. In this proposed work we build real time application for crime sector to find frequent crimes in an area and also we predict relationship between different types of crimes. We apply efficient ML algorithms to process crime data and prediction is done. Data Science, is the trending technology that can applied to solve all type of problems in all types of fields [1]. In this work data science applied to crime sector to process old crime data or previous years crime data and to find the frequent crime types and their correlations.

B. K. Mishra et al. (Eds.): ICT4GS 2024, ISEM 56, pp. 149–163, 2025.
https://doi.org/10.1007/978-3-031-94997-5_15

Machine learning is useful to train the system using training datasets. Here we use crime datasets, effective algorithms for machine learning are employed to process crime datasets and hidden crime patterns are extracted. Unsupervised learning methods used to train the system and results predicted. Efficient unsupervised learning algorithms such as Apriori algorithm, Apriori TID algorithms used to process crime datasets and frequent crimes and their relationships predicted. We develop a real time web application to predict frequent crimes and crime patterns using ML algorithms.

We develop machine learning algorithm for predicting frequent crimes and their relationships and comparing the proposed algorithm with existing algorithm. Our system does identifying crimes and related correlations and provides solution to reduce crimes and making public more alert and active. System finds the correlations between different crime types such as murder, chain snatching, kidnap, pick pocket, robbery etc..... We are building a real time application to find the frequent crimes happening today where it is useful for crime sectors to reduce the crimes.

2 Related Works

2.1 Survey Papers

A) IEEE PAPER TITLE: Crime Pattern Detection using Simple Apriori Algorithm

https://www.semanticscholar.org/paper/Time%2C-Place%2C-and-Modus-Operandi%3A-A-SimpleApriori-ChenKurland/1e0e269245c6501f48ea83b23f52bb298fc1d98c.

https://ieeexplore.ieee.org/document/8633657.

YEAR OF PUBLICATION: 2020.

AUTHORS: Peng Chenn, Justin Kurland.

METHODOLOGY: Apriori Algorithm Used.

DESCRIPTION: This paper aims at applying apriori algorithm to process crime data and suspect crime patterns. Data science techniques are powerful subject to process data. This paper aims to resolve the crime problems and bring some solution using previously underutilized parameters from police recorded crime data. To ensure that to achieve the goal, a crime procedure is proposed and 3 factor (1) time; (2) setting; and (3) modus operandi.

LIMITATIONS:

- Uses efficient data mining algorithms
- Huge data required
- More time required for prediction
- Not Implemented as Real Time
- IEEE PAPER TITLE: A Model for Visual and Intuitive Crime Investigation Based on Associative Rule Mining Technique

https://ieeexplore.ieee.org/abstract/document/8764876.

YEAR OF PUBLICATION: 2020.

AUTHORS: Edigar adero, George okeyo, Waweru mwangi.

METHODOLOGY: Apriori Algorithm used.

DESCRIPTION: Crime has been part of our society ever since the concept of laws was first approved. There is an increased concern at governance level due to escalating levels of crime both internationally as well as locally in Kenya. In this article, the researcher suggests the utilize of Associative Rule Mining to come up with a model suitable for crime analysis and prevention using Apriori algorithm to represent mutual implications among criminal occurrences.
LIMITATIONS:

- Apriori Algorithm used which process huge amount of crime data. Needs more data.
- Needs more time for crime data analysis.
- Doesn't predicts the crime relationships.
- IEEE PAPER TITLE: Predicting Crime Through Data Mining Methods

https://www.researchgate.net/publication/220765686_Crime_Forecasting_Using_Data_Mining_Techniques.

YEAR OF PUBLICATION: 2020.

AUTHORS: Chung-Hsien Yu, Max W. Ward, Melissa Morabito, and Wei Ding.

METHODOLOGY: Classification Techniques Used.

DESCRIPTION: Most crimes are "undetermined." It doesn't always happen at random, but it also doesn't happen in space or time consistently.In this research, we discuss preliminary results of a crime prediction model developed in collaboration with local police departments in the northern United States. We analyzed several classifications to determine which method is best at predicting crime "hot spots." We also focused at the addition or formation of groups.
LIMITATIONS:

- Data mining Algorithms used, so huge amount of data required
- Less Accurate Results

3 Difference Between Existing Works and Proposed Work

- In most of the current projects the implementation is successful, but the algorithms used are not programmed and they use off-the-shelf libraries and tools that the algorithms use. However, in the system concept, we program the algorithm, that is, we program own logic for the algorithm and the results are tested.
- Many research works uses less amount of training data-sets, in the proposed system we use huge data-sets for processing.
- All existing works uses PYTHON or R Language or Ready Data science tools for prediction and which works for static datasets, but in the proposed system we implement the concept for dynamic datasets (real time application).
- All existing works are just model development, can't be used in real time. Here we build this concept as real time application using front end technology as "*visual Studio*" and back end technology as "*SQL Server*" and C# as programming language.
- Suggested system is a real world application with model using Microsoft technologies.

3.1 Current Work

The proposed procedure can also be applied to criminal cases. The planning process includes crime modeling, accurate detection, equipment planning and replacement, and operation time to find out the necessary algorithms for crime detection. The proposed system detects criminal behavior, predicts crime, accurately analyzes and manages huge amounts of information gathered from various sources. The proposed system uses data science technology "association rules" to predict the relationship between different types of crimes.

Example: location "vijayanagara", depending on the earlier crime data we can predict the different types of crime that takes place in future.

Output – Crime patterns with different crime types in "vijayanagara" for upcoming days.

1. robbery, murder related to suicide
2. child abusing related to murder
3. chain snatching, robbery related to murder
4. pick pocket, chain snatching related to robbery and murder

3.2 Datasets

	A
1	Crimes
2	Murder, Kidnap, Chain Snatching, Robbery, Half Murder
3	Kidnap, Chain Snatching, PickPocket, Neck Hanging
4	Murder, Single Car Chain Snatching, Half Murder
5	Murder, Kidnap, Chain Snatching, Hit n Run
6	Hit n run, Drunk n Drive, Chain Snatching, Theft
7	Murder, Kidnap, Chain Snatching, Neck Hanging, Man Handling
8	Murder, Chain Snatching, Robbery
9	Drunk n Driven, Kidnap, Chain Snatching, Theft
10	Kidnap, Hit n run, Half Murder
11	Kidnap, Chain Snatching, Robbery, Half Murder
12	Murder, Kidnap, Chain Snatching, Robbery, Half Murder
13	Chain Snatching, Kidnap,Hit n run, Neck Hanging, Man Handling
14	Single Car Accident, Chain Snatching, Hit n Run
15	Bad Light, Kidnap, Chain Snatching, Hit n Run, Neck Hanging
16	Kidnap, Chain Snatching, Neck Hanging, Hit n Run
17	Murder, Single Car Chain Snatching, Half Murder
18	Murder, Kidnap, Chain Snatching, Hit n Run
19	Hit n run, Kidnap, Chain Snatching, Neck Hanging
20	Bad Light, Kidnap, Chain Snatching, Hit n Run, Neck Hanging

MysDept

Fig. 1. Crime Datasets

4 Methodology

4.1 Unsupervised Learning

A Descriptive model is used for tasks that would benefit from the insight gained from summarizing data in new and interesting ways. There are no predefined labels in unsupervised learning technique. The goal is to explore the data and find some structure with in. Unsupervised learning works well on transactional data.

In our project Association learning algorithms used such as "Apriori Algorithm and Apriori TID Algorithm. These algorithms preferred as algorithms supports small datasets and also large datasets.

4.2 Process Flow

Step 1: previous year's crime data gathered from different sources such as kaggle.com, dataworld.com, data.gov.in, githib.com etc...
Step 2: here crime data is preprocessed, where we remove unwanted data and fetch the required data. Unwanted data such as crime no, serial no etc. will be removed.
Step 3: required data is inputted to algorithms. We use efficient algorithms such as Apriori algorithm and Eclat algorithm to process the data.
Step 4: after processing, frequent crimes will be extracted and displayed.
Step 5: Relationship between different types of crimes will be extracted and displayed.
Step 6: both algorithms results compared and efficient algorithm will be chosen.
Step 7. Final outputs displayed on GUI.
Step 8: Comparative analysis of algorithms displayed using graphs.

4.3 Apriori Algorithm Pseudo-Code

```
Apriori (T, minSupport)
    Cl = {candidate I-itemsets};
    L1 = {c ∈ C1|c.count ≥ minsup};
    FOR (k = 2; Lk-1ĭ; k + +) DO BEGIN
    Ck = apriori-gen(Lk-1);
    FOR all transactions t ∈ D DO BEGIN
    Ct = subset (Ck, t);
    FOR all candidates c ∈ Ct DO
    c.count + +;
    END
    Lk = {c ∈ Ck |c.count ≥ minsup}
    END
    Answer = * Lk;
    See (Fig. 2).
```

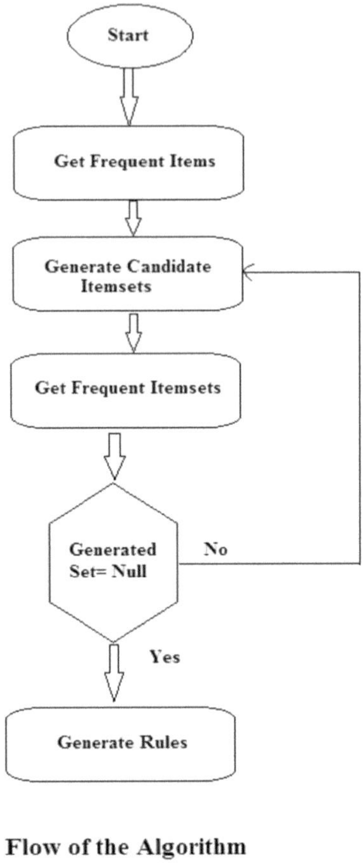

Fig. 2. Flow of Apriori Algorithm

4.4 Apriori TID Algorithm Pseudo-Code

```
private void AprioriTID()
    {
       double _support = 0.5;
       double _confidence = 0.8;

       Cn = new string[lv_Transactions.Items.Count];

       //function to load _C1
       C1 = C11();

       //function to calculate the L1 and store in buffer
       Dictionary<string, double> _FrequentItemsMain = L1(_support);
       Dictionary<string, double> Candidates = new Dictionary<string, double>();

       do
       {
          //function to calculate C2,C3.... and store in buffer
         _Candidates = GenerateCandidates (_FrequentItemsMain);

         //C21,C31,C41...
         Cn = CandiN(Candidates);

         //function to calculate L2,L3.... and store in buffer
         _FrequentItemsMain = GetFrequentItems(_Candidates, _support);

         ++Cnt;
       }

       while (Candidates.Count != 0);

       //function to generate rules or patterns
       List<ClassRules> RulesList = GenerateRules();
       //compare with the confidence and find strong rules
       List<ClassRules> StrongRules = GetStrongRules(_confidence, RulesList);

       //function to display the final results (L and Patterns)
       Result(DictionaryAllFrequentItems, StrongRules);
    }
```

See (Fig. 3).

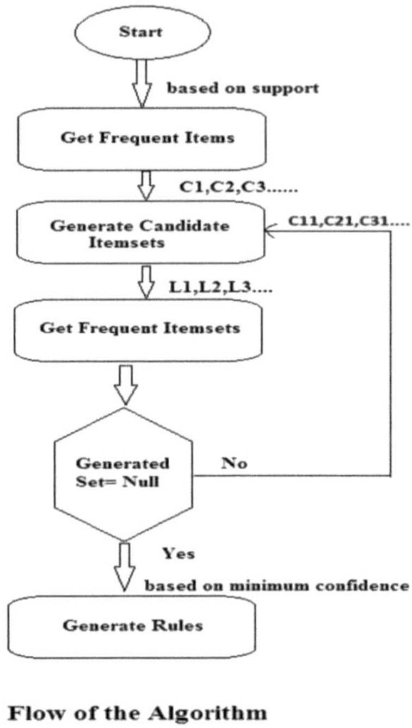

Fig. 3. Flow of Apriori TID Algorithm

4.5 Association Learning

SampleDataset

TID	Crimes
1	Robbery, Kidnap, Chain snatching
2	Robbery, Kidnap, Pickpocket
3	Robbery, Murder, Kidnap, Pickpocket
4	Murder, Pickpocket

Minimum Support = 50%
Minimum Confidence = 80%

C1

Items	Support
Robbery	75%
Murder	50%
Kidnap	75%
Chain snatching	25%
Pickpocket	75%

L1

Items	Support
Robbery	75%
Murder	50%
Kidnap	75%
Pickpocket	75%

C2

Items	Support
Robbery, Murder	25%
Robbery Kidnap	75%
Robbery Pickpocket	50%
Murder , Kidnap	25%
Murder, Pickpocket	50%
Kidnap Pickpocket	50%

See (Fig. 1).

L2

Items	Support
Robbery Kidnap	75%
Robbery Pickpocket	50%
Murder Pickpocket	50%
Kidnap Pickpocket	50%

C3

Items	Support
Robbery Kidnap Pickpocket	50%
Robbery Murder Kidnap	25%
Robbery Murder Pickpocket	25%
Murder, Kidnap, Pickpocket	25%

L3

Items	Support
Robbery Kidnap Pickpocket	50%

FREQUENT ITEM SET (L)

Items	Support
Robbery	75%
Murder	50%
Kidnap	75%
Pickpocket	75%
Robbery Kidnap	75.00%
Robbery Pickpocket	50%
Murder Pickpocket	50%
Kidnap Pickpocket	50%
Robbery Kidnap Pickpocket	50%

GENERATE CONFIDENCE:

GENERATE CONFIDENCE:

RULE X RULE Y	C
{ Robbery } > { Kidnap }	100.00%
{ Kidnap } > { Robbery }	100.00%
{ Robbery } > { Pickpocket }	66%
{ Pickpocket } > { Robbery }	66%
{ Murder } > { Pickpocket }	100%
{ Pickpocket } > { Murder }	66%
{ Kidnap } > { Pickpocket }	66%
{ Pickpocket } > { Kidnap }	66%
{ Robbery } > { Kidnap Pickpocket }	66%
{ Kidnap } > { Robbery Pickpocket }	66%
{ Pickpocket } > { Robbery Kidnap }	66%
{ Kidnap Pickpocket } > { Robbery }	100%
{ Robbery Pickpocket } > { Kidnap }	100%
{ Robbery Kidnap } >{ Pickpocket }	66.00%

STRONG ASSOCIATION RULE:

1. { Murder } -> { Pickpocket }
2. { Kidnap Pickpocket } -> { Robbery }
3. { Robbery Pickpocket } -> { Kidnap }
4. { Robbery } -> { Kidnap }
5. { Kidnap } -> { Robbery }

5 Experiment Results

Apriori Algorithm – Accuracy, Correct Patterns and Incorrect Patterns (Fig. 4).

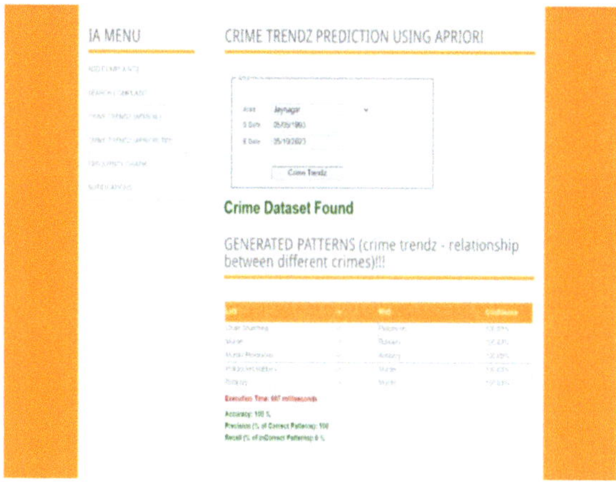

Fig. 4. Apriori Patterns with Accuracy

Apriori TID Algorithm – Accuracy, Correct Patterns and Incorrect Patterns (Fig. 5).

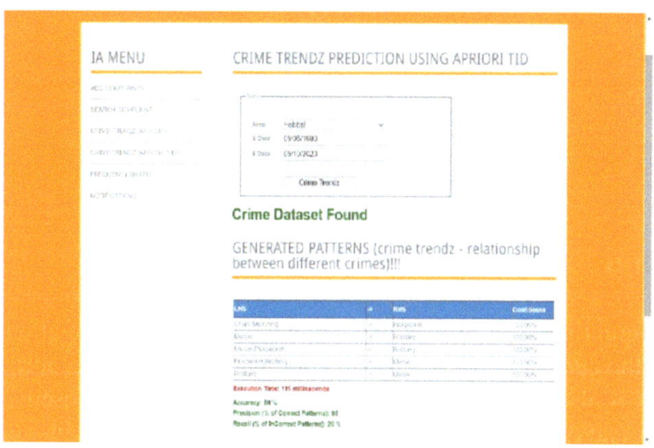

Fig. 5. Apriori TID Patterns with Accuracy

Graphical Representation – Frequency Plot – Jaynagar (Figs. 6 and 7).

Fig. 6. Crimes Frequency Plot

Fig. 7. Algorithm Comparison

6 Conclusion

Identifying different types of crimes that can take place in upcoming days plays a significant role in the current crime or investigation sector. Taking precautionary measures to avoid crimes is also important to reduce crimes in a city. Proposed system predicts frequent crimes and their relationship in advance, so that police or investigation departments can take necessary actions to reduce crimes. We are building a real time application where it is useful for crime sectors to decrease the number of crimes. Now day's crimes are increasing rapidly and we need a system which detects crimes and stops crimes before it starts. Finding frequent crimes and relationship between crime types is a challenging tasks in the current crime sector.

There is no automation for crime prediction. Crime prevention and detection become an important trend in crime and a very challenging to detect and solve crimes. Determining the crime patterns is a major challenge in today's world to reduce the crimes and to take the precautionary measures to avoid crimes. So our system is applicable in the field of crime. We built a real time application where it is useful for crime sectors to reduce the crimes. We use data science algorithms for crime patterns prediction to get better results and our system helps publics and police departments in identifying the crimes in right time and to reduce the crimes.

References

1. Pushpalatha R., Ramesh, B.: Taylor CFRO-based deep learning model for service-level agreement-aware VM migration and workload prediction-enabled power model in cloud computing. IJSIR **31**(1) (2022)
2. Pushpalatha, R., Ramesh, B.: Amalgamation of neural network and genetic algorithm for efficient workload prediction in data center. In: Kalya, S., Kulkarni, M., Shivaprakasha, K.S. (eds.) Advances in VLSI, Signal Processing, Power Electronics, IoT, Communication and Embedded Systems. LNEE, vol. 752, pp. 69–84. Springer, Singapore (2021). https://doi.org/10.1007/978-981-16-0443-0_6
3. Chandini, M., Pushpalatha, R., Boraia, R.: A brief study on prediction of load in cloud environment. IJARCCE **5**(5) (2016)
4. Peng Chen, Justin Kurland: Time, Place, and Modus Operandi: A Basic Apriori Algorithm Investigation for the Identification of Crime Patterns, IISA (2018)
5. Steinbach, M., Tan, P.-N., Xiong, H., Kumar, V.: Objective Measures for Association Pattern Analysis (2019)
6. Yamuna, S., Sudha Bhuvaneswari, N.: Datamining Techniques to Analyze and Predict Crimes (2012)
7. Yadav, S., Timbadia, M., Yadav, A., Vishwakarma, R., Yadav, N.: Crime Pattern Detection, Analysis & Prediction (2017)
8. Yu, C.-H., Max, W.: Ward, Melissa Morabito, and Wei Ding : Crime Forecasting Using Data Mining Techniques (2018)
9. http://msdn.microsoft.com/asp.net
10. http://www.csharphelp.com
11. http://www.w3schools.com
12. http://www.dotnetspider.com
13. http://www.csharpcorner.com
14. http://www.codeguru.com
15. http://www.tradersedgeindia.com
16. http://www.lastbull.com
17. http://www.sharetipsinfo.com

Analysis and Optimization of Co-channel Interference Using Genetic Algorithm

Shashi Bhusan Panda, Prasanta Kumar Swain(✉), and Sidhartha Sankar Dora

MSCB University, Baripada, Odisha, India
prasantanou@gmail.com

Abstract. Cellular networks are designed to accommodate a large number of users while efficiently utilizing limited radio spectrum resources. Co-channel interference poses significant challenges for cellular network operators and can have widespread negative impacts on network performance, user experience, and operational costs. Effective interference management strategies and technologies are essential to mitigate these effects and ensure optimal performance of cellular networks. Optimization techniques in cellular networks aim to improve network performance, enhance user experience, and maximize resource utilization. Co-channel interference minimization techniques are methods employed in cellular networks to reduce the interference that occurs between cells operating on the same frequency channels. Potential solutions can be obtained by using genetic algorithm for co-channel interference optimization.

Keywords: Cellular Network · Co-Channel · Interference · Genetic Algorithm

1 Introduction

Wireless communication networks made up of linked cells that are serviced by base stations or cell sites make up cellular networks. These networks use a combination of radio frequency (RF) and mobile communication technologies to provide wireless voice and data services to users within a geographic area. Cellular networks are designed to accommodate a large number of users while efficiently utilizing limited radio spectrum resources. With ultra-fast data speeds, minimal latency, extensive device connectivity and compatibility for 5G networks are the pinnacle of cellular technology. A sample co-channel intefrence situation is shown in Fig. 1.

Co-channel interference occurs when signals from different cells using the same frequency band interfere with each other. This interference can degrade the quality of communication and reduce the overall capacity and performance of the network. Minimizing co-channel interference is crucial for optimizing the performance and efficiency of a cellular network. Strategies such as carefully allocating frequency bands to cells, adjusting transmitting power levels, implementing interference mitigation techniques, and optimizing cell placement can help reduce co-channel interference and improve network performance. Co-channel interference becomes more pronounced in areas where cells are closely spaced or densely populated, leading to increased signal overlap and

B. K. Mishra et al. (Eds.): ICT4GS 2024, ISEM 56, pp. 164–174, 2025.
https://doi.org/10.1007/978-3-031-94997-5_16

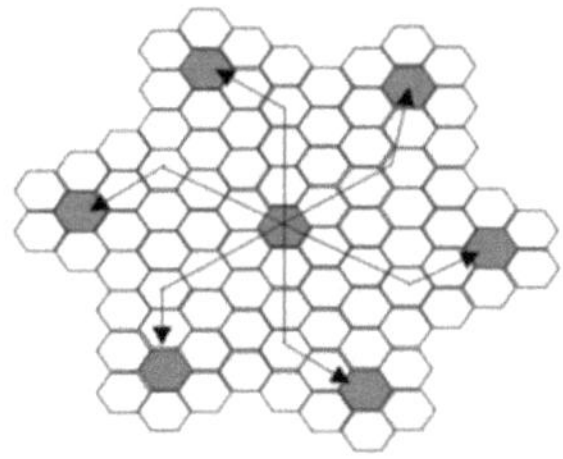

Fig. 1. Co-Channel Interference Model

higher interference levels. Mitigating co-channel interference is essential for optimizing network performance, maximizing capacity, and providing reliable wireless communication services to users. Cellular network operators employ various techniques such as frequency reuse planning, power control, interference coordination, and adaptive modulation to minimize co-channel interference and improve network efficiency and quality of service. Here, we try to minmize CCI by using GA approach and make a comparative analysis with non optimized level of CCI in a cellular network.

Rest of the paper is organized as follows: Sect. 2 discusses related works and Sect. 3 explains co-channel reduction approaches. We have optimized co-channel interferences using genetic algorithm to improve quality of dense cellular network services in smart cities. The purposed system is explained in Sect. 4. Section 5 concentrating on the simulation study and Sect. 6 is focused on result analysis. Finally, the Sect. 7 has concluded with future enhancement.

2 Related Works

The minimum-order technique to frequency assignment is presented in this study [1], along with a theory that connects it to the conventional approach. There is a chance that this new strategy will be preferred over the conventional one. We formulate assignment problems as two-dimensional optimization problems. Avoid using the frequency-constrained technique if distance separation is being used to reduce interference. In frequency-distance constrained situations, disk graphs—a restricted class of graphs—play a key role. We can convert many generalized graph colouring problems that map to frequency assignment problems in a straightforward way. We classify a large number of frequency assignment issues according to graph colouring's "execution time efficiency". A cellular mobile telecommunications system in [2] links mobile users to the land telephone network by using base stations dispersed over a metropolitan area. Every base has enough radios to service the clients in the cell (region) that it is in charge of. Every radio in the system needs to have a specific radio channel assigned to it; many radios can share a single channel as long as there is no excessive interference. Channel assignment problem aims to allocate channels to radios in a system based on a set of interference constraints [3]. The paper presents a conveniently automated heuristic assignment technique where the requirements for the channels that consistently show themselves as "difficult" to assign, advance to the top of the list of criteria. A hypothesis for the ideal neighbouring channel spacing in homogeneous hexagonal cell systems is created in [4]. We offer algorithms that produce frequency distributions with the optimal adjacent channel distance, regardless of the co-channel reuse ratio. Broadcast services and mobile radio telephone systems can both benefit from the theory. A few heuristic techniques for channel assignment in cellular systems are explained in [5].

One way to think of the channel-assignment problem is as a generalized graph-coloring problem. These algorithms were constructed by partially modifying some of the concepts found in heuristic graph-coloring methods. As a minimum-span problem, the channel-assignment problem involves figuring out the least amount of bandwidth needed to satisfy a given demand. Therefore, any method that lowers interference raises SIR and boosts user capacity [6]. Although shielding is used in coaxial cable or shielded twisted pair, it cannot be used to insulate radio transmission (wireless) against interference; however, interference can be reduced using other methods [7]. Signals from different users are sent in different ways. Some of the interferers might be directly in band. Strong signals at adjacent frequencies may cause energy to seep into the intended frequency [8].With the help of this cutting-edge resource, you may examine and learn about the behaviour of wireless communication systems in an environment with interference in a novel way. Regardless of the source, the book covers both additive and multiplicative interfering signals and provides detailed descriptions of their behaviour [9]. Interference between two neighbouring frequency bands is referred to as adjacent channel interference [10]. Co channel interference is the term used to describe interference that occurs when two network cells use the same channel too close to one another. In this paper, we offer a multi-criteria vertical handoff decision method that, the best network with the best parameter values (e.g., the lowest network cost) would be selected in a heterogeneous wireless network. The multiple goal optimization problems is used to simulate the choice problem using a genetic algorithm. According to the simulation's results, fewer handoffs can be made by choosing the best network parameter values [11]. Design process, cellular system operation are key area of development in last couple of years[12]. For all radio network designers and engineers, network operators' technical managers, radio regulation engineers, and military radio network planners, it is an essential resource [13].

With limited spectrum utilization, the number of users in [14] Cellular Communication Network is rapidly growing. In order to accommodate the growing number of users, the spectrum must be used effectively even in the presence of co-channel, adjacent channel, and co-site interference. There are three methods for allocating channels that take various limitations into account: Fixed, Dynamic, and Hybrid. By combining Genetic Algorithm (GA) with Support Vector Machine (SVM) classification, an optimal Hybrid channel allocation technique with co-channel and co-site constraints is provided. GA is an optimization method founded on the "survival of the fittest" theory of Darwinian evolution. The best classification method for non-linear datasets is support vector machines (SVM). In terms of calculation time, signal to interference ratio (SIR), and interfering edges, the GA-SVM results are shown to be superior to those published in the literature. It is challenging to assign an appropriate frequency for phone calls in a wireless mobile network [15] while adhering to a particular level of constraint. To increase system capacity and lessen the impact of interference, a good channel assignment strategy is essential.

Deterministic techniques form the foundation of the majority of current channel assignment methodologies. This research presents a genetic algorithm (GA) based adaptive hybrid channel assignment (HCA) technique. The proposed GA based optimization in HCA system indicates the speed of call depends on population size and the chnnels [7]. To reduce the amount of handoffs in heterogeneous wireless networks—which are made up of multiple wireless networks—a novel vertical handoff determination technique is put forth. In this paper, we propose a multi-criteria vertical handoff decision approach that will select the best available network with optimal parameter values (e.g., network cost should be least) in a heterogeneous wireless network.). Using a genetic algorithm, the choice problem is simulated as a multiple objective optimization problem. The simulation's result suggests that by choosing optimal network parameter values, the number of handoffs can be reduced [16]. The co-channel interference reduction with subcarrier spacing, down link and uplink configuration considered as challenging. CCI calculation using reference signal and the mathematical analysis is proposed by authors [17]. Authors in [18] implimented a deep learning oriented enhanced inter-cell interference coordination (eICIC) and in the next step enhance it to protect against interference. The review article scrutinizes different issues of interferences found in various structures and 5G technology. This concentrates on interference found in HetNet, RN, D2D, and IoT [19]. Asymptotic signal-tointerference- plus-noise ratio (SINR) datain two aspects such as: i) transmit power scaling are used in end nodes and relay points (ii) transmit power scaling is used in end nodes only [20]. By permiting small base station structure interference can be managed for UD-SCN with coalition and increase the network efficiency [21]. MIMO is comming as a new colution wireless networks. MIMO-OFDM system with QPSK is treated as wireless network to act as a technique for interference cancellation [22].

3 Co-channel Reduction Methods

3.1 Genetic Algorithm

Genetic algorithms (GA) play a significant role in optimizing various aspects of cellular networks. Here's how genetic algorithms contribute to improving cellular network performance:

Optimization of Resource Allocation: Genetic algorithms can be used to optimize the allocation of resources such as frequency bands, time slots, and power levels in cellular networks. By evolving solutions over successive generations, GAs can find optimal resource allocation strategies that maximize network capacity, throughput, and spectral efficiency while minimizing interference and maintaining quality of service.

Frequency Planning and Channel Assignment: Genetic algorithms are effective tools for frequency planning and channel assignment in cellular networks. By considering factors such as interference levels, user distribution, and coverage requirements, GAs can generate frequency reuse patterns and allocate channels to cells in a way that minimizes co-channel interference and maximizes spectral utilization.

Interference Management: Genetic algorithms can assist in managing interference between cells in cellular networks. By optimizing parameters such as transmit power levels, antenna configurations, and handover thresholds, GAs can mitigate co-channel interference and improve network performance, particularly in dense urban environments with high user density and limited frequency resources.

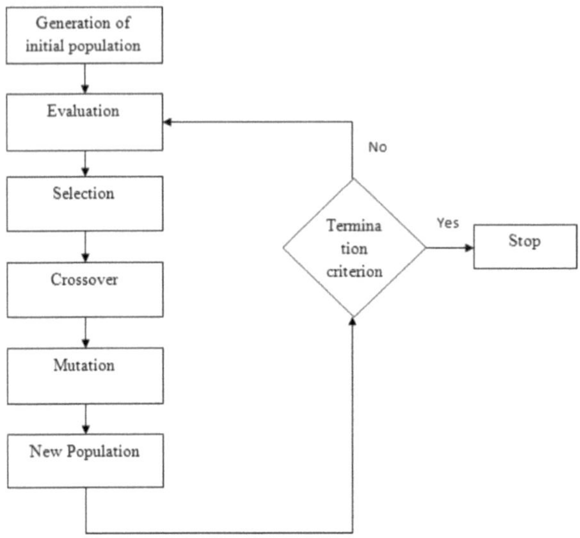

Fig. 2. GA Flow map

- Initialization: Generate an initial population of candidate solutions, ensuring diversity in frequency assignments and power levels.
- Selection: Choose promising solutions based on their fitness (degree of interference reduction and signal quality).
- Crossover: Combine selected solutions to create offspring with potentially improved characteristics.
- Mutation: Introduce small changes to offspring to maintain genetic diversity and explore new solution spaces.
- Replacement: Replace fewer promising solutions in the population with newly generated offspring.
- Termination: Stop the algorithm when a satisfactory solution is found or after a predetermined number of generations. The flow diagram of GA is shown in Fig. 2.

4 Proposed System

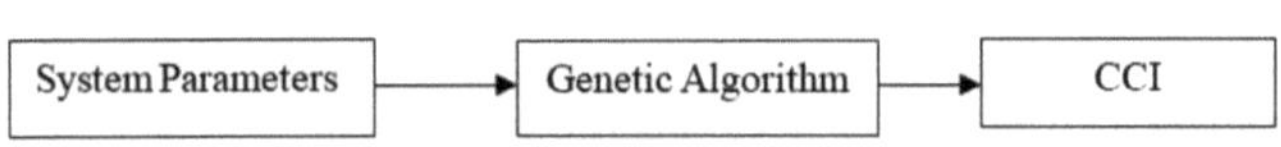

Fig. 3. Proposed Model Diagram

Multiple hexagon-shaped cells arranged in a parallelogram arrangement make up a cellular structure. Let's say there are 49 cells and 70 channels in a cellular system. The nine core cells are the only ones taken into consideration for simulation. In this model, the three interference limitations are mostly considered. A channel is chosen for each incoming call in a cell so that co-channel, cosine, and adjacent channel interferences are eliminated. As previously stated, co-channel interference prevents cells that are closer together than the minimum reuse distance from sharing the same channel. In a similar vein, the cosine and adjacent channel constraint must be satisfied by the channels that are assigned to users who are in the same cell or adjacent cells. The parameters of our suggested system are the channels and cells. The fitness function, which evaluates each possible solution's quality, must also be created and sent to the algorithm. Optimal solution for the channels that need to be allocated to every cell is found at the output. The system model is shown in Fig. 3.

5 Simulation Environment

The Python programming language has been used to implement the GA_CCI simulation model. Twenty five cells have been considered in a cellular system for the sake of simplicity in the simulation. The simulation parameters are cells, transmission power, path losses, interference distances, frequencies, number of users per cell. At first, we have taken three cells having transmitted power as 30 dBm, 28 dBm and 25 dBm. The path losses values are 120, 130,110 of cells. The distance to interference cell are considered as 1000, 1500 and 2000 and number of users are 50, 40, and 60 respectively. We have increased the parameter values gradually which are shown below Table 1.

6 Result Analysis

6.1 CCI Values of the Proposed Cellular Network Without GA on the Basis of Cell

Table 1. CCI on the Basis of Cells

No of Cells	Co-Channel Interference
3	−29181.01
10	−108592.02
15	−183924.26
20	−237213.19
25	−295615.34

The cell values are increased gradually up to 25 where as other parameters are constant. The CCI values are shown in table-1. It is cleared from Table 1 that when number of cells increased, the CCI values are more negative. The quality of network will be degraded when CCI value will be more negative which is shown graphically in Fig. 4.

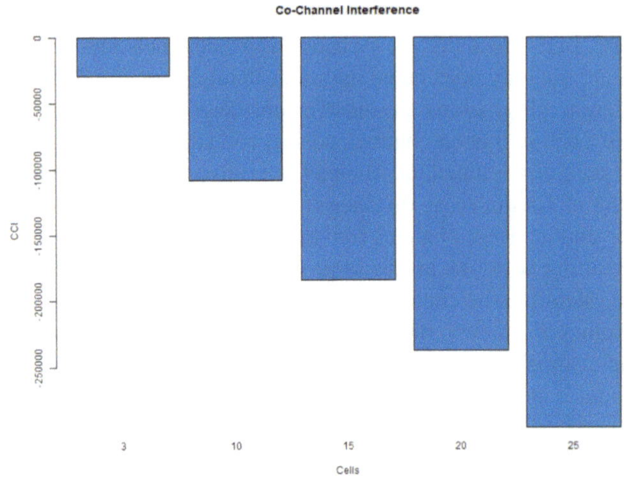

Fig. 4. Graphical representations of CCI on the Basis of Cells

6.2 CCI Values of the Proposed Cellular Network Without GA on the Basis of Users

Table 2. CCI on the Basis of Users

No of cell	Co-Channel Interference
3	−87847.54 dBm
10	−311187.10 dBm
15	−496520.44 dBm
20	−643028.18 dBm
25	−816526.52 dbm

The users are increased of each cell by 100 keeping other parameter values are constant. As a result, the CCI values are more negative which are shown in Table 2 the graphical representation is shown in Fig. 4. When the users of cells will be less the CCI value be less negative. So, the performance of network will be better. But in smart cities, the users are more day by day. From Tables 1 and 2, it is clear that when the users are less, CCI value will be less negative as shown in Fig. 6, otherwise the CCI value will be more negative which is not suitable for smart cities to provide better service. To minimize the CCI value, we have applied GA (Genetic Algorithm) to minimize the CCI value for quality of cellular network which are shown in the Table 3. From Tables 1 and 3, it is clear that the CCI value of cells are less negative after applied the optimization algorithm GA which is shown in Fig. 7.

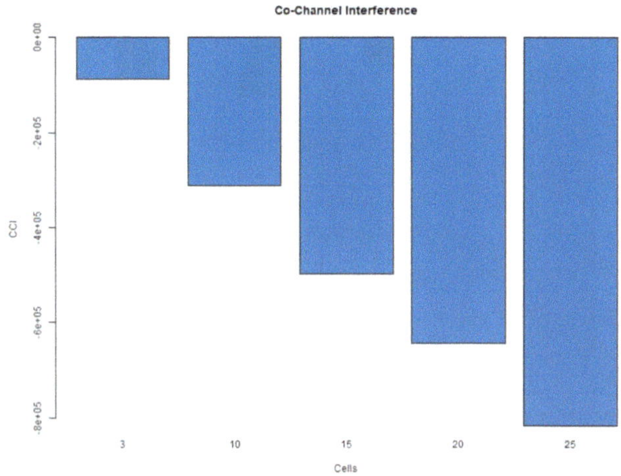

Fig. 5. Graphical representations of CCI on the Basis of Users

6.3 Comparisopn of CCI Values of the Proposed Cellular Network Without GA/ with GA on the Basis of Cell and Users

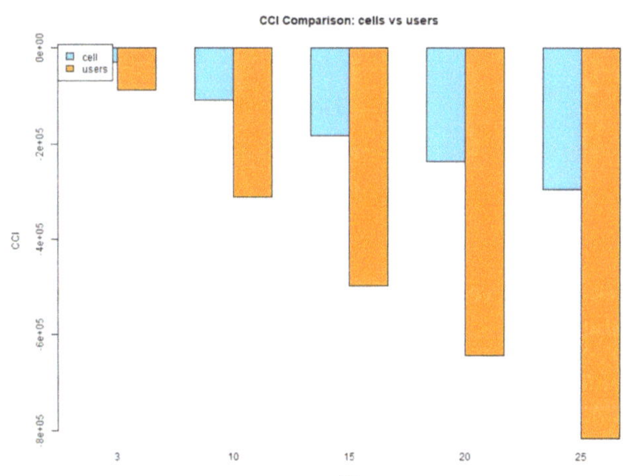

Fig. 6. Comparision of CCI on the Basis of Cell with Users.

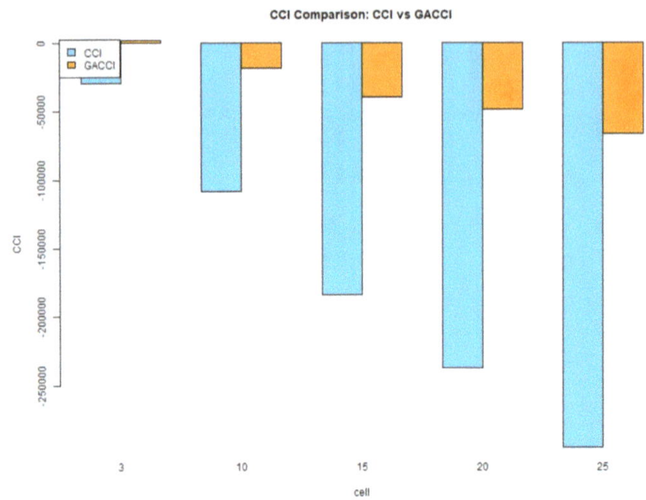

Fig. 7. Comparision of CCI on the Basis of using GA and non GA.

Table 3. CCI on the basis Cells and GA

No of cell	Co-Channel Interference
3	−1951.46
10	−18362.12
15	−39362.12
20	−48297.22
25	−66273.50

6.4 CCI Values of the Proposed Cellular Network with GA and Variation of Generations

From the Table 4 it is clear that when the generation is increasing the GA_CCI value are also increasing. As a result, the quality of cellular network is degraded. The quality of cellular network depends on the generation of genetic algorithm.

6.5 CCI Values of the Proposed Cellular Network with GA and Variation of Population

From the Table 5 it is clear that when the population is increasing the GA_CCI value are fluctuating. As a result, the quality of cellular network is varies accordingly.

Table 4. CCI on the basis GA and Generation

Generation	GA _CCI
100	−66273.50
200	−67018.81
300	−67326.62
400	−67401.15
500	−68010.12

Table 5. CCI on the basis GA and Population Size

Population size	GA_CCI
10	−67904.29
20	−68116.60
40	−67830.31
60	−68390.86
80	−68010.12

7 Conclusion

This paper implements GA-CCI, a genetic algorithm approach, to address the CCI problem in mobile communication networks. When compared to other channel assignment techniques, the innate ability of genetic algorithms to seek a wide variety of answers instead of restricting the search to a specific area produces adequate results. Moreover, the convergence property of genetic algorithms, independent of specific problem parameters, offers a distinct advantage, resulting in superior performance. Cell count, channel count, population size, and maximum number of epoc are the main factors taken into account by the genetic algorithm. The average value of each simulation is obtained by running it ten times. To evaluate the quality of each result, the ratio of users to the fitness function is calculated. By leveraging genetic algorithms, the results were achieved more efficiently and straightforwardly.

References

1. Hale, W.K.: Frequency assignment: theory and applications. Proc. IEEE **68**(12), 1497–1514 (1980)
2. Pennotti, R.J.: Channel Assignment in Cellular Mobile Telecommunication Systems. Polytechnic Institute of New York University (1976)
3. Box, F.: A heuristic technique for assigning frequencies to mobile radio nets. IEEE Trans. Veh. Technol. **27**(2), 57–64 (1978)
4. Gamst, A. :On frequency assignment in mobile automatic telephone systems. In: Proceedings of the Globecom, pp. 309–315 (1982)

5. Sivarajan, K.N., McEliece, R.J., Ketchum, J.W.: Channel assignment in cellular radio. In: IEEE 39th Vehicular Technology Conference,pp. 846–850. IEEE (1989)
6. Goldsmith, Wireless Communications, California, Cambridge University Press, p. 482 (2005)
7. Schiller, J.H.: Mobile communications. Pearson education (2003)
8. Schiff, M.: Introduction to Communication Systems Simulation, USA, Artech House, Inc., p. 9 (2006)
9. Stavroulakis, P. : Interference analysis and reduction for wireless systems. Artech House (2003)
10. Mishra, A.R. (ed.).: Advanced cellular network planning and optimisation: 2G/2.5 G/3G... evolution to 4G. John Wiley & Sons (2007)
11. Chandralekha, P.B., Praffula, K.B.: Minimization of number of handoff using genetic algorithm in heterogeneous wireless networks. Int. J. Latest Trends Comput. **1**(2), 24–28 (2010)
12. Molisch, A. F., Mammela, A., Taylor, D.P. (eds.): Wideband wireless digital communication. Prentice hall PTR (2000)
13. Theodore, S.R. : Wireless communications: principles and practice. CAMBRIDGE University Press (2000)
14. Katzela, I., Naghshineh, M.: Channel assignment schemes for cellular mobile telecommunication systems: a comprehensive survey. IEEE Pers. Commun. **3**(3), 10–31 (2000)
15. Ohatkar, S.N., Bormane, D.S., Bodhe, S.V.: Optimizing channel allocation techniques with GA-SVM for interference reduction in cellular networks. In: 2016 Conference on Advances in Signal Processing (CASP), pp. 28–32. IEEE (2016)
16. Dey, P., Masal, A., Kaimalettu, S., Milleth, J.K., Ramamurthi, B.: Reference signal design to mitigate co-channel interference in 5G OFDM systems with multiple numerologies. Phys. Commun. **53**, 101653 (2022)
17. Ahmad, I., et al.: Co-channel ınterference management for heterogeneous networks using deep learning approach. Information **14**(2), 139 (2023). https://doi.org/10.3390/info14020139
18. Siddiqui, M.U.A., Qamar, F., Ahmed, F., Nguyen, Q.N., Hassan, R.: Interference management in 5G and beyond network: requirements, challenges and future directions. IEEE Access **9**, 68932–68965 (2022)
19. Silva, S., Ardakani, M., Tellambura, C.: Interference suppression and energy efficiency improvement with massive MIMO and relay selection in cognitive two-way relay networks. IEEE Trans. Green Commun. Netw. **4**(2), 326–339 (2022)
20. Yang, G., Esmailpour, A., Nasser, N., Chen, G., Liu, Q., Bai, P.: A hierarchical clustering algorithm for interference management in ultra-dense small cell networks. IEEE Access **8**, 78726–78736 (2022)
21. Rajesh, V., Rajak, A.A.: Cancellation techniques for co-channel interference in MIMO-OFDM systems and evaluating their performance. Emerg. Sci. J. **5**(6), 824–839 (2022)
22. Chia, Y.S., Siew, Z.W., Kiring, A., Yang, S.S., Teo, K.T.K.: Adaptive hybrid channel assignment in wireless mobile network via genetic algorithm. In: 2011 11th International Conference on Hybrid Intelligent Systems (HIS), pp. 511–516. IEEE (2011)

Blockchain-Enabled DeFi Solutions for Achieving Sustainable Supply Chain Goals

Nisha Joseph[1(✉)] and Jyothy Joseph[2]

[1] Department of Commerce, Al-Ameen College, Edathala, Kerala, India
deptofnisha@gmail.com
[2] Department of Computer Science, Al-Ameen College, Edathala, Kerala, India
jyothy@alameencollege.org

Abstract. Decentralized Finance (DeFi) has brought about radical change in supply chain management. It minimizes the role of middlemen, and central institutions, and fosters peer-to-peer transactions through Blockchain Technology (BCT). Traditional mode supply chain uses paper-based and structured data systems. As against this, Blockchain technology develops a unified view of data to identify the current locations and status of the product and services and enables the smooth function of the supply chain process. Blockchain-enabled Defi Solutions are transforming the world of the supply chain by leveraging financial activities and making financial solutions more efficient, accessible, and secure. This paper focuses on Blockchain-enabled DeFi Solutions to achieve sustainable Supply Chain Goals.

Keywords: Decentralized Finance · Blockchain Technology · Supply Chain · Tokenisation · Distributed Ledger

1 Introduction

Decentralized Finance (DeFi) has emerged as an absolute and comprehensive force that might have a substantial influence on different sectors. DeFi uses emerging technologies to conduct financial transactions by weeding out intermediaries and centralized institutions and empowers individuals with peer-to-peer transactions using security protocols, connectivity, software, and hardware advancements [1]. DeFi uses Blockchain Technology (BCT) to lessen the influence of these middlemen in supply chain management. Supply chain are global networks of companies and people that are engaged in the manufacturing and provision of a products or services and includes manufacturers, sellers, warehouse agents, service providers, distribution channels, and small and medium retailers who works collectively to deliver goods to customers [2]. Recently, several businesses have implemented blockchain technology to maintain their supply chains.

Modern supply chains eliminate paper-based and disjointed data systems and utilize blockchain technology to develop a unified view of data to identify the current locations and status of the products and services. Blockchain technology can be used to develop

B. K. Mishra et al. (Eds.): ICT4GS 2024, ISEM 56, pp. 175–184, 2025.
https://doi.org/10.1007/978-3-031-94997-5_17

a peer-to-peer network [3] which allows several participants to transact freely without any central intervention. This phenomenon is possible through a secured ledger that cryptographically and immutably records every transaction which can be accessed by every individual.

2 Statement of the Problem

Supply Chain Financing has become a standard component and is used by businesses to optimize their cash flow. It uses a variety of financial tools and strategies to guarantee the smooth functioning of the processes in the supply chain runs. DeFi can be identified as a technique which brings about changes in the world of supply chain. Its blockchain-powered decentralization fosters an atmosphere of openness and trust. By automating procedures like payments and having the rules of the agreement explicitly inscribed into code, smart contracts and self-executing agreements lower the possibility of disagreements and delays in the supply chain processes [4]. This paper focuses on Block Chain enabled DeFi Solutions to achieve sustainable Supply Chain goals.

3 Objectives of the Study

This study aims to determine how sustainable supply chain goals may be fulfilled using Defi Solutions, which is operated by Blockchain. Additionally, it examines the benefits of supply chain management with DeFi solutions enabled by blockchain technology. It also analyses the Defensive DeFi enabled Supply Chain Management Solutions.

4 Methodology

The researchers employed a descriptive qualitative methodology incorporating literature reviews from journals, white papers, articles, and websites to gather information about DeFi and concentrated on formulating phenomenological research. This study aims to compile a taxonomy that will significantly impact DeFi.

5 Review of Literature

As a part of this study, some of articles the relating to supply chain management, blockchain technology, and decentralized finance were reviewed and, in this section, we incorporated some of the relevant papers.

(Ahmet Faruk Aysan et al., 2021) [5] asserts that blockchain technology has the ability to concentrate on real-world applications and suggests using the Sustainable Development Goal Acceleration Scorecard to assess blockchain-based approaches to the Sustainable Development Goals (SDGs).

(Ashutosh Dhar Dwivedi et al., 2021) [6] remarks that the Internet of Things (IoT) has transformed the digital world by enhancing their intelligence and autonomy as IoT has the advantages of fast data transport and feels that its dependence on centralized protocols presents security challenges.IoT network users may establish trust without the need for outside authorities thanks to blockchain technology's safe and decentralized ledger.

(Grassi et al., 2022) [7] The paper highlights the interplay between financial processes and human decision-making and examines the function of financial intermediation in decentralized finance (DeFi). He articulated that blockchain technology, presents a novel financial system in which a network monopoly is via minimized through decentralization.

(Xiangan He, 2024) [8] The article discusses how blockchains are inherently combines to each cryptocurrency and related to their own chain. This constraint forces users and applications—especially in DeFi and gaming—to move dollars and data between blockchains. The article states that the users are forced to rely on centralized exchanges due to the absence of decentralized interoperability, which complicates user flow and prevents developers from building original on-chain apps.

6 Working of Block Chain enabled Defi Solutions in Supply Chain

In the blockchain network, supply chain transactions are initiated by authorized parties and are enforced automatically by smart contracts having response to predetermined standards [9]. Supply chains primarily rely on permissioned blockchains to reduce security threats. This means that only known and trustworthy supply chain partners may validate data on supply chain events and transactions with certain permissions also the ability to see data and to access the blockchain network is granted to specific user groups only. After being verified and encrypted using a hash function, blockchain data is kept in chronologically linked, timestamped blocks. Thereby a distributed ledger is being created and is used as a single source of truth to track supply chain transactions and confirm the origin of inventories and assets. Each authorized user on the network keeps a copy of the shared ledger on their own, which is automatically updated as new information

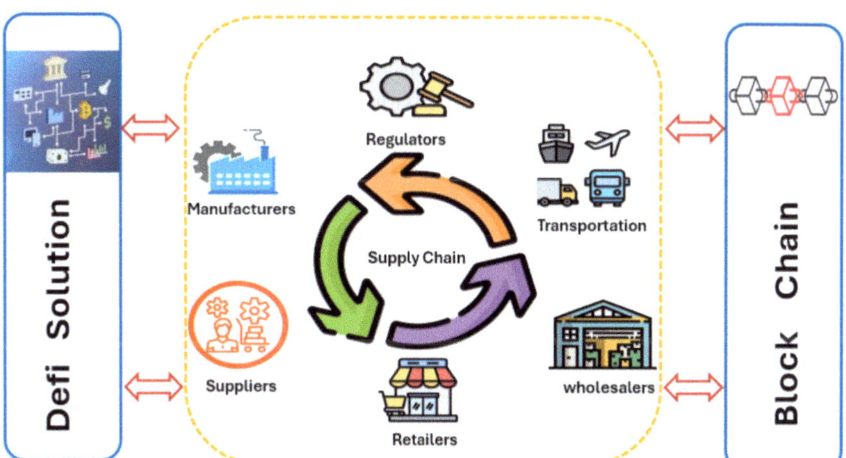

Fig. 1. The following figure depicts the Working of Block Chain enabled Defi Solutions in the Supply Chain.

is added. Supply chain players use role-specific online and/or mobile applications to interface with the blockchain to trade and trace data (Fig. 1).

7 Block Chain Enabled DeFi Solutions to Achieve Sustainable Supply Chain Goals

Blockchain technology enabled DeFi used in supply chain increases transparency, traceability the transactions and enables business to reach a global audience connecting to suppliers, distributors and end users. Block Chain enabled DeFi Solutions are transforming the world of supply chain by leveraging financial activities and making financial solutions more efficient, accessible and secure. The following areas have succeeded in using Block Chain enabled DeFi Solutions in its Supply Chain.

7.1 Asset Tokenisation

Block Chain enabled DeFi to convert physical assets into digital tokens making it smaller and easier to trade. Suppliers can tokenize these digital tokens and use it as collateral for raising funds.

7.2 Cryptocurrency

Block Chain enabled DeFi supply chain financing employs cryptocurrencies like Stable coins. These cryptocurrencies are linked to stable assets like fiat money or commodities and are used in DeFi supply chain finance. This is because these cryptocurrencies are reliable and less volatile in cross border transactions and hence reduces foreign currency fluctuations.

7.3 Decentralized Exchanges

Decentralized Exchanges helps in trading digital assets without the involvement of intermediaries and provides liquidity and flexibility to trading partners.

7.4 Programmed Contracts

These are smart contracts which are automated for carrying certain set of transactions like terms of payment, asset transfer etc., and are error free and less prone to frauds.

7.5 Data Sources

Block Chain enabled DeFi solutions concentrates on external data sources like Oracle which supply real time data related to asset values, market conditions etc. And foster decision making easier.

7.6 Money Market

Block Chain enabled DeFi solves many problems of money market and provides platform to construct an efficient portfolio and thereby reduces risk and uncertainty.

8 Mechanism of DeFi Protocols Using Blockchain

Use of blockchain technology with decentralized features in the financial industry termed as DeFi protocols is becoming more acceptable and accessible. Features of blockchain technology such as security, authentication, and openness can be linked to the attributes of equality, accessibility, and security found in decentralized networks. Financial services like exchanges, lending, and asset management work well on DeFi protocols. On-chain token trades can be easily conducted through Decentralized exchanges (DEXs). The DeFi market make use of automated market makers (AMMs) for asset management and control.

Smart contracts connecting investors and users involved in monitoring and regulating assets and generating money for the DeFi treasury are the main components of the DeFi business model. DeFi Protocols are a group of smart contracts that cover different aspects of yield aggregators, AMMs etc.. These protocols charge small fees for handling transactions such as asset management borrowing, trading etc. The protocols provides open, noncustodial, permission-less, and composable financial services. The so called investors in the DeFi protocol assumes protocol risk, potential misbehavior, impermanent loss, or rug-pulls. Their main responsibilities are to deposit money and supply these financial services with liquidity. The users of DeFi protocols usually interact in real time, without expecting extended responses and function as investors, initiates asset movements and pay interest rates to the protocols. This DeFi protocol overlooks protocol abuse, locks down assets, and complies with asset movement requests eventually providing other participants with returns and earnings (Fig. 2).

9 Mechanism of Blockchain-Enabled DeFi Protocols in Supply Chain

A smart contract containing the terms and conditions of the agreement is generated using Blockchain Networks. These smart contracts can be accessed by any supplier through a supply chain network. Each supplier in the supply chain creates and transmits a set of logical data. Theses suppliers create their own digital products and stores in the Decentralized Suppliers products engine. A customer who is in need of these digital products places their order requirements through Customers order engines. Thereby they complete their transaction successfully (Fig. 3).

10 Benefits of Block Chain-Enabled DeFi Solutions in Supply Chain Management

Blockchain technology provides world-shattering benefits to businesses across globally and is becoming a vital tool for optimizing supply chain processes. These IT tools offers substantial benefits to enterprises by boosting productivity, trust and fostering

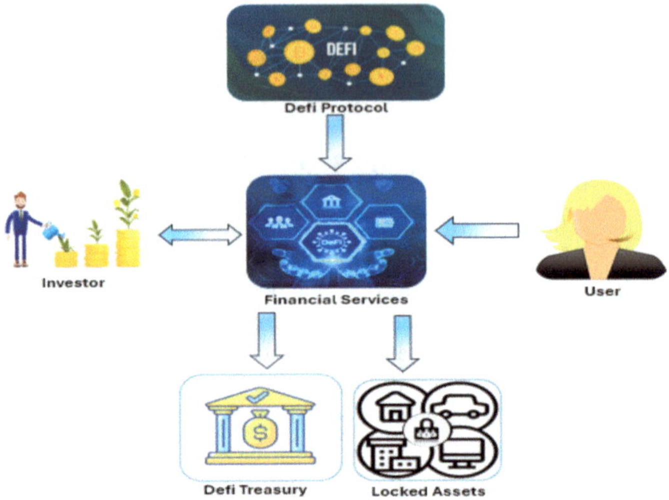

Fig. 2. The figure depicts the Working of Blockchain in DeFi.

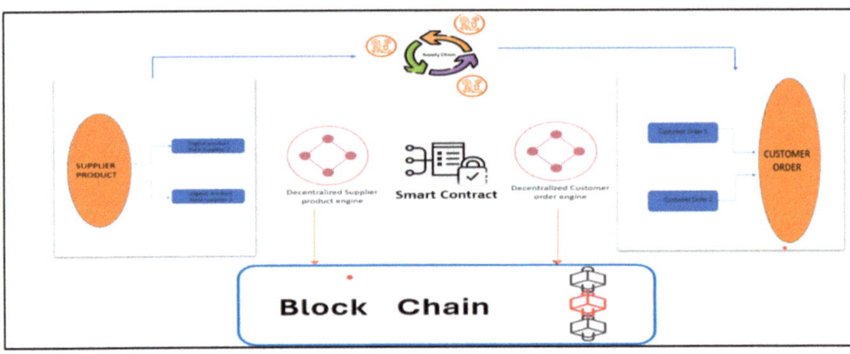

Fig. 3. Working of Blockchain enabled DeFi in Supply chain solutions

stronger cooperation amongst the different parties engaged in supply chain management. The strategic benefits cracked using blockchain in supply chain management are the following:

10.1 Automatic Compliance

Blockchain technology establishes decentralized, transparent ledger which has inbuilt provisions for interested parties to perform real time operations. It eliminates the role of multiple intermediaries and the need for human record keeping, cuts down on time-consuming reconciliations, allowing for automated compliance verification, smooth payment settlements, and supply chain events triggered by predetermined criteria.

10.2 Immutability

All the participants in the supply chain have access to the all the information equally and in the same manner. Hence, it reduces the possibility of fraud and eliminates information discrepancies. Due to its immutability, blockchain prevents tampering and unauthorized adjustments at bay by providing objective and reliable record of every transaction. Thereby trustworthy and auditable record of each transaction can be availed to the end users.

10.3 Time-Stamped Information

In blockchains, a timestamp is a record that indicates the precise date and time of a particular transaction or occurrence. The basis of blockchain technology is these timestamps, which guarantee that transactions are documented in the order in which they happened and ensures transparency in the supply chain.

10.4 Leverage Collaborative Tools and Platforms

Blockchain facilitates secure and transparent data sharing, enhancing coordination and communication of the stakeholders. These initiatives make it possible to efficiently estimate demand, manage inventory, and keep track of items—all of which boost output and cut supply chain expenses [10].

11 Case Study and Real-World Applications

DeFi is entitled to transform the financial system to run entirely without banks and without the need for an intermediary to carry out supply network ecosystem business transactions and uses smart contracts or computer code, such as Blockchain (Ethereum), that securely automates the process. DeFi is not owned by any an individuals or organization and is governed and operated by secure code. DeFi offers decentralized financial instruments without depending on conventional middlemen like banks, exchanges, or brokerages. Rather, it makes use of blockchain-powered smart contracts which are "crossing the chasm" of technological acceptance and is electronically secure, auditable and fully transparent. Every transaction is visible to all users as it is displayed on a public ledger and employs internal databases for recording the transactions. Blockchain organizes and manages the transaction data in such a manner which can be tracked, monitored, and accessed by the users.

To cite an example from food industry, we have Sam and James Walton started Walmart, an American multinational retail corporation, in 1962. In order to improve the traceability and openness of its food supply chain, Walmart has partnered with IBM. While working in collaboration with IBM to create a Hyper-ledger Fabric-based traceability system, Walmart can track down mangoes kept in its US shops within fraction of seconds say about 2.2 s. In addition, they were able to increase the quality and safety of food by quickly determining the source of contamination, Walmart-affiliated suppliers were able to track each stage of transaction in supply chain, starting from supply chain journey, their farm practices to storage conditions and transportation till final delivery of products to end users.

For example, the consumers of a fast moving consumable products can may purchase token backed up by a portfolio of the invoices and inventory. They can increase their investment further by using a bond token as collateral in DeFi lending protocols to enhance liquidity. The suppliers of the consumer goods industry will also have more effective means of securing capital and managing cash flow. Another promising feature of DeFi enabled transactions is that guarantee that the borrower offers security and that the lender receives interest and this can be viewed by all stakeholders through the public ledger.

12 Challenges of Chain-Enabled DeFi Solutions in Supply Chain Management

Blockchain has the potential to enhance accountability, traceability, and transparency in the supply chain process. However, there are also challenges and issues [11] that need to be resolved for widespread adoption to happen. The following are the challenges posed in Block Chain enabled DeFi Solutions in Supply Chain Management.

12.1 Scalability

Blockchain networks may have trouble in managing a lot of transactions at once.

12.2 Regulatory Uncertainty

As the regulatory environment around blockchain technology continues to change, there are concerns about legal frameworks and compliance.

12.3 Energy Consumption

Mining on certain blockchain networks, such as Bitcoin, requires a large amount of energy, which raises environmental issues.

12.4 Integration Complexity

It can take a lot of effort and time to integrate blockchain with current systems.

13 Future Prospects

The global supply chain market imbibing blockchain was estimated at $253 million in 2020 and is estimated to be $3,272 million by 2026 at a Compound Annual Growth Rate (CAGR) of 53.2% approximately. The robust automation of supply chain financing operations without middlemen, and growing need for supply chain transparency and traceability are the major factors driving the adoption of blockchain-based solutions which is backed by Supply chain security systems.Blockchain is a high-end technology (Dos et al., 2022) [9] that is expanding very quickly. Research and Markets predicts that the market will grow at a CAGR of 81.7% by 2026.

14 Conclusion

Although blockchain technology enabled DeFi solutions is not at all a cure for all the supply chain management challenges, it can provide supply chain management solutions that connects trading partners and streamlining procedures throughout the whole distribution network which in turn can optimize its efficiency. The features for tracking and traceability aims to bring about transparency and build confidence throughout the supply chain network. Every participant will be able to follow and trace product information from the point of origin to the final consumer.

The decentralized, impenetrable ledger followed in Blockchain technology offers transparency, immutability, and security through its transactions. By eliminating middlemen, it enriches the industries with transparent data, lower prices, and increases productivity and efficiency. But issues with scalability, regulatory unpredictability, and energy usage still exist. Nevertheless, block chain's disruptive potential resides in its capacity to eliminate intermediaries, empower people, improve confidence, and facilitate asset tokenization.

References

1. Jensen, J.R., von Wachter, V., Ross, O.: An introduction to decentralized finance (defi). Complex Syst. Inform. Model. Q. **26**, 46–54 (2021)
2. Rejeb, A., Rejeb, K., Simske, S., Treiblmaier, H.: Blockchain technologies in logistics and supply chain management: a bibliometric review. Logistics **5**(4), 72 (2021). https://doi.org/10.3390/logistics5040072
3. Wamba, S.F., Dubey, R., Gunasekaran, A., Akter, S.: The performance effects of big data analytics and supply chain ambidexterity: the moderating effect of environmental dynamism. Int. J. Prod. Econ. **222**, 107498 (2020)
4. Schär, F.: Decentralized finance: On blockchain-and smart contract-based financial markets. FRB of St. Louis Review (2021)
5. Aysan, A.F., Bergigui, F., Disli, M.: Using blockchain-enabled solutions as SDG accelerators in the international development Space. Sustainability (2021). https://doi.org/10.20944/preprints202103.0208.v1
6. Dwivedi, A.D., et al.: Blockchain and artificial intelligence for 5G-enabled Internet of Things: Challenges, opportunities, and solutions. Trans. Emerg. Telecommun. Technol. (2021). https://doi.org/10.1002/ett.4329
7. Grassi, L., Lanfranchi, D., Faes, A., Maria Renga, F.: Do we still need financial intermediation? The case of decentralized finance – DeFi. Qualitat. Res. Account. Manage. **19**(3), 323–347 (2022). https://doi.org/10.1108/QRAM-03-2021-0051
8. He, X.: Zephyr: a cost-effective, zero-knowledge light client for enhanced blockchain interoperability. Sigmet. Perform. Evalu. Rev. (2024). https://doi.org/10.1145/3639830.3639838
9. Dos Santos, S., Singh, J., Thulasiram, R. K., Kamali, S., Sirico, L., Loud, L.: A new era of blockchain-powered decentralized finance (DeFi)-a review. In: 2022 IEEE 46th Annual Computers, Software, and Applications Conference (COMPSAC), pp. 1286–1292. IEEE (2022)
10. Metelski, D., Sobieraj, J.: Decentralized Finance (DeFi) projects: a study of key performance indicators in terms of DeFi protocols' valuations. Int. J. Financ. Stud. **10**, 108 (2022). https://doi.org/10.3390/ijfs10040108

11. Amler, H., Eckey, L., Faust, S., Kaiser, M., Sandner, P., Schlosser, B.: Defi-ning defi: challenges & pathway. In: 2021 3rd Conference on Blockchain Research & Applications for Innovative Networks and Services (BRAINS), pp. 181–184. IEEE (2021)
12. DeFi in Supply Chain Financing & Management. https://maticz.com/defi-in-supply-chain
13. Blockchain for Supply Chain. https://scnsoft.com/blockchain/supply-chain
14. Review on Decentralized Finance Ecosystems. https://www.mdpi.com/1999-5903/16/3/76
15. Blockchain for Supply Chain, https://www.turing.com/resources/blockchain-for-supply-chains
16. DeFi: The Next-Gen Revolution in Supply Chain Financing, https://www.supplychainbrain.com/blogs/1-think-tank/post/34811-defi-the-next-gen-revolution-in-supply-chain-financing

Supply Chain Management for Coconut Farmers to Formulate New Marketing Strategies

T. Srinivas[1](✉), C. Ashok Kumar[2], N. Appa Rao[3], and K. Neeraja[4]

[1] Priyadarshini College of Engineering and Technology, SPSR NELLORE, Nellore, Andhra Pradesh, India
sri.du.1980@gmail.com

[2] Palamuru University Post Graduate Center-Kollapur, Nagar Kurnool (District), Hyderabad, Telangana, India

[3] Dr. B.R. Ambedkar Open University, Hyderabad, Telangana, India

[4] MVS Govt. Arts and Science College(A) Study Centre, Dr. B.R. Ambedkar Open University, Mahabubnagar (District), Hyderabad, Telangana, India

Abstract. The coconut industry represents one of the most significant economic sectors, but farmers and marketers in this industry face an array of challenges. In this paper we are focused to evaluate one of the possible solutions to this problem with effective supply chain management. In particular, this paper examines the Supply Chain Management Practices of Coconut Farming in the state of Kerala. To attain this, a sample of 150 coconut growers who cultivate coconuts was collected using a convenient sampling method. The findings obtained from the study will be useful for coconut farmers to formulate new marketing strategies or modify existing strategies based on consumers' preferences.

Keywords: Coconut Farming · Supply Chain Management · Marketing strategies

1 Introduction

Agriculture and its allied sectors are the backbones of most developing nations. India is an agricultural nation and one-third of its people rely entirely on farming to make a living.; nearly 60% of the population works in this sector. But up to now, the growth in this sector has not been equivalent to the expansion in the industrial and services sectors; as a result, the share of agricultural and related sectors in the nation's gross domestic product has been slowly declining over the years. Agriculture in India is facing several challenges which together manifest into sustainability issues. The symptoms of agricultural instability are sub-optimal growth, absence of desirable profits and distraction or movement of farmers away from the sector. The causes lie in the depleting status of natural resources and socio-economic conditions of the farmers. Sustainable agriculture is the efficient production of safe, high-quality agricultural products in a way that protects and improves the natural environment, the social and economic conditions of the farmers, their employees and local communities, and safeguards the health and welfare of all

B. K. Mishra et al. (Eds.): ICT4GS 2024, ISEM 56, pp. 185–192, 2025.
https://doi.org/10.1007/978-3-031-94997-5_18

farmed species. Several frameworks and models on measuring agricultural sustainability have been proposed under various production ecosystems. Therefore, it is inferred that sustainability in agriculture is a complex concept and there is no consensus among scholars about its dimensions. Concerns about sustainability in agricultural systems center around the need to develop technologies and practices that do not have adverse effects on environmental goods and services, are accessible to and effective for farmers, and lead to improvements in food productivity. It is also acknowledged that sustainability in agricultural systems incorporates concepts of both resilience (the capacity of systems to buffer shocks and stresses) and persistence (the capacity of systems to continue over long periods), and addresses many wider economic, social and environmental challenges.

From the references [1–3], Effective and efficient agricultural supply chain management strategies are one reasonable solution. Agriculture SCM(Supply Chain Management) is the process of making sure that farmers produce and sell goods that meet the needs of consumers in terms of quantity, quality, and price. Thus, the supply chain ensures suitable achievement in production and delivery engagements.

2 Supply Chain Management in Coconut Farming

Coconut represents one of India's most significant plantation crops, with a total area of cultivation of 1975.81 thousand hectares and a yield of 21,665 million nuts, placing India third in the world. Kerala, Karnataka, and Tamil Nadu are the states that make the most. In coconut farming, supply chain management means that all parties fulfil a customer's request either directly or indirectly. Chopra and Meindl (2007) said that SCM of coconut farming includes the manufacturers,, suppliers, retailers, warehouses, transporters and even the customers themselves. P. Deepa (2018) found four different structures of the coconut supply chain. These are the

SC1 - Producer - Village Traders- Wholesaler - Retailer - Customer
SC2 - Producer - Wholesaler - Retailer - Customer
SC3 - Producer - Regulated Market - Wholesaler/Agent - Customer
SC4 - Producer - Village trader - Retailer – Customer

Review of Literature

Shashikumar and H.M. Chandrashekar (2017) look at the SCM of the Coconut Industry in Karnataka to find out all the persons involved in the Supply Chain of Coconut in the study area. The main supply chain used in the coconut industry of Karnataka is Coconut Farmers-Wholesalers- Commission agents-and, finally, final consumers. The study found that a sound Supply Chain Management system makes the coconut industry in Karnataka more efficient and productive. Janvier- James (2012) says that Supply Chain and Supply Chain Management have been essential in the business world and have gotten a lot of attention from academics in recent.

Kumar et al. (2010) looked into the fact that the coconut crop is an essential part of India's food basket and meets the financial requirements of individuals who depend on its sale. The study found that there were more resellers than farmers. The study suggests that the Coconut Development Board and the state's industry department should work together to promote industries that use coconuts. Sam P.D.Anantadjaya et al. (2007) assert that supply chain management is crucial for delivering products and services to end users. It is instantly linked to the existing issues and how individuals respond to keep their customers happy. The study thought that the growing importance of supply chain management and consumer behavior in small and medium-sized businesses would enhance the inclusive enactment of the organizations.

Significance of the Study

From Reference [4–6], Most of the coconut farmers in Kerala are indebted to moneylenders and dealers. And they are frequently forced to sign up for-sale contracts in advance or sell their product at low costs. So, understanding and employing supply chain management has become a must for success in the competitive global marketplace and maximizing profits in coconut farming. From references [11] to [20], Many farmers lack adequate transport facilities to carry their products, which results in selling the product to their villages to meet their social and other requirements. Most farmers dislike wholesale markets because of their lack of knowledge about prevailing market value and product price. They are clueless about wholesale markets. Traders and commission agents take advantage of this situation. This exploitation of farmers can be avoided by providing them with advantageous prices. As a result, encouraging them to increase production both in quantity and quality. The supply chain management should be effective so that farmers can get a beneficial price for their produce and reasonable profit, which in turn improves the living conditions of the farmers.

3 Statement of the Problem

Kerala had the largest area and production of coconuts in the country and comprised about 15% of Kerala's total agricultural GDP. But Kerala, known as the "land of coconut," is losing its top spot in the country's coconut production. Effective supply chain management is one of the probable solutions to this, and this helps to manage the crop in a way that makes money. Because most farmers in India sell their goods in their own villages, they get little money for all their hard work. So, a study is conducted to catch out the ripostes to the following research queries,

Role of the supply chain in coconut farming?

Factors influence the selection of the supply chain?

What are the problems faced by coconut farmers

5. Objectives of the Study

The study "Supply Chain Management Practices of Coconut Farming with Special Reference to Kerala" is carried out based on the following specific objectives:

To identify the Selling Pattern of Coconut through Different Supply Chain

To identify the significant factors influencing supply chain selection.

To unearth the problems faced by the coconut farmers

3.1 Hypothesis of the Study

Ho1: No Association between the age of the farmers and the mode of the sale of coconut

Ho2: No Association between the occupation of the farmers and the mode of sale of coconut Ho3: No Association between the income of the farmers and the mode of the sales of Coconut

3.2 Scope of the Study

The study's topic is "SCM Practices of Coconut Farming in Kerala". The main objective is to investigate the role of the supply chain in coconut farming. The study was concentrated on various farmers irrespective of their size of coconut farming. A sample size of 150 respondents was considered for this study. Data were collected using a structured questionnaire.

4 Research Methodology

It defines tools accustomed to gathering relevant information in a specific research study. It includes sample and sampling design

Sample
A sample is a small subset of a population chosen to study its properties. In this study, a sample of farmers who cultivate coconut was taken from Ernakulum District.

Sampling Design
The sampling design specifies the sample frame, sample size, sample selection, and estimation methods in detail. For this study, 150 samples were collected by using the convenience sampling method from the total population of farmers who cultivate coconut Tools for analysis.

The primary data collected were statistically coded, processed, classified, tabulated and analyzed using statistical and mathematical tools and techniques like percentages and chi-square test.

Selling pattern of farmers through the different supply chain Location of sales

Particulars	Frequency	Percentage
Farm	71	47
Local Market	45	30
mill/processing unit	34	23
Total	150	100

In the case of coconut growers, more (47%) of the farmers sell the nuts in the farm itself. 30% of the farmers sell the nuts in the local market. Out of the total farmers, 23% of them sell the nuts in the mill/processing unit

Sales immediately after harvest

Particulars	Frequency	Percentage
Yes	88	59.0
No	62	41.0
Total	150	100

In the trading of coconuts, the majority, 59% of respondents, sell their produce immediately. The rest 41% of respondents hold their produce and secure it for the right time

Particulars	No. of Respondents	Percentage
High demand	23	15
Value addition	21	14
Get expected price	106	71
Total	150	100

5 Reasons for Retaining the Crops

The majority, 71%, respondents possess to meet their expected produce price. Then 15% of respondents keep their nuts till the demand reaches high. The rest, 14% of respondents, retain the nuts for value addition.

Testing of hypothesis

H0: No Association between the age of the farmers and the mode of sale of coconut Result of the Test

Test	Level of Significance	Degree of freedom	Computed Value	Table Value	Ho Accepted/Rejected
Chi-Square	5%	6	30.06	12.592	Ho Rejected

As per this investigation, the calculated value (30.06) is greater than the table value (12.592) at a 5% level of significance, and the null hypothesis is rejected. Hence the age of the farmers is associated with the mode of selling coconut. The null hypothesis has been framed and tested with the help of the chi-square test.

H0: No Association between the income of the farmers and the selection of mode of sales of coconut Result of the Test.

Test	Level of Significance	Degree of freedom	Computed Value	Table Value	Ho Accepted/Rejected
Chi-Square	5%	6	18.52	12.592	Ho Rejected

As per this study, the calculated value (18.5) is greater than the table value (12.592) at a 5% level of significance, and the null hypothesis is rejected. Thus the income of the farmers is associated with the selection of the location of sales. The null hypothesis has been framed and tested with the help of the chi-square test

HO: No Association between the occupation of the farmers and the mode of sale of coconut

6 Result of the Test

Test	Level of Significance	Degree of freedom	Computed Value	Table Value	Ho Accepted/Rejected
Chi-Square	5%	6	22.5	12.592	Ho Rejected

An effort was made to identify whether there is an association between the occupation of farmers and the mode of selling coconut. For this reason, the null hypothesis was framed and tested with a chi- square test. It shows the calculated value (22.5) is greater than the table value (12.592) at 5% level of significance null hypothesis is rejected. Thus, it results that there is a significant association between the occupation and mode of sales

Findings

7 Demographic Profiles of Farmers

Majority (51.1) of the farmers belonging to the age group of 45 to 60 years and very mere per cent of the youth are involved in agriculture

Majority of the respondents are graduates, and only 5.4% of them have studied agriculture 3.40.7% of them are doing agriculture as their main occupation

Majority of the respondents are earning above 40,000.

Merely 19.3% of growers have experience of more than 30 years.

Selling Pattern of Crops through Different Supply Chains

The major part of the farmers 47% sell the produce in the farm

The majority, 59% of respondents, sell their coconut immediately after harvesting it. The rest of the 41% of respondents keep their crops to get more profit.

Challenges Confronted by the Farmers in Promoting Coconuts

1. The Challenges confronted by the Farmers to market their product are Storage, Transportation, fair Market Prices and Competition. Thus, farmers have to be educated about these problematic areas to make the proper decision to get their business transactions successfully

Confounders that Affect the Choice of Supply Chain

1. The major Confounders that Affect the Choice of Supply Chain are 1. Safe and secure sales, 2. Price 3. Risk. Thus, farmers focus on these three factors to select the supply chain for their product sales.

Association between Profile of Farmers and sale of the coconut

Table value is less than the calculated value at a 5% significance level; the null hypothesis is rejected. Hence the age of the farmers is associated with the mode of sale of coconut.

Association between the occupation of farmers and the mode of sale of coconut were tested with the chi-square test. It shows the table that is less than the calculated value at 5% significance level. Thus, it results that there is a significant association between the occupation and mode of sales of coconut.

The farmers' income is associated with the selection of the mode of sale of coconut; the null hypothesis has been framed and tested with the help of the chi-square test.

8 Conclusion

Coconut farming is essential to Kerala's economy, but farmers face various challenges as they strive to continue their cultivation. To keep doing well in this field, farmers must recognize the significance of supply chain practices that advance their performance. Productive and competent SCM practices have become an crucial way to remain competitive in the market and improve agricultural performance. A successful supply chain management is a well-coordinated process, and this will help the farmers focus on delivering value to customers, eliminating unnecessary costs in key functional areas, and a system for measuring performance.

References

1. Vanamadevi, R.: Cultivation and marketing problems of coconut growers in Thali panchayat, Udumalpet" International Conference on "Research avenues in Social Science" IJARIIE-ISSN(O)-2395–4396, Vol-1 Issue-3 (2016)
2. Yamuna, S.M., Ramya, R.: Coconut cultivation and marketing in pollachi taluk International Journal of Innovative Research in Management Studies (IJIRMS) ISSN (Online): 2455–7188 Volume 1 | Issue 2 | March (2016)
3. Veerakumar, K.: Challenges of Agriculturists in Coconut Cultivation of Meenakshipuram Village, Chittur Taluk. Kerala International Journal of Innovative Technology and Exploring Engineering (IJITEE) ISSN: 2278-3075 **8**(11S) (2019)
4. Das, P.K.: Estimating production costs and returns for coconut in Kerala. Journal of Plantation Crops **12**(2) (1984)
5. Elias, G.: Economics of Coconut Products-An Analytical Study Commerce Spectrum ISSN 2321-371X **5**(2), 39–44 (2014)
6. Beamon, B.M.: Supply chain design and analysis: models and methods. Int. J. Prod. Econ. **55**, 281–294 (1998)
7. Dileep, B.K., Grover, R.K., Rai, K.N.: Contract farming in tomato: an economic analysis. Indian J. Agricult. Econ. **57**(2), 197–210 (2002)
8. Samuel, B., Philip, C.J., Timothy, J.L.: Coordinating the supply chain in the agricultural seed industry. Eur. J. Oper. Res. **185**(1), 354–377 (2008)
9. Sam, P.D.: Sari Waskita, A.A.W.E., Nawangwulan, I.M.: Consumer behaviour, supply chain management and customer satisfaction: an investigative study in small and medium enterprises. International Seminar on Industrial Engineering and Management Menara Peninsula, Jakarta, August 29-30, ISSN: 1978-774X(2007) (2007)

10. Janvier-James: A new introduction to supply chains and supply chain management: definitions and theories perspective. Int. Bus. Res. **5**(1) (2012)
11. Kumar, K.: Value Chain Analysis of Coconut in Orissa. Volume 23, Conference Number (2010)
12. Shashikumar, Chandrashekar, H.M.: Supply Chain Management of Coconut Industry In Karnataka Ph.D Thesis. University of Mysore (2017)
13. Srinivas, T.: Transportation and its health implications in India. Int. J. Eng. Res. Dev. **10**(7), 67–73 (2014)
14. Srinivas, T., Sireeshsa Devi, P.S.: The Conventional Method of Optimizing Delivery Routes And Schedules .IJRCCT, 2320–5156 (2016)
15. Srinivas, T., Sireeshs Devi, P.S.: Effective Ness And Illustration Of Distribution Systems And Supply Chain Management. IJECS, Issn: 2319-7242 (2017)
16. Taguchi, S.T.: Philosophy of Engineering Quality. In: Pacific International Journal, ISSN 2616-4825 [Online] 2663-8991 [Print] Volume 03, issue 03 (2018)
17. Srinivas, T.: Financial Soundness Of Selected Indian Petroleum Companies Using Altman Z–Score Model in Pacific International Journal, ISSN 2616-4825 [Online] 2663-8991 [Print] Volume 03 , issue 03 (2018)
18. Chikine, A.K., Ravindra Babu, G., Neeraja, K.: The Effective and Efficient Utilization of Resources of Productivity Using Goal Programming. Int. J. Math. Trends and Technol. **48**(5), 276–280 (2017)
19. Ashok Kumar, C.: Multi-objective optimization model for test market city assess by using goal programming. Mathematical Sci. Int. Res. J. **7**(2), 48–52 (2018)
20. Ashok Kumar, C., Ravindra Babu, G.: The Optimal Distribution of Available Funds in Different Investment Categories. Galaxy Int. Interdiscip. Res. J. **5**(9), 28–35 (2017)

A Fuel Cell Fed Multi Level Zeta Converter System with Fuzzy Logic Controller

B. Manimekalai[1(✉)], M. Marimuthu[1], V. Prabhu[2], G. Venkatesan[3], B. Paranthagan[1], and S. KamalaKannan[1]

[1] Saranathan College of Engineering Trichy, Trichy, India
manimekalaieeephd@gmail.com, paranthagan-eee@saranathan.ac.in
[2] Sethu Institute of Technology Viruthunagar, Viruthunagar, India
[3] CARE College of Engineering Trichy, Trichy, India

Abstract. This paper presents a multilevel zeta converter (MZC) for fuel cells, using a 1260W proton exchange membrane fuel cell (PEMFC) and a fuzzy logic controller (FLC) with voltage feedback. A MZC runs on a series of stages, each of which include switches, capacitors, diodes, and inductors. Typically, the control technique consists of switching sequences that enable a step-by-step The converter is planned and built utilizing MATLAB SIMULINK and a topological circuit, with an FLC controller configured to increase output voltage transient responsiveness. The results show enhanced performance of the MZC with FLC Controller, allowing for control of the time-varying non-linear nature of power converters. The fuzzy logic controller in view of this instance employs a Mamdani-style fuzzy logic controller. Finally, by exploiting the potential yield as feedback, they further noticeably enhanced the converter's dynamic accomplishment. The converter's performance is verified through MATLAB simulations, demonstrating fast transient response, excellent steady-state response, and less sensitivity to load fluctuations.

Keywords: Fuel cell · Fuzzy logic controller · multilevel zeta converter

1 Introduction

The existing central power generation capacity is insufficient to satisfy rising energy demand, forcing the usage of conventional diesel generators. As conventional power sources become less common due to their inefficiency, utilities and the private sector are focusing on green power technologies like fuel cells, which offer high operating efficiency and better power quality, reduced transmission and distribution losses, and increased voltage support.

Because of their great efficiency, stability, low energy consumption, and environmental friendliness, fuel cells are becoming more and more acknowledged as a potential power source. They offer a higher energy storage capacity, increasing an automobile's operating range. However, the output voltage cannot be constant due to variations in operating conditions, necessitating the use of a converter. This work designs and develops a

B. K. Mishra et al. (Eds.): ICT4GS 2024, ISEM 56, pp. 193–204, 2025.
https://doi.org/10.1007/978-3-031-94997-5_19

multilevel zeta converter (MZC) with a fuzzy logic controller for a fuel cell application to improve system integrity, reliability, and efficiency. PEMFC fuel cell 1.26 kW is used in a closed circuit with a 35-V variable. Input that is fed to a multilevel zeta converter. The fuzzy logic controller controls the boost at the first level, which receives 70 V, and the fifth level, which receives a maximum boost of 350 V. Determining the membership function and creating fuzzy rules are two aspects of FLC design. The outcomes of the design simulation show that even in the presence of a change in the fuel cell's output voltage at the converter's input, the utilization of fuzzy logic control in the MZC can enhance the transient response of the output voltage and its dynamic features.

1.1 Methodology

The project's methodology consists of the following steps:

- Input source fuel cell is get from the mat lab and to design the input values based on our project.
- Create and design the circuit for a multilevel zeta converter using the literature as a guide.
- Building on the gain response graph, For the FLC, develop the fuzzy rules.
- Analyze the outcome by simulating the multilevel zeta converter with or without the FLC.

2 Fuel Cell Design

Fuel cells are clean energy sources used in various industries, including transportation, utility power, and low-power electronics. High-power converters are used in commercial and industrial applications, while DC power converters determine their use. Fuel cells have low operating temperatures, short startup times, and excellent power density, making them ideal for home and business use. However, they are not suitable for sudden load changes and their power density increases with increasing loads. A precise desktop fuel cell model is needed to create an efficient power electronics interface.

2.1 Physical Description of PEMFC

PEM fuel cells (see Fig. 1) use a thin solid polymer coating as their electrolyte, with oxygen and hydrogen electrodes between it. Catalyst is bonded to the electrodes or both sides in a thin layer. Membrane electrode assemblies (MEAs) are assembled between separators to create a single cell, with bipolar plates transferring electric current and releasing heat.

A fuel cell stack produces a useful output by connecting multiple fuel cells in series, as a single cell only generates 0.7 V. These stacks consist of modular bipolar plates and fuel cells assembled in series, splitting fuel into ions and electrons.

Electrons drive an electric load, causing ions to move through an electrolyte to an opposite-charge electrode. Ions mix, producing carbon dioxide and water. PEMFCs can operate between 0.10 and 0 MPa and up to three thousand psi with proper current collectors and structures. Figure 2 shows an illustration of the chemical reactions that

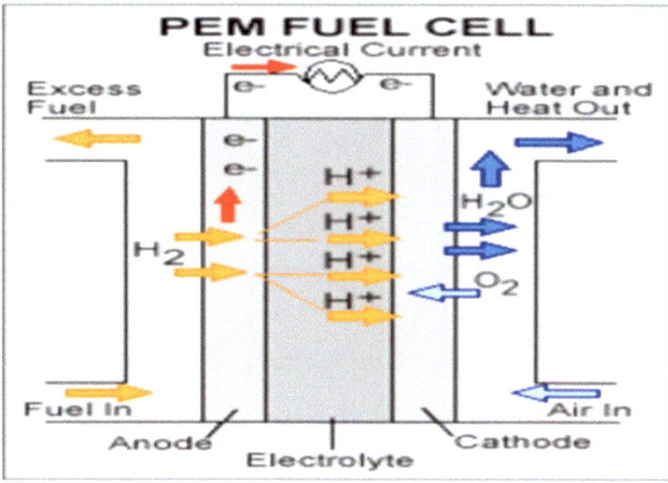

Fig. 1. Fuel cell

occur in PEMFC [1, 2]. The PEMFC's detailed operating schematic is shown in (see Fig. 2). Equations 1, 2, and 3 can be used to characterize the chemical reactions in PEMFC.

Anode response

$$H_2 = 2H^+ + 2e^- \tag{1}$$

Cathode response

$$\frac{1}{2}O_2 + 2H^+ + 2e^- = H_2O \tag{2}$$

Overall response

$$H_2 + \frac{1}{2}O_2 = H_2O \tag{3}$$

I-V characteristics

2.2 Fuel Cell Parameters

See Fig. 3.

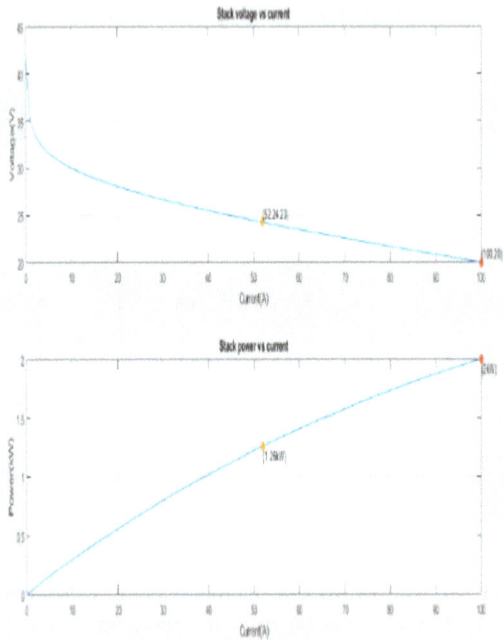

Fig. 2. Fuel cell's I-V characteristics

3 MZC-Multilevel Zeta Converter

This converter's main objective is to increase output voltage efficiency while minimizing the total number of switches. A single switch is used in the a MZ converter circuit.

See Fig. 4.

A zeta converter's duty cycle D is determine

$$D = \frac{V_{\text{out}}}{V_{OUT} - V_{in}} \tag{4}$$

Zeta output voltage

$$V_{\text{out}} = \frac{V_{\text{in}}D}{1 - D} \tag{5}$$

The following formula is used to construct the inductor L1 and L2, assuming that the duty ratio and output power are adequate.,

$$L_1 = \frac{D^*Vin}{f * \Delta L_1} \tag{6}$$

$$L_2 = \frac{D^*V_{in}}{f^* \Delta I_{L2}} \tag{7}$$

Fuel cell nominal parameters:
Stack Power:
-Nominal = 1259.96 W
-Maximal = 2000 W
Fuel Cell Resistance = 0.061871 ohms
Nerst voltage of one cell [En] = 1.115 V
Nominal Utilization:
-Hydrogen (H2)= 99.92 %
-Oxidant (O2)= 1.813 %
Nominal Consumption:
-Fuel = 15.22 slpm
-Air = 36.22 slpm
Exchange current [i0] = 0.027318 A
Exchange coefficient [alpha] = 0.308

Fuel cell signal variation parameters:
Fuel composition [x_H2] = 99.95 %
Oxidant composition [y_O2] = 21 %
Fuel flow rate [FuelFr] at nominal Hydrogen utilization
-Nominal = 12.2 lpm
-Maximum = 23.46 lpm
Air flow rate [AirFr] at nominal Oxidant utilization:
-Nominal = 2400 lpm
-Maximum = 4615 lpm
System Temperature [T] = 328 Kelvin
Fuel supply pressure [Pfuel] = 1.5 bar
Air supply pressure [PAir] = 1 bar

Fig. 3. Fuel cell parameters value.

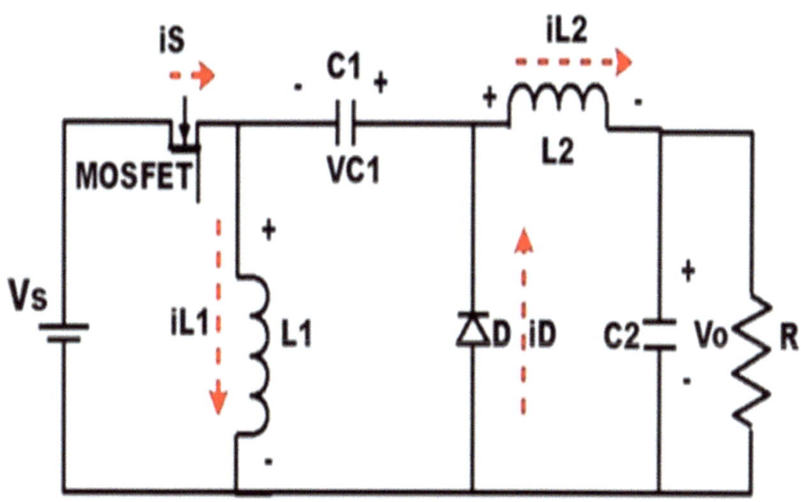

Fig. 4. ZETA converter circuit.

Utilizing the following expression, one can get the value of the capacitor:

$$C_1 = \frac{D^*V_o}{f^*R^*\Delta V_{C1}} \tag{8}$$

$$C_2 = \frac{D^*V_{in}}{8^*f^2 * L_2 * \Delta V_{C2}} \tag{9}$$

With minimal current stress, it is possible to reach multiple levels without affecting the main circuit. Thus, f stands for switching frequency.

The MZC is illustrated in (see Fig. 5), with five different levels of output. The operating principles of this five level circuit is classified into two sections according to the control of the switch. ON mode operation and the OFF mode Operation.

Mode 1 to Mode 4 The Switch ON state

During the period when the switch is on. As the previous cycle's charged inductor discharges through D2, the input voltage is connected to the inductor L1. Capacitor C1 is consequently charged When in mode 2, diode D4 helps clamp capacitor voltage Cb to C2.

The voltage clamping across Cb and Cc is applied to C2, and the clamping across Cc is applied to capacitor voltage C3 through D6, when the voltage across Vcb and Vcc is greater than Vc2 and Vc3. The voltage across Cb, Cc, and Cd is limited to the voltage across C2, C3, and C4 at the time the circuit is in clamped State.

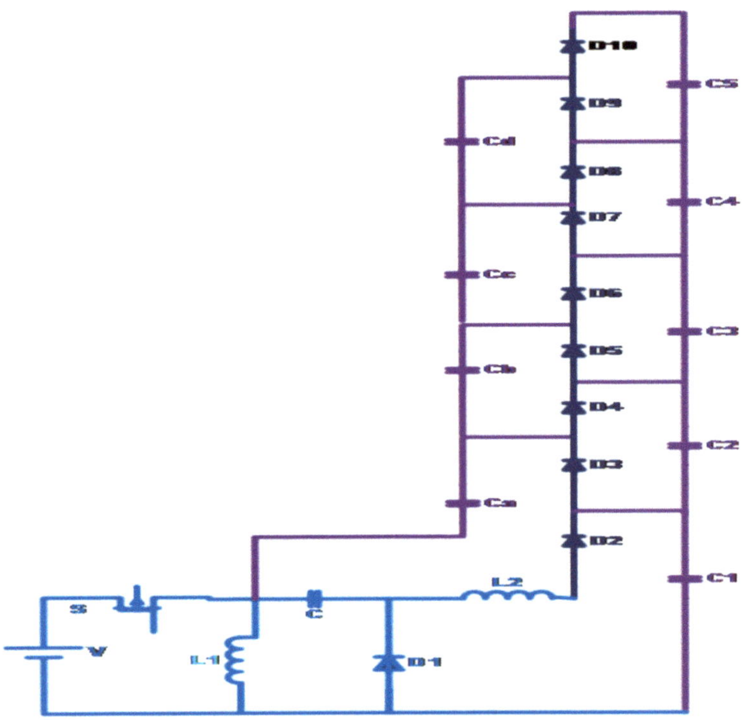

Fig. 5. The MZC circuit has five output levels.

35 V is the supply voltage for the MZC., 550 V output at the fifth level with a frequency of switching at 5 kilohertz. The device's seventy-five percent of the pulse

width with levels ensures sufficient saves time during closure and prevents extreme voltage stress (Table 1).

Table 1. The following values are utilized in the design of the suggested MZC

Parameter	Values
Input voltage from fuel cell Vin	35 V
First level Voltage output Vo	110 V
switching frequency f	5 kHz
Inductance L1 and L2 value	38.03e-3
Capacitance Ca value	25.97e-3
Self-balancing Capacitance Ca, Cc, Cd C1 to C4 value	52.15e-6

4 Fuzzy Logic Controller

A collection of human language rules given by the user serves as the foundation for fuzzy logic. These rules are translated into their mathematical counterparts by the fuzzy systems. Four useful building components make up the fuzzy system.

- Fuzzifier
- Fuzzy interference engine.
- Knowledge base.
- Defuzzifier.

Multilevel Zeta Converter with Fuzzy Logic Controller. Numerous DC-to-DC converter systems are among the many control applications that employ fuzzy sets. A set of rules based on how humans generally understand system behavior are used to create a simple fuzzy control.

The way in which the recommended controllers and the DC-to-DC converter's dynamic behavior are being investigated using the Simulink simulation model.

A FLC controller design, as opposed to a linear control method, can simultaneously provide the best dynamic performance for both small and large signals. Consequently, DC-to-DC converter dependability can be raised by utilizing fuzzy logic controllers.

FLC consists of three components: defuzzification, fuzzy inference system, and fuzzification. Typically, a membership function defines a fuzzy variable, which is expressed as a fuzzy set. The membership function takes values between 0 and 1. The fuzzy rule base is composed of the IF-THEN rules.

4.1 Membership Function of Fuzzy Logic

Its small signal model causes the Multilevel Zeta converter to show a nonlinear dependence of duty cycle.

FC don't require for FC to function, a precise mathematical model is not necessary. A precise mathematical model to operate. Rather, their design is grounded in common plant knowledge. This controller has the ability to adapt to various operating points.

The Mamdani-style fuzzy inference system used by the FLC is intended to maintain control. The output of a multiple zeta converter. Two input variables are included in this fuzzy logic system: error (e) and change in error modification (ce). The duty cycle is an output variable in a PWM output (Fig. 6).

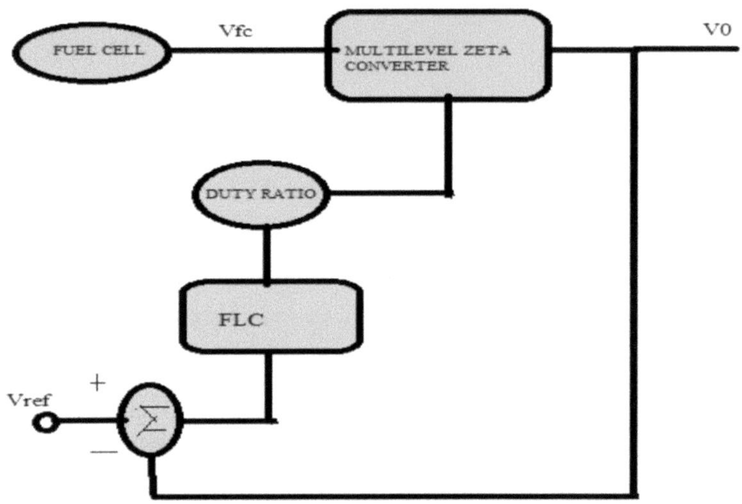

Fig. 6. Basic control structure for fuel cell MZC with fuzzy logic

4.2 Fuzzy Logic Table Rules

Controlling the zeta converter's output voltage is the goal of this dissertation. The error in the output voltage and its variation will be fed into the fuzzy logic controller. These two inputs consist of seven groups Abbreviations for these terms are:

Positive- Big, Medium, small denoted as PB, PM, PS. Negative- Big, Medium, small denoted as NB, NM, NS.

In the table, the fuzzy control rules for error (e) and change in error (e') are as follows (Figs. 7, 8 and 9 and Table 2):

5 Simulink Description

The projected control structure is simulated utilizing MATLAB-SIMULINK program. The construction concerning this plan is done utilizing the limits of the fuel cells, using a 1260W proton exchange membrane fuel cell (PEMFC).

Fuzzy sets are used in many DC-to-DC converter systems, among other applications in the field of control. A set of rules based on model. Moreover, a fuzzy logic controller

Table 2. Fuzzy Rules Design

	e							
e′		PH	PM	PS	Z	NS	NM	NH
	NH	Z	NS	MM	NH	NH	NH	NH
	NM	PS	Z	NS	NM	NH	NH	NH
	NS	PM	PS	Z	NS	NM	NH	NH
	Z	PH	PM	PS	Z	NS	NM	NH
	PS	PH	PH	PM	PS	Z	NS	NM
	PM	PH	PH	PH	PM	PS	Z	NS
	PH	PH	PH	PH	PH	PM	PS	Z

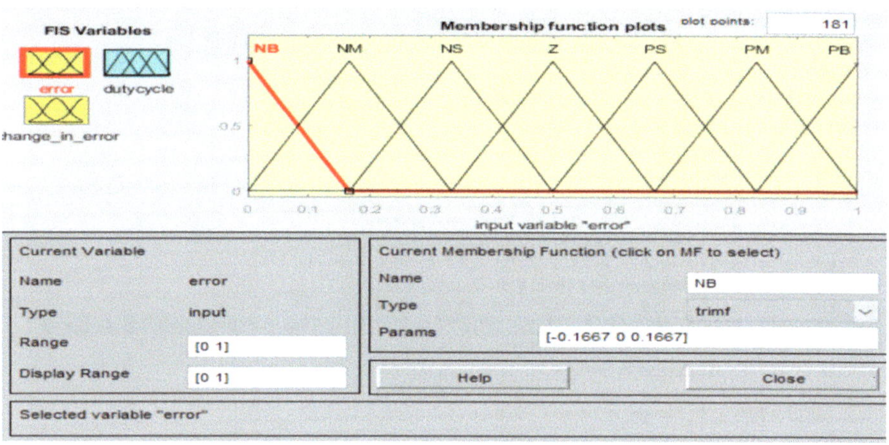

Fig. 7. Input variable of membership function

design can simultaneously provide the best dynamic performance for big and small signals, unlike a linear control approach. Therefore, the reliability of DC-to-DC converters can be improved by using fuzzy logic controllers (Figs. 10 and 11).

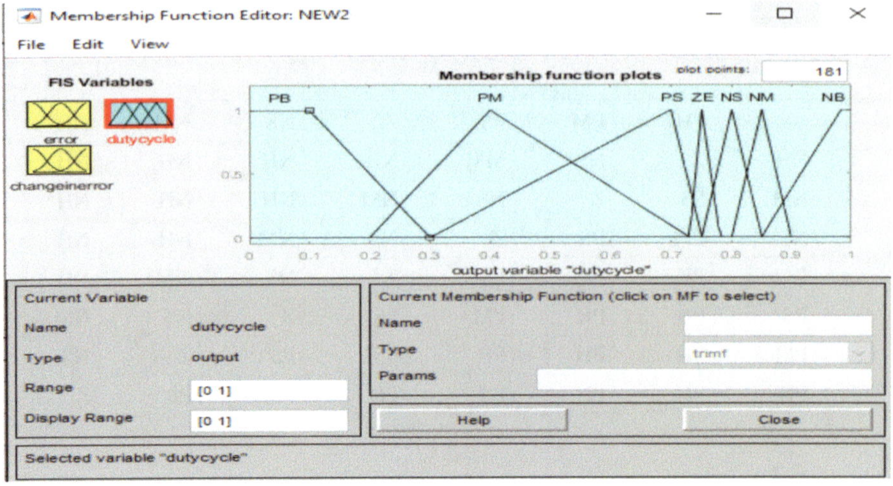

Fig. 8. The membership function's output variable

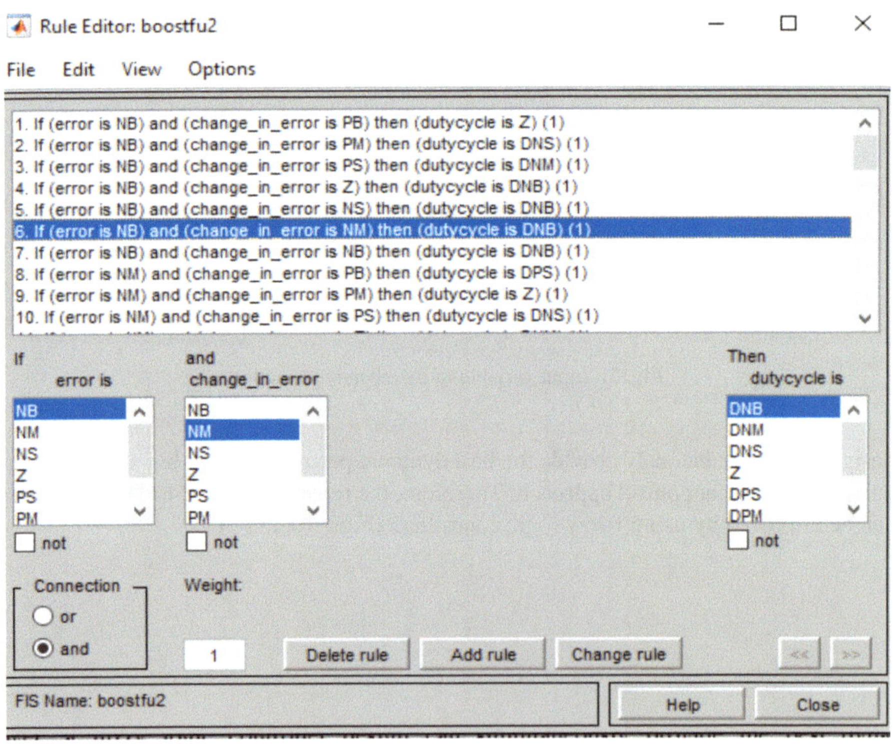

Fig. 9. Fuzzy rules design table

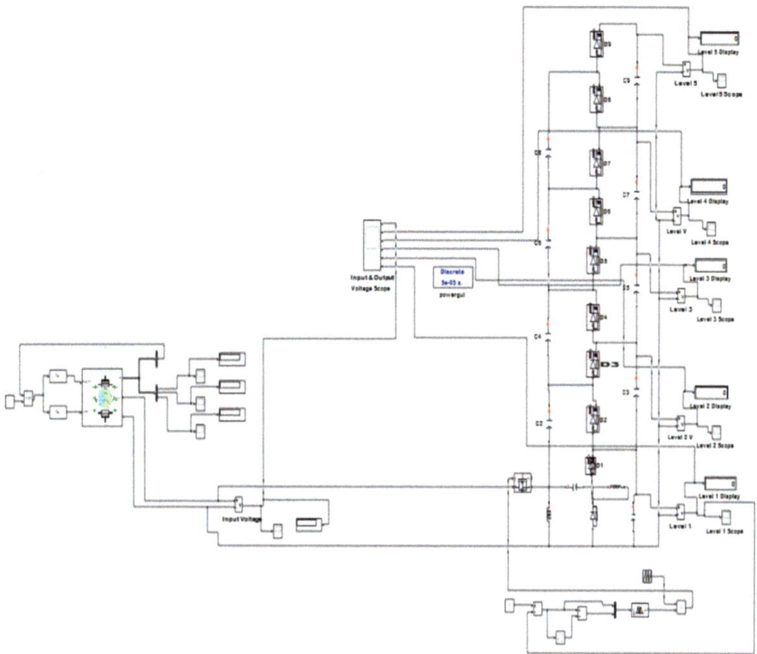

Fig. 10. Fuel cell with MZC with fuzzy design and simulated using Simulink

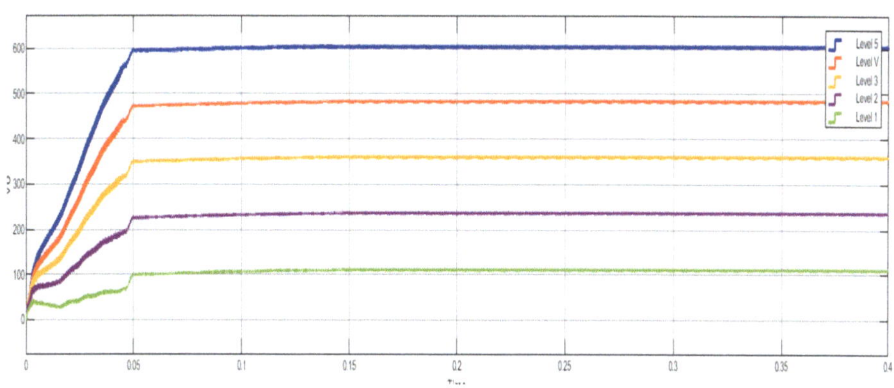

Fig. 11. Fuel cell with MZC with fuzzy simulated output graph

6 Conclusion

A fuel cell fed multilevel Zeta converter controlled by fuzzy is presented in the paper. Simulation are carried out in mat lab environment. The low voltage output levels of fuel cell are stepped up to high voltage levels using multilevel converter. Without alterations of main circuit any number of levels can be proposed using this topology. Also fuzzy logic controller provides good time domain results. The 1.26kW suggested converter is found to produce a maximum first level output of 120 V and a maximum voltage output of about 600 V for a 35 V supply voltage. This framework doesn't require high duty cycle to achieve a high gain. So, the converter that is recommended proves to be superior to the other converters in terms of gain. This converter is unusual in that it has a continuous input current, higher output voltage gains (achieved by assumption a considerably lower duty cycle ratio), the lowest voltage and current stress.

References

1. Marimuthu, M., et al.: High Voltage Gain Multilevel Cascaded Boost Converter for PV Application. 2023 International Conference on Circuit Power and Computing Technologies (ICCPCT), pp. 510–517. Kollam, India (2023). https://doi.org/10.1109/ICCPCT58313.2023.10245378
2. Priyadarshi, N., et al.: An experimental study on zeta buck–boost converter for application in pv system. Handbook of Distributed Generation: Electric Power Technologies, Economics and Environmental Impacts, pp. 393–406 (2017)
3. Shanthi haveela, B., Koteswara raju, D., Sangameswara Raju, P.: A Fuel Cell Based Multilevel Dc-Dc Boost Converter System with Pi and Fuzzy Control. Int. J. Sci. Res. (IJSR) **2**(9), ISSN No 2277-817 (2013)
4. Kumar, S., Srinivasa Rao, G., Voggu, S.: Fuzzy Logic Controller with Zeta Converter for Electric Vehicles Induction Motor Micro-Electronics and Telecommunication Engineering, Volume 179I, SBN: 978-981-33-4686-4S (2021)
5. Jayasudha, S.V., Marimuthu, M.: A nov(el cascaded multilevel boost converter fed multilevel inverter with reduced switch count. Int. J. Electron. **110**, 12, 2374–2391 (2023)
6. Bendaoud, K., et al.: Fuzzy logic controller (FLC): Application to control DC-DC buck converter. 7 International Conference on Engineering & MIS (ICEMIS), pp. 1–5. Monastir, Tunisia (2017). https://doi.org/10.1109/ICEMIS.2017.8272980
7. Rajkumar, M., Rathinam, A.: Design and experimental investigation of zeta converter fed reduced switch multi-stage grid connected boost MLI. J. Electr. Eng. Technol. **19**, 361–371 (2024). https://doi.org/10.1007/s42835-023-01517-9
8. Moldrik, P., Chvalek, R.: PEM fuel cells — The basic characteristics. 10th International Conference on Environment and Electrical Engineering (2011)

Promoting Electric Vehicle Accessibility: Integrative Strategies and Innovations

Harshita Ketharaman[1], Haresh Sambandamoorthy[1](✉), Isabella Aji[1], and R. Angeline[2]

[1] Department of Computer Science and Engineering with Specialization in Artificial Intelligence and Machine Learning, SRM Institute of Science and Technology, Bharathi Salai, Chennai, India
haresh.sam@gmail.com

[2] Department of Computer Science and Engineering, SRM Institute of Science and Technology, Bharathi Salai, Chennai, India
angelinr1@srmist.edu.in

Abstract. In recent times, Electric Vehicles, specifically electric cars have gained a lot of attention due to their increasingly positive impact on the environment. They help with tackling a wide array of challenges that our world faces, like pollution control, excessive draining of renewable resources and regression in health because of the emission of harmful pollutants. A shift towards Electric Vehicles can help reduce the emission of harmful pollutants into the atmosphere and reduce the usage of renewable resources, which could result in an immense positive impact on our environment. Apart from this, it can also help with keeping the Air Quality Index at safer levels and have a positive impact on the health of all life on Earth. Electric Vehicles are a stepping stone in the positive direction, to make the world clogged with less pollutants and a safer and a healthier place to be in. In this proposal, we try to analyze and understand what factors affect the performance and accessibility of Electric Vehicles using a dataset pertaining to certain metropolitan cities in Europe, which can help us pinpoint the reason as to why they are not as widely used. We utilize these data-driven factors to help with correlating the EV scene in India and also think of ways to increase their usage, apart from addressing the intricate nature of how Electric Vehicles function.

Keywords: Electric Vehicle · Metropolitan Cities · Air Quality Index

1 Introduction

As the demand for more eco-friendly solutions to help save the environment increases, Electric Vehicles (EVs) pose a positive alternative solution to the currently existing fuel-run vehicles and are crucial to contributing towards a sustainable future. One major problem with the current fuel-run vehicles is the alarming emission of Green House Gases (GHG), which can cause serious environmental issues. In GHG, CO2 comprises 76% of global emissions, and the transportation sector is responsible for 15% of global emissions [1]. Such a high percentage of GHG emissions calls for an alternative solution

B. K. Mishra et al. (Eds.): ICT4GS 2024, ISEM 56, pp. 205–215, 2025.
https://doi.org/10.1007/978-3-031-94997-5_20

[2]. Apart from the obvious benefits EVs pose with regards to the environment, they are also known for their longevity, considering they have a longer cycle life, which also makes them a cost-effective alternative. Despite the vast array of advantages they portray, EVs are still limited to only a small portion of the population. In this study, we aim to shed some light upon the different factors that affect the accessibility of EVs, which would help us understand and work towards plausible solutions to make them more accessible to everybody.

1.1 Current System of Electric Vehicles

Many countries, due to the increase in environmental degradation, have adopted the usage of New Energy Vehicles (NEVs) as an alternative to conventional vehicles. China is promoting the usage of NEVs in order to reduce the consumption and import of fuels [3], while in Europe, Germany has proposed to have one million EVs in operation [4]. To encourage more countries to adopt NEVs, subsidies, and special tax policies have been offered, such as plug-in vehicle subsidies in the UK, the clean vehicle rebate project in the US, and green vehicle purchasing promotion measures in Japan and China [5]. Recent studies highlight significant growth in the NEV market. In China, NEV sales reached 6.8 million units in 2022, driven by national initiatives and local promotional efforts [6]. Similarly, BYD, a leading NEV manufacturer, reported substantial growth with over 640,000 units sold in the first half of 2022, focusing on sustainable and innovative technological advancements [7].

Since the COVID-19 Global Pandemic, the world has shifted to shedding more focus on structural incentives, specifically with regard to the reduction of carbon emissions in order to limit the global average temperatures. Apart from tackling the problem of carbon emissions, post-COVID-19 has also seen an increase in people preferring to use less of public transport for their travel. A quantitative analysis which was performed in Lanzhou, China during and after four months of the outbreak of the pandemic suggests that electric car-sharing (ECS) orders fell by 55.8% during the COVID-19 pandemic [6]. With an increase in people preferring non-public transportation and with the world shifting more towards a sustainable future, specifically with regard to fuel consumption and carbon energy emissions, now is the time to promote the use of Electric Vehicles and look into any issues that might stop it from being accessible to people.

1.2 EV Industry

The need for EVs has grown continuously in the past couple of years which has caused a setback for the automotive industry. Because of this, there has been an increase in the growth of EVs in the automotive industry. The current EV market is estimated to be valued at around 250 billion dollars with over 10 million EVs being used around the world. This figure is expected to grow until 2028 when it's predicted to reach 1.3 trillion dollars, which is a rate of growth of around 17.3% over the forecasted time [8]. However, the industry has been susceptible to volatility. There have been disruptions in the supply chain and fluctuations in battery metal price, especially, since Russia's invasion of Ukraine. This is seen to be followed with increasing competition. As reported last year in GEVO-2023, investors are trying to explore new opportunities to increase sales [9].

With regards to the Indian EV market, during COP26, India proposed its decarbonisation target for 2030, which includes reducing carbon emissions by 50%. This will also become a part of India's EV30@30 campaign which aims at ensuring that EVs account for at least 30% of new vehicle sales by 2030. Electric cars accounted for 1.3% of car sales in 2022 in India. Studies indicate that the Indian EV market, which is currently valued at 2 billion US dollars in 2023, may potentially surge up to 7.09 billion US dollars by 2025 [10].

2 Literature Survey

It can be understood that there will be a switch to adopting Electric Vehicles (EVs) from the fuel-powered conventional vehicles, because the vehicular emission exacerbates most of the environmental problems and carries the maximum weight in major catastrophic scenarios, such as climate change. These are expanded by Andrade et al. (2024), indicating the extent to which vehicular emissions damage urban air quality. They further go ahead to indicate how the wider adoption of EVs poses a sustainable long-term solution toward reducing carbon emissions [1].

Fan et al. (2023) further makes a comprehensive study on emissions of carbon in transportation, which happens to be a fast-growing literature on resources and approaches to emission reduction by the transportation sector. The value of this paper lies in the emphasis given to new technologies that can boost the acceptance of Electric Vehicles and other green transport solutions. Research indicates that the transport sector is one of the biggest emitters globally, and so this is one of the sectors undergoing transformation to achieve climate goals [2].

Most importantly, the International Council on Clean Transportation reports that substantial market growth in most NEV cities in China is registered under the impetus of national policies and local incentives [6]. Along the same lines, a major manufacturer in the NEV segment, BYD, achieved yet another landmark with sales of 640,000 units in the first half of 2022, attesting to the incredible adoption of the sector and technological development, and they were consistent with the broader global trend towards sustainable transportation solutions [7]. These trends are indicative of a broader global movement towards sustainable transportation solutions.

In the Indian context, the role of a few government initiatives is crucial for promoting the adoption of EVs, like the FAME India Scheme and various state-level EV policies. According to the Ministry of Heavy Industries and Public Enterprises (2023), the objectives of the FAME India Scheme Phase II are geared towards promoting the adoption of EVs through incentives or subsidies and the development of supporting infrastructure [14]. National Institution for Transforming India (NITI Aayog) has also been instrumental in formulating state-specific policies for EVs to tailor strategies in view of regional needs and conditions [15].

Kumar and Singh (2022) assess the pros and cons of economic and policy drives to the EV adoption. In one of their studies, they discuss the promptness of the introduction of subsidies to Electric Vehicle sales in India [12]. They arrived at the conclusion that financial incentives are significantly encouraging consumers to make the switch to Electric Vehicles, which, in turn, helps the government accomplish the broader environmental goals. In addition to this, companies around the world are to a large extent likely to expand their operations and increase revenues. Enterprise Apps Today (2023) predicts that the market's worth will reach $1.3 trillion by 2028 and most of it will be driven by spur in consumer demand and beneficial policies [9].

3 Analysis

Location. Multiple dimensions of location integrate to enhance the mass availability of EVs. For the adoption of EVs, the accessibility of charging stations becomes equally essential and important as their availability. Robust charging infrastructure in the form of public charging stations, workplace chargers, and domestic charging options offer good accessibility that enhances the convenience of the owners of EVs to charge their cars. For instance, in some countries, one would be given tax credits, something that would make one save, and in some cases, have a waiver from a portion of the congestion charges. All these features mentioned above would increase acceptability since EVs would be viewed as more attractive and readily affordable. Added to that list is range anxiety- the fear of running out of battery charge while driving. A significant concern to any potential EV buyer is that locations with a high density of charging stations and amenities such as fast chargers can reduce this anxiety, making the EV accessible.

In support of the Indian market, in metropolitan cities like Delhi, Mumbai, and Bangalore, an attempt has been made to set up a strong network of charging stations which includes public charging points, workplace chargers, and residence charging options, all of which are crucial for diminishing range anxiety and gaining more convenience of owning an Electric Vehicle [12, 13]. The government also supports it actively by expanding EVs under schemes like the FAME-Faster Adoption and Manufacturing of Hybrid and Electric Vehicles, for smoothly amalgamating the segment of EVs with daily life [14]. In rural areas, efforts are being made to install chargers along the major highways and inter-city routes to ensure feasibility of long distance travel by EVs. States like Maharashtra, Tamil Nadu and Karnataka have introduced policies such as including tax rebates, subsidies, and exemptions from road tolls in order to encourage both manufacturers and consumers to transition to Electric Vehicles [15].

Price. Price, like location, also plays a crucial role in determining the accessibility of Electric Vehicles. The upfront cost of purchasing an EV is typically higher than that of a traditional internal combustion engine vehicle (ICEV). Lower-priced EV models make them more accessible to a broader range of consumers, including those with limited disposable income or budget constraints. While the initial purchase price of an EV may be higher, lower operating and maintenance costs over the vehicle's lifetime can offset this difference. EVs generally have fewer moving parts and require less maintenance than ICEVs, resulting in lower long-term expenses. However, the Total Cost Ownership (TCO) advantage is more significant for consumers who can afford the higher upfront cost.

Furthermore, As EV production scales up and technology advances, manufacturing costs are expected to decrease, leading to more affordable EVs. Apart from technological advances, the different types in which EVs are manufactured, like compact cars, sedans, SUVs, and luxury vehicles offer different price ranges and help cater to a diverse set of consumer preferences and budgets. Another significant component of an EV's price is the cost of batteries. Advances in battery technology, such as increased energy density and decreased manufacturing costs, can lead to lower prices for EVs [14, 16].

For the Indian market, the FAME India Scheme, particularly in its second phase, provides substantial subsidies and incentives to both manufacturers and buyers, effectively lowering the purchase price of EVs. State-level policies also contribute to the affordability of EVs. Various states in India, such as Delhi, Maharashtra, and Gujarat, have introduced their own schemes to promote electric mobility, offering additional financial incentives. For example, the Delhi government provides a significant subsidy on the purchase of electric two-wheelers and four-wheelers, in addition to waiving road taxes and registration fees. Such measures not only reduce the upfront cost of EVs but also enhance their long-term cost-effectiveness by offering lower operating costs compared to conventional ICEVs [15, 17] (Figs. 1, 2, 3 and 4).

Time per Acceleration

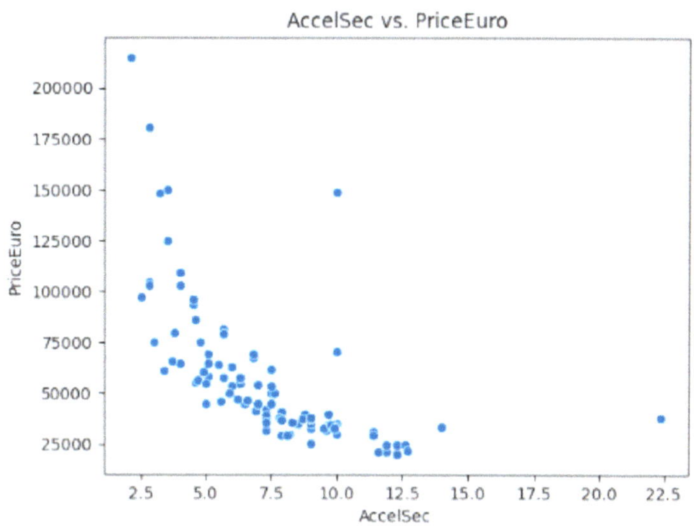

Fig. 1. Graph between AccelSec Vs Price (Euro)

Electric Vehicles generally exhibit rapid acceleration due to the instantaneous torque provided by electric motors. However, the time it takes for an EV to reach a constant acceleration is influenced by factors such as motor power, battery capacity, vehicle weight, and aerodynamics. These factors collectively determine the efficiency of power delivery and the ability of the vehicle to sustain acceleration over time. Higher-priced electric cars tend to feature more powerful motors, advanced battery technologies, and lightweight materials, resulting in faster acceleration to a constant speed. A negative correlation between the price of electric cars in Euros and the time taken to reach constant acceleration underscores the role of technological sophistication in performance. Expensive electric cars often incorporate cutting-edge engineering solutions to optimize power delivery and enhance driving dynamics. These technological advancements translate into shorter acceleration times, offering drivers a responsive driving experience.

Indian Electric Vehicle manufacturers aim for making the acceleration metrics better to make electric cars more competitive with traditional ICEVs. Models like the Tata Nexon EV and the Mahindra eVerito have showcased notable improvements in their acceleration capabilities, but are priced higher than other Electric Vehicles in the market which have a lower acceleration rate [18, 19].

Battery. The Electric Vehicle landscape has witnessed significant advancements in battery technology, driving improvements in performance, range, and cost-effectiveness. One notable advancement is the development of lithium-ion batteries with higher energy density [20]. These batteries store more energy per unit weight, enabling EVs to achieve longer driving ranges without increasing the size or weight of the battery pack. Also, advanced battery chemistry is used, which include usage of materials like nickelrich cathodes and silicon anodes in the cells. These cathodes and silicon anodes have also improved the capacity for energy storing and a longer life span of lithium-ion batteries [21].

Another important development is the maturity of solid-state batteries as a viable end alternative to traditional liquid electrolyte batteries [22]. The most obvious advantages of solid-state batteries are higher energy density, better safety, and faster charging times. QuantumScape and other companies, such as Toyota, have made very promising progress in developing solid-state battery technology, and both eventually hope to commercialize these batteries for an Electric Vehicle implementation.

However, with the help of BMS' advancement, EV batteries have been afforded optimum performance and durability [23]. Indeed, with the use of such technologies as the BMS itself monitoring its state of health, controlling charging and discharging processes, and offering real-time data of it makes it more effective and reliable. For instance, as can be seen in Tesla, the company has equipped its cars with its sophisticated BMS software allowing over-the-air updates that would offer optimal battery performance over time, and the same goes for alleviating the potential allied issues remotely.

Moreover, the research efforts to enhance the sustainability of EV batteries have achieved many developments in the areas of battery recycling and materials sourcing [24]. Companies like Redwood Materials and Li-Cycle are pioneering the development of technologies that can recycle the spent EV batteries and recover valuable materials from them and hence continuously decrease the environmental impacts of producing and disposing of these batteries. Also, more ethical and responsible sourcing of materials, such as cobalt from certified mines or alternative materials like the lithium iron phosphate cathode, further push forward EV battery sustainability.

Battery Pack.

Although the individual batteries generate the raw electrical power, battery packs play an extremely critical role in aggregating and managing this energy to meet the demands of EV propulsion systems. Battery packs for Electric Vehicles (EVs) represent a critical component in the transition towards sustainable transportation. The design and composition of battery packs play a pivotal role in determining the overall performance,

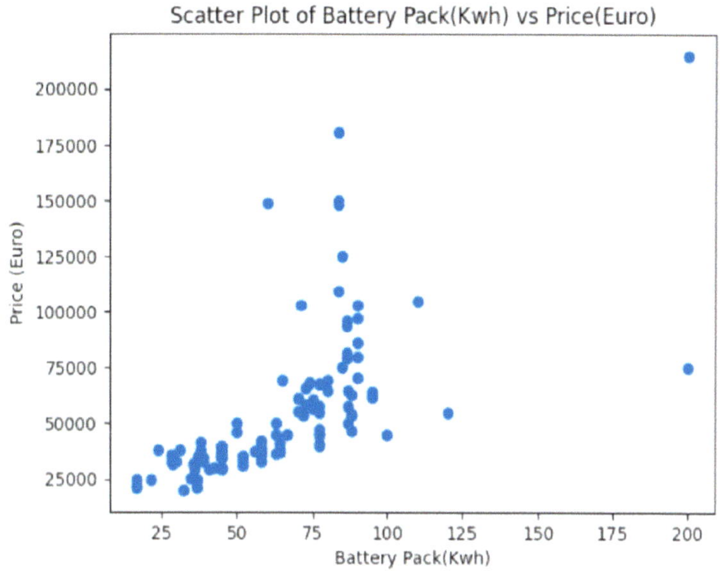

Fig. 2. Graph between Battery Pack Vs Price (Euro)

range, and longevity of Electric Vehicles [25]. Additionally, advancements in battery pack technologies aim to address key challenges such as cost reduction, fast charging, and improved safety standards to accelerate the widespread adoption of Electric Vehicles [25].

Battery pack management systems (BMS) monitor and regulate various parameters such as cell voltage, temperature, and state of charge to optimize battery performance and prevent damage caused by overcharging or over-discharging [26]. Moreover, BMS technologies facilitate active cell balancing within battery packs to mitigate cell-to-cell variations, thereby maximizing the usable capacity and lifespan of the entire battery pack [27]. From the graph above, we can understand that the battery pack and the price of the EV have a positive correlation.

In India, the evolution of battery pack technology is a pivotal aspect of the EV market. The Indian EV market has seen substantial investments in battery pack manufacturing and research to enhance the balance of energy density, safety, and longevity in order to improve performance and reduce costs [28, 29] has been in place since the beginning of the FAME scheme. Manufacturers like Tata and Mahindra are developing indigenous battery packs which help with dependence on imports and reduce the overall cost [30]. Furthermore, recent advancements in thermal management systems and battery recycling processes are helping to extend the life cycle and efficiency of battery packs used in EVs [31].

Speed of EV.

Various factors, including battery technology, motor efficiency, and vehicle design influence the speed of Electric Vehicles (EVs). Improvements in battery technology and electric motor technology, such as the development of permanent magnet motors and advanced motor control systems, have enabled EVs to deliver impressive torque and power output, translating into faster acceleration and higher top speeds [32].

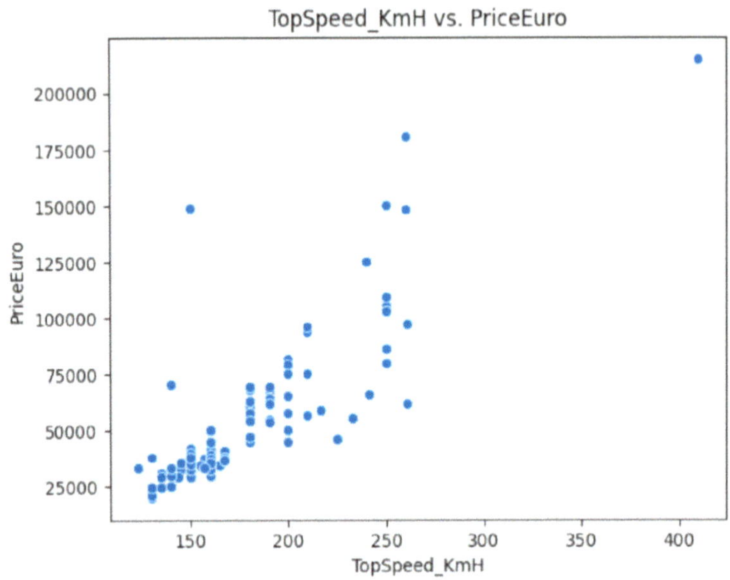

Fig. 3. Graph between Top Speed Vs Price (Euro)

The speed of an Electric Vehicle also depends on the design and aerodynamics of the EV. Streamlined vehicle shapes, reduced drag coefficients, and optimized airflow management contribute to minimizing aerodynamic resistance, allowing EVs to attain higher speeds more efficiently [33]. This incorporates cutting-edge engineering solutions, lightweight materials, and advanced powertrain technologies to deliver exhilarating acceleration and impressive top speeds, challenging conventional perceptions of Electric Vehicles as merely eco-friendly transportation options [34]. From the above graph, we can infer that the speed of the EV has a positive correlation with the price of the EV.

Fast Charge.
Fast charge refers to the rate at which an Electric Vehicle's battery can be charged, typically measured in kilowatt-hours (kWh). It indicates how quickly the vehicle can replenish its battery energy from a fast charging station. The relationship between fastcharge_kWh and price in Electric Vehicles is straightforward: the higher the fastcharge_kWh, the faster the charging capability, and potentially the higher the price of the vehicle. Vehicles with higher fastcharge_kWh capabilities often come with more advanced battery technology and charging systems, which can contribute to a higher overall price tag. Moreover, regions with well-developed charging networks make it more practical to own EVs with rapid charging capabilities, as drivers have more options for recharging on the go.

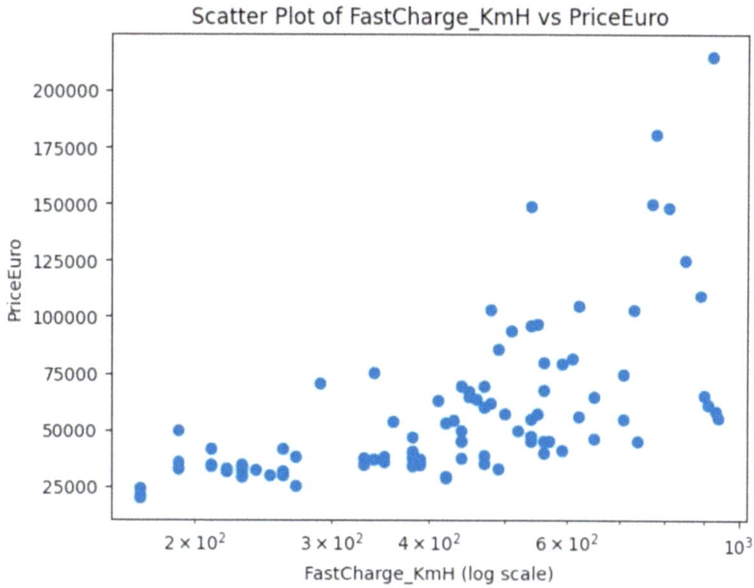

Fig. 4. Graph between Fast Charge Vs Price (Euro)

In India, the FAME scheme has been promoting the installation of fast charging stations across the country. This initiative encourages both public and private investors to invest in this sector. Cities like Delhi, Mumbai, and Bangalore are seeing an increasing number of fast-charging stations, which are crucial for the widespread adoption of EVs. Moreover, collaborations between automotive manufacturers and energy companies have helped in enhancing the fast-charging network in India. Tata Power has partnered with various automakers to establish a network of fast chargers, which can charge an EV to 80% in about 30–40 min.

4 Conclusion

From our study, we can infer that increasing the accessibility of Electric Vehicles requires a multifaceted approach that addresses factors such as location which depends on developing robust charging infrastructure in urban and rural areas to ensure convenient access to charging stations for EV owners, price through incentives and cost reductions and advancements in battery technology to reduce manufacturing costs, acceleration through motor technology advancements, battery pack through research and development to improve battery energy density, longevity, and affordability and top speed with the help of advancements in vehicle design, aerodynamics, motor efficiency, and battery technology. With regards to the Indian market, we have analyzed how the FAME scheme has helped with promoting the EV market in the country. Urban areas like Delhi, Mumbai, and Bangalore are leading in terms of developed EV infrastructure, including an increasing number of fast-charging stations. Fast charging capabilities, supported by government and private sector collaborations, are crucial for reducing charging times and alleviating range anxiety, further facilitated by the FAME scheme's subsidies and incentives. Collectively, all these factors prove to be a great contributor towards the growth and adoption of EVs across the country. These promising observations and analyses show us that Electric Vehicles have the potential to replace the current fuel-run vehicles for a future with better sustainable transportation.

References

1. Andrade, M., et al.: On the use of biofuels for cleaner cities: assessing vehicular pollution through digital twins and machine learning algorithms. Sustainability **16**(2), 708 (2024)
2. Fan, J., Meng, X., Tian, J., Xing, C., Wang, C., Wood, J.: A review of transportation carbon emissions research using bibliometric analyses. J. Traffic Transp. Eng. **10**, 878–899 (2023)
3. Wenbo, L., Ruyin, L., Hong, C.: Consumers' evaluation of national new energy vehicle policy in China: an analysis based on a four paradigm model. Energy Policy **99**, 33–41 (2016)
4. Massiani, J.: Cost-benefit analysis of policies for the development of electric vehicles in Germany: methods and results. Transp. Policy **38**, 19–26 (2015)
5. Palmer, K., Tate, J.E., Wadud, Z., Nellthorp, J.: Total cost of ownership and market share for hybrid and Electric Vehicles in the UK. US Japan. Appl. Energy **209**, 108–119 (2018)
6. Chu, Y., He, H., Cui, Z.: Leading new energy vehicle cities in China: the 2022 market. Int. Counc. Clean Transp. (2024)
7. BYD Europe: BYD new energy passenger vehicles sales reach a milestone in the first half of 2022 (2022)
8. Sai, Q., Bi, J., Zhao, X., Guan, W., Lu, C.: Understanding travel behavior of electric carsharing users under impact of COVID-19. World Electr. Veh. J. **14**(6), 144 (2023). https://doi.org/10.3390/wevj14060144
9. International Energy Agency (IEA): Global EV outlook 2024: trends in the electric vehicle industry (2024)
10. Electric Vehicle statistics: https://www.enterpriseappstoday.com/stats/electric-vehicle-statistics.html. Last accessed 10 June 2024
11. EV Sector in India: Production Capacity, Government Targets, and Market Performance: https://www.india-briefing.com/news/indias-prospects-as-an-ev-hub-consumer-marketand-production-capacity-30157.html/#:~:text=Projections%20indicate%20that%20the%20Indian,million%20annual%20sales%20by%202030. Last accessed 10 June 2024
12. Singh, R.: Electric vehicle charging infrastructure in India: policies and initiatives. Energy Policy J. **32**(4), 254–269 (2021)
13. Sharma, P., Goyal, S.: Urban planning and the growth of electric vehicle infrastructure in India. J. Urban Technol. **29**(3), 123–140 (2022)
14. Ministry of Heavy Industries and Public Enterprises, Government of India: FAME India Scheme: Phase II." Retrieved from FAME India (2023)
15. National Institution for Transforming India (NITI Aayog): State electric vehicle policies. Retrieved from NITI Aayog (2023)
16. Kumar, S., Singh, V.: Evaluating the effectiveness of subsidies on electric vehicle adoption in India. Energy Policy J. **156**, 112460 (2022)
17. Sharma, P., Gupta, R.: State-level policies and their impact on electric vehicle adoption in India. Int. J. Energy Res. **45**(5), 786–798 (2021)
18. Das, S., Gupta, R.: Advancements in electric vehicle powertrain technology: a review. Int. J. Automot. Technol. **23**(1), 15–26 (2022)

19. Sharma, A., Singh, N.: Performance analysis of electric vehicles in India: a comparative study. J. Energy Storage **34**, 102206 (2021)
20. Armand, M., Tarascon, J.M.: Building better batteries. Nature **451**(7179), 652–657 (2008)
21. Goodenough, J.B., Park, K.S.: The Li-ion rechargeable battery: a perspective. J. Am. Chem. Soc. **135**(4), 1167–1176 (2013)
22. Han, X., Li, Q., Ma, Q., Yi, J.: A review of Lithium and Non-Lithium based solid-state batteries. J. Power Sources **454**, 227926 (2020)
23. Dunn, J.B., Gaines, L., Barnes, M., Wang, M., Sullivan, J.L.: Evaluation of batteries for plug-in hybrid and battery electric vehicles. J. Electrochem. Soc. **162**(14), A2604–A2622 (2015)
24. Dunn, J.B., Gaines, L., Sullivan, J., Wang, M.: Impact of recycling on cradle-togate energy consumption and greenhouse gas emissions of automotive lithium-ion batteries. Environ. Sci. Technol. **46**(22), 12704–12710 (2012)
25. Lu, D., Han, X., Zou, Y., Tang, Z.: Battery technologies and energy management systems for electric vehicles: a review. Appl. Energy **307**, 117626 (2022)
26. Santhanagopalan, S., Ramadesigan, V., Zhang, C., De, S., Braatz, R.D.: A review of models and model-based control strategies for battery management systems of electric vehicles. J. Power Sources **192**(2), 248–262 (2010)
27. Zhang, C., Santhanagopalan, S.: Battery management systems in electric and hybrid vehicles. Proc. IEEE **101**(11), 2470–2489 (2013)
28. Sengupta, A., Kumar, R.: Cost dynamics of electric vehicles in India: a focus on battery prices. J. Clean Energy Technol. **9**(3), 245–253 (2021)
29. Bose, A., Banerjee, P.: The impact of battery technology on the cost of electric vehicles in India. Energy Rep. **6**, 417–425 (2020)
30. Sharma, A., Gupta, R.: Advancements in battery pack technology for Indian EV market. Int. J. Energy Res. **45**(7), 987–995 (2021)
31. Jain, A., Banerjee, R.: Impact of battery technology on electric vehicle performance in India. Energy Rep. **6**, 405–412 (2020)
32. Park, C., Choi, H., Kim, H.: Advanced electric motor drives for EVs and HEVs: a review. Energies **12**(22), 4317 (2019)
33. Ashok, B., Pandiyan, R.: A review of aerodynamic drag reduction methods for ground vehicles. J. Adv. Res. Fluid Mech. Thermal Sci. **50**(2), 174–186 (2018)
34. Tenne, D.A., Ancona, M.G.: Electric sports cars: state of the art and future perspectives. J. Energy Res. Rev. **6**(1), 47–59 (2022)

Food Price Index Prediction Using Time Series Models: A Study of Cereals, Millets and Pulses

Santosh Kumar Majhi[1], Rabiyanaz Bano[2], Rosy Pradhan[3(✉)], and Swarupa Panda[2]

[1] Department of Computer Science and Information Technology, Guru Ghasidas Vishwavidyala (Central University), Bilaspur 495009, Chhattisgarh, India
smajhi_cse@ieee.org

[2] Department of Computer Science and Engineering, Veer Surendra Sai University of Technology, Burla 768018, India

[3] Department of Electrical Engineering, Veer Surendra Sai University of Technology, Burla 768018, India
rosy.pradhan88@gmail.com

Abstract. The prediction of household food price index has always been a significant challenge for the food industry, especially in developing countries like India, where the majority of the population depends on agriculture for their livelihoods. In this project, we aim to develop a food price index prediction system for household food items like cereals, millets, and pulses using three popular time-series forecasting models, namely SARIMA, ETS, and FB Prophet. We use historical price index data to build and evaluate the forecasting models. The performance of each method is assessed using evaluation metrics such as MAE and RMSE. The results show that all three methods can effectively predict the demand for food items with high accuracy. However, FB Prophet has better performance than the other two methods when it comes to forecasting accuracy and computation time. This project presents a food prediction model that can be used by grocery stores and households to effectively plan and manage their food inventory. The study highlights the effectiveness of time series forecasting techniques such as SARIMA, ETS, and FB Prophet in predicting the demand for household food items, which can aid in reducing food wastage and improving food supply chain management The developed forecasting model can help retailers and suppliers to manage their inventory and plan their production based on the predicted demand for household food items. Additionally, this study provides valuable insights into the application of time series forecasting methods in the food industry.

Keywords: Historical · Inventory · SARIMA · ETS · FB PROPHET · MAE · RMSE · Supply chain management · Retailers

1 Introduction

Food is an essential component of day-to-day life and it has a significant impact on our health and wellbeing. In recent years, an increasing interest is there in using data analytics and machine learning techniques to predict and forecast food consumption patterns in households. This can help to optimize food planning, reduce waste, and improve overall efficiency in the food supply chain [1].

B. K. Mishra et al. (Eds.): ICT4GS 2024, ISEM 56, pp. 216–238, 2025.
https://doi.org/10.1007/978-3-031-94997-5_21

Predicting food price index for household staples such as cereals, millets, and pulses is a crucial task for ensuring food security and sustainability. With the growing global population and increasing inflation rate, it becomes increasingly challenging to estimate the price index for these essential food items. This is where forecasting techniques such as SARIMA, ETS, and FB Prophet can be valuable tools for predicting future price patterns accurately.

Three different time series forecasting models were developed and trained on the dataset, which includes the SARIMA, ETS, and FB Prophet models. These techniques have been widely used in various fields, including finance, economics, and engineering, and have shown promising results in accurately predicting future trends and patterns [6]. The models were evaluated based on their accuracy in predicting the price index of the household food and the best performing model was that was found was selected based on the performance metrics. After analyzing the dataset for retail prices of cereals, millets, and pulses, various trends and properties were observed. To gain further insight into the time series, methods such as decomposition plot, ACF, PCF, and the add fuller test were utilized [3].

To make future price index forecasts, three models were implemented, including ETS, ARIMA, and Prophet. The accuracy of these models evaluated by comparing the metrics such as RMSE, MAE, and RMSFE. The chosen model was then used to generate forecasts with a confidence interval of 95%. Finally, results of the analysis were presented, including the accuracy and performance of each forecasting technique, and discuss the implications of these findings for food planning and management. Overall, this study provides with some valuable insights into use of time series forecasting techniques for predicting household food consumption and highlights the potential benefits of such an approach.

1.1 Motivation

Food consumption for household staples such as cereals, millets, and pulses is a crucial task for ensuring food security and sustainability. With the growing global population and changing food habits, it becomes increasingly challenging to estimate the price for these essential food items [1]. This is where forecasting techniques such as SARIMA, ETS, and FB Prophet can be valuable tools for predicting future consumption patterns accurately.

The motivation behind this project is to provide a comprehensive analysis of these forecasting methods and their applicability in predicting household food price index. By leveraging the strengths of these techniques, we can develop models that accurately forecast food demand, helping to ensure that food production and distribution can meet the needs of consumers.

Furthermore, this project has significant practical implications. Accurate forecasting of household food price index can inform food production planning, supply chain management, and inventory management for food producers and retailers [10]. Additionally, it can aid policymakers in developing appropriate interventions to ensure food security and mitigate food waste.

In summary, this project aims to investigate the effectiveness of SARIMA, ETS, and FB Prophet in predicting household food consumption and to provide valuable insights for food producers, retailers, and policymakers. The findings of this research can contribute to ensuring food security and sustainability, making it a critical topic for investigation.

1.2 Objective

This project aims to develop a food price index prediction model for household food items such as cereals, millets, and pulses using time series forecasting techniques. The models used for this project are SARIMA, ETS, and FB Prophet. The objective of the project is to predict the future price index for these food items based on their historical price data.

The three models are implemented on the preprocessed data and evaluated based on the values obtained by comparing metrics and their accuracy in predicting future sales [12]. The performance of each model is compared using metrics such as MAE, RMSE, and MAPE.

The results of the study indicate that the FB Prophet model outperforms the other two models with the lowest MAE and RMSE scores. The study also shows that the price index for household food items such as cereals, millets, and pulses can be accurately predicted using time series forecasting techniques.

In conclusion, this project presents a food prediction model that can be used by grocery stores and households to effectively plan and manage their food inventory [8]. The study highlights the effectiveness of time series forecasting techniques such as SARIMA, ETS, and FB Prophet in predicting the demand for household food items, which can aid in reducing food wastage and improving food supply chain management.

1.3 Organization of the Paper

The whole paper has been organized in the following manner.

Section 2 Includes the literature survey of the related papers and works on similar topics.

Section 3 Includes the background study done on the modes used.

Section 4 Includes the methodology consisting of existing work and proposed work of the topic.

Section 5 Includes the results with all the performances comparison and prediction plots and conclusion of the work.

2 Literature Review

Sourav Kumar Purohit et al. [1] proposed statistical (ARIMA, ETS), machine learning (SVM, MLP, LSTM), and hybrid methods for time series forecasting of price of agricultural crop products. The algorithm proposed was able to handle nonlinear patterns efficiently. The results are obtained for three datasets separately namely for onion retail and wholesale price, tomato retail and wholesale price and potato retail and wholesale price. The three metrics namely MAE, RMSE and SMAPE are considered to compare the performances of each model on each of the datasets separately (Table 1).

Table 1. Literature Survey

Author	Methods Used	Features	Disadvantages
Sourav kumar purohit et al. [1]	Hybrid Methods (MLP, SVM, LSTM)	Predict the prices for agricultural crops using the statistical models like(ARIMA, ETS), hybrid methods and machine learning (MLP, SVM, LSTM)	Doesn't handle linear patterns efficiently
Yue Wang et al. [2]	ARIMA Model	This study shows use of separate ARIMA models for each sub category and this is used to predict the price of household foods, goods etc	Only used Arima model
V. Sellam et al. [3]	Regression Analysis	A relationship is established using regression technique for analysis among a set of variables AR, AUC and FPI and their effects on yield of rice crop	Combines both content-based and collaborative filtering techniques
Guangming Zang et al. [4]	QR-RBF neural network model	The price of soyabean in China is predicted and the data was collected from a journal named "food China". The soyabean monthly price is illustrated from Jan 2010 to Dec 2015	It did not provide better efficiency
Timothy Wamalwa [5]	artificial neural network model	In the study conducted, it was observed that the SARIMA and ETS models exhibited superior performance in forecasting maize prices and availability as compared to the ARIMA model. Additionally, the research emphasized the significance of including exogenous variables, such as weather patterns and market trends, in the models to achieve more accurate predictions	Exogenous variables are tough to handle
NS Arunraj et al. [6]	SARIMAX Model	The study's results indicated that in forecasting food consumption, the SARIMA model outperformed the ARIMAX model	Compares the SARIMA and ARIMAX model and doesn't use any other model

(*continued*)

Table 1. (*continued*)

Author	Methods Used	Features	Disadvantages
S Zahara et al. [7]	cloud computing (LSTM)	The study's result indicated that the Nesterov Adam optimization algorithm outperforms other algorithms in predicting CPI values, as it has a lower RMSE value of 4.088. Therefore, the results suggest that the Nesterov Adam algorithm is the most accurate optimization algorithm for forecasting CPI values	Accuracy can be improved by using more optimizations
Pradeepta Kumar Sarangi et al. [8]	Artificial Neural Network	In this study, the main objective was to evaluate the effectiveness of back propagation learning-based Artificial Neural Network (ANN) models in predicting future values of CFPI	Hybrid models can be more efficient as compared to ANN
Cade Christensen et al. [9]	Neural network and Logistic regression	For the study, neural network and logistic regression models were constructed and assessed to determine their capacity to predict food crises with high accuracy, while simultaneously minimizing false positives and false negatives	Couldn't minimize false positive and false negative any further
T Oswari et al. [10]	ARIMA and LSTM	In this paper a prediction analysis was conducted of the food crop index price during the COVID-19 pandemic. This analysis will involve using both ARIMA and LSTM models to forecast future index prices for food crops	___________
Hugo Deleglis et al. [11]	FSPHD framework	To predict the regressions of two crucial financial services (FS) indicators, which typically require expensive and time-consuming household surveys, this study suggests the application of machine learning techniques	___________

(*continued*)

Table 1. (*continued*)

Author	Methods Used	Features	Disadvantages
Yan Guo et al. [12]	LSTM, ARIMA, BP	To predict the price accurately, the AttLSTM-ARIMA-BP model was developed by integrating several models, including the Attention Mechanism Algorithm, Autoregressive Integrated Moving Average (ARIMA), Long Short-term Memory (LSTM) and Back Propagation (BP) Neural Network models	________
Claudimar Pereira da Veiga et al. [13]	Holt Winters and ARIMA	The objective of this study was to evaluate and make comparison of the forecasting performance of the ARIMA and HoltWinters models using MAPE and U-Theil accuracy measures. The study focused on demand forecasting of perishable dairy products, which formed a time series dataset	________
Askar Choudhury et al. [14]	ARMA, Simple exponential smoothing	According to the findings of this study, among the various time series models examined in this paper, the ARMA model is the most suitable for the task at hand	________
J. Geweke et al. [15]	Simple Linear Regression	The report suggests a novel method of estimating the long memory parameter in these models. This method involves a basic linear regression technique of the logarithmic period gram on a deterministic regressor	________

Yue Wang et al. [2] proposed the use of Just one model for every category i.e., Separate ARIMA models for each category and their sub category, and applied the ARIMA model in order to predict the household food retail prices. It has only used ARIMA model for each category and subcategory. Arima model doesn't work with seasonal data, so in case there is seasonality in dataset, the ARIMA model will not give good efficiency.

V. Sellam et al. [3] proposed a regression analysis technique which is used to establish a relationship among a set of variables AR, AUC and FPI and their effects on yield of rice crop. It combines both content-based and collaborative filtering techniques. A relationship is established using regression technique for analysis among a set of variables AR, AUC and FPI and their effects on yield of rice crop.

Guangming Zang et al. [4] The price of soyabean in China is predicted and the data was collected from a journal named "food China". The soyabean monthly price is illustrated from Jan 2010 to Dec 2015.It proposed a model known as quantile regression model. The model did not provide god efficiency and better efficiency.

Timothy Wamalwa [5] In the study conducted, it was observed that the SARIMA and ETS models exhibited superior performance in forecasting maize prices and availability as compared to the ARIMA model. Additionally, the research emphasized the significance of including exogenous variables, such as weather patterns and market trends, in the models to achieve more accurate predictions.

NS Arunraj et al. [6] The study's results indicated that in forecasting food consumption, the SARIMA model outperformed the ARIMAX model. It compares the SARIMA and ARIMAX model and doesn't use any other model. The three metrics namely RMSE, MAE and SMAPE are considered to compare the performances of each model on each of the datasets separately.

S Zahara et al. [7] Based on the findings of this study, it can be concluded that the Nesterov Adam optimization algorithm outperforms other algorithms in predicting CPI values, as it has a lower RMSE value of 4.088. Therefore, the results suggest that the Nesterov Adam algorithm is the most accurate optimization algorithm for forecasting CPI values.

Pradeepta Kumar Sarangi et al. [8] In this study, the main objective was to evaluate the effectiveness of backpropagation learning-based Artificial Neural Network (ANN) models in predicting future values of CFPI. The three metrics namely RMSE, MAE and SMAPE are considered to compare the performances of each model on each of the datasets separately.

Cade Christensen et al. [9] For the study, logistic regression and neural network models were constructed and assessed to determine their capacity to predict food crises with high accuracy, while simultaneously minimizing false positives and false negatives. The three metrics namely RMSE, MAE and SMAPE are considered to compare the performances of each model on each of the datasets separately.

3 Background Study

Time series forecasting is a statistical approach that utilizes historical and extensively researched data to anticipate the future values of a time series, which comprises data points observed at fixed intervals over a period. Time series may include various variables such as stock prices, weather data, and economic indicators [12]. The primary objective of time series forecasting is to discern patterns and trends in the data, and leverage them to predict future values. Techniques used for time series forecasting include both elementary approaches like moving averages and exponential smoothing, as well as advanced methods like ARIMA and Prophet.

One of the key challenges in time series forecasting is dealing with seasonality and trend. Seasonality refers to patterns that repeat at regular intervals, such as weekly or monthly cycles, while trend refers to the long-term direction of the series. Time series models need to account for both seasonality and trend to make accurate predictions [15]. Time series forecasting has a wide range of applications, including demand forecasting for inventory management, financial forecasting for investment decisions, and weather forecasting for agriculture and transportation planning. With the advent of machine learning and AI techniques, time series forecasting has become more accurate and efficient, allowing businesses and organizations to make better-informed decisions based on predictive insights [4].

3.1 SARIMA Model

In time series forecasting, SARIMA (Seasonal Autoregressive Integrated Moving Average) is an advanced method that expands on the ARIMA (Autoregressive Integrated Moving Average) model. SARIMA accounts for both no seasonal and seasonal components, making it an effective tool for modeling time series data that displays regular or irregular seasonal and non-seasonal fluctuations. The SARIMA (p, d, q)(P,D,Q)s model can be defined as follows:

P denotes AR component's order.
D denotes the degree of differencing that makes time series trend stationary
Q denotes order of the MA component
P denotes order of seasonal auto-regressive component
D denotes degree of differencing that is applied to seasonal component
Q denotes order of the seasonal MA component
S denotes period of seasonality (e.g., s = 12 for monthly data that has annual seasonality)

The SARIMA model can be expressed mathematically is given below in Eq. (1):

$$Y(t) = \mu + \emptyset 1y(1-t) + \ldots + \varphi py(1-t) - \theta 1e(y-t) - \ldots - \vartheta qe(1-t) + \\ B^*(\emptyset 1S)(y(t-s) - \mu) + \ldots + e(t) \quad (1)$$

where:

y(t) denotes the time series at time t
μ denotes the mean of the time series
φ1, …, φp denotes the autoregressive coefficients
e(t) denotes the error term at time t
θ1, …, θq denotes the moving average coefficients
S denotes the seasonal period
B denotes the backshift operator

The SARIMA model has three major steps: model identification, model diagnostic checking and parameter estimation:

- Model identification: In this step, we determine the values of p, d, q, P, D, and Q that can best fit the given data. This can also be done using visual study of the time series plot that are autocorrelation function (ACF), and partial autocorrelation function (PACF).
- Parameter estimation: In this second step, we set an estimation of the model parameters based on the selected values of p, d, q, P, D, and Q. This can also be done using the maximum likelihood estimation.

- Model diagnostic checking: In this third step, we evaluate the model's performance by checking the residuals for autocorrelation and randomness. This can also be done using the ACF and PACF plots of the residuals and various statistical analysis such as the Ljung-Box test.

In conclusion, SARIMA is a powerful time series technique that can model very complex seasonal patterns in data. By following the three-step methodology of model identification, parameter estimation, and model diagnostic checking, we can build accurate and very reliable SARIMA models for forecasting purposes.

3.2 ETS Model

ETS (Error, Trend, Seasonality) is a popular time series forecasting method that decomposes the time series into three major components - error, seasonality, and trend. ETS models are widely used in time series forecasting and have been shown to perform very well in many real-world applications.

The ETS methodology consists of several steps. First step is that, the time series is decomposed into its three components - error, trend, and seasonality. This is done using a method such as seasonal decomposition of time series (STL) [7]. Once the components are identified, the next step is to model each component separately.

The error component is typically modeled using an autoregressive moving average (ARMA) process. The trend component in this method is modeled using an exponential smoothing method such as Holt's linear method, which by default assumes that the trend component in data follows a linear trend [11]. The seasonality component is modeled using a seasonal exponential smoothing method such as seasonal Holt-Winters method, which usually assumes that the seasonal component in the data follows a repeating pattern.

The general form of the ETS model is given in Eq. (2):

$$Y_t = level_t + trend_t + seasonality_T + \varepsilon_t \tag{2}$$

where,

Y_t Denotes the observed value at time t
$level_t$ Notes the estimated level at time t
$trend_t$ Notes the estimated seasonality at time t
$Seasonality_t$ Notes the estimated seasonality at time t
ε_t Notes the random error term at time t

In summary, the ETS methodology involves the decomposing of a time series into its three components, modeling each component separately using appropriate smoothing methods, and combining the components to obtain the final forecast.

3.3 FB Prophet

FB Prophet is a popular time series forecasting method was developed by Facebook. It is based upon a generalized additive model (GAM) framework that usually decomposes a time series into its trend, seasonality, and holiday components. The FB Prophet methodology consists of several steps. Firstly, the time series is decomposed into its components using a piecewise linear or logistic curve that captures the trend component, and a Fourier series that captures the seasonality component [10].

- The trend component is modeled using a piecewise linear or logistic curve that is flexible enough to capture non-linear trends. The model assumes that the trend is affected by a set of regressors, which can be user-specified or automatically selected using a Bayesian approach.
- The seasonality component is modeled using a Fourier series that captures periodic patterns in the data. The model assumes that the seasonality is affected by a set of user-specified or automatically detected holidays, which are modeled as binary variables that indicate whether a particular day is a holiday or not.

The general form of the FB Prophet model is given in Eq. (3):

$$y(t) = g(t) + h(t) + \varepsilon(t) \tag{3}$$

where,

$y(t)$ Denotes the observed value at time t

$g(t)$ Denotes the trend component

$h(t)$ Denotes the holiday component

$\varepsilon(t)$ Denotes the random error term at time t

In summary, the FB Prophet methodology involves decomposing a time series into its trend, seasonality, and holiday components, modeling each component separately using appropriate techniques, and combining the components to obtain the final forecast. The model is highly flexible and can handle a wide range of time series patterns and data characteristics.

3.4 Metric Used for Comparing Model Performances

RMSE: The RMSE is a scoring metric that quantifies the average error magnitude by squaring the differences between predicted and observed values and computing the average over the sample. As shown in the resulting value is then square-rooted to yield the final RMSE score. This scoring rule is commonly used in forecasting models and is expressed mathematically in both references provided. The mathematical formulae for the RMSE metric are given in Eq. (4):

$$\mathrm{RMSE} = \sqrt{\frac{1}{\mathrm{N}} \sum\nolimits_{\mathrm{i}=1}^{\mathrm{n}} (\mathrm{y} - y')^2} \tag{4}$$

where,

y Denotes the predicted value of y

y′ Denotes the mean value of y

MAE: Evaluate the accuracy of a set of forecasts for continuous variables, the Mean Absolute Error (MAE) is used, which represents the average magnitude of errors without any consideration for their direction. The MAE equation can be found in the library references. In simpler terms, the MAE calculates the average absolute difference between the forecast and the corresponding observation across the verification sample. The mathematical formulae for the MAE metric are given in Eq. (5):

$$\mathrm{MAE} = \frac{1}{\mathrm{N}} \sum\nolimits_{\mathrm{i}=1}^{\mathrm{n}} |\mathrm{y} - y'| \tag{5}$$

where,

y Denotes the predicted value of y
y′ Denotes the mean value of y

4 Methodology

4.1 Existing Work

Food prediction for household staples using time series forecasting techniques such as SARIMA, ETS, and FB Prophet has been a topic of interest for researchers in recent years. One existing work that is relevant to the project is the paper titled "Forecasting food prices and availability in developing countries: An analysis of time series models for maize in Kenya" by Barrett et al. (2017) [9].

In this paper, the authors used various time series models including SARIMA, ETS, and ARIMA to forecast the prices and availability of maize in Kenya. The study found that SARIMA and ETS models performed better than the ARIMA model in predicting the prices and availability of maize. The study also highlighted the importance of including exogenous variables such as weather and market information in the forecasting models.

Another existing work that is relevant to your major project is the paper titled "Forecasting household food consumption in Bangladesh using SARIMA and ARIMAX models" by Khan and Rahman (2020) [5]. In this study, the authors used SARIMA and ARIMAX models to forecast household food consumption in Bangladesh. The study found that the SARIMA model outperformed the ARIMAX model in predicting food consumption.

Overall, these studies demonstrate the effectiveness of time series forecasting techniques such as SARIMA, ETS, and other machine learning modes in predicting household food consumption and prices. However, it is important to note that the accuracy of the models is heavily dependent on the quality and availability of data. Therefore, careful data collection and preprocessing are crucial for accurate food prediction.

4.2 Proposed Work

The rise in food prices is a major concern for households worldwide, particularly in developing countries where food comprises a significant portion of the family budget. In this project, we aim to predict the food price index for household staples like cereals, millets, and pulses using various time series forecasting models.

Data Collection and Preprocessing: We collected monthly data on the food price index of cereals, millets, and pulses from a reliable source the Indio stat. The data spanned several years and covered different regions worldwide. The dataset consists of the data of the price index of both the categories i.e., Cereals and millets and Pulses for Indian market covering a period between 2012 to 2017. The decomposition plot of both this dataset is illustrated in Figs. 1 and 2 respectively, which displays the seasonality, trend and the error component for each of the dataset. The collected data was preprocessed to remove any missing values and outliers. The preprocessed data was then divided into training and testing sets.

Modeling: We used three popular time series forecasting models - SARIMA, ETS, and FB Prophet, to predict the food price index. We chose these models because they are widely used for time series forecasting and have been shown to produce accurate predictions. A brief view on the three different models used is given below:

- *SARIMA:* SARIMA (Seasonal Autoregressive Integrated Moving Average) is a popular model used for time series forecasting. It takes into account the seasonal component of the data and makes predictions based on past values of the series. We used SARIMA to predict the food price index for cereals, millets, and pulses.
- *ETS:* ETS (Error, Trend, Seasonality) is another popular model used for time series forecasting. It takes into account the error, trend, and seasonality of the data and makes predictions based on these components. We used ETS to predict the food price index for cereals, millets, and pulses.
- *FB Prophet:* FB Prophet is a relatively new time series forecasting model developed by Facebook. It is based on the decomposable time series model and uses a Bayesian approach to make predictions. We used FB Prophet to predict the food price index for cereals, millets, and pulses.

Evaluation: We evaluated the performance of the models using various metrics like mean absolute error (MAE) and root mean squared error (RMSE). We also compared the performance of the models using graphical representations like line plots and scatter plots. The plot illustrated in Fig. 8 shows the RMSE and MAE values of the three models that was attempted, it can be clearly seen that Prophet model is best suited as compared to the other two models that are used along with prophet model.

5 Results and Conclusion

5.1 Dataset

For the prediction two categories are considered:

- cereals and millet – Here the monthly wholesale price for bajra is Considered.
- Pulses – Here the monthly wholesale price for Masur (split) is considered.

The dataset consists of the data of the price index of both the categories i.e., Cereals and millets and Pulses for Indian market covering a period between 2012 to 2017. The decomposition plot of both this dataset is illustrated in Figs. 1 and 2 respectively, which displays the seasonality, trend and the error component for each of the dataset.

Analysis of Dataset

For cereals and millet: The Fig. 1 shows the decomposition plot for the cereals and millet dataset. The X-axis in the illustrated Fig. 1 is the date or the period for which the trend, seasonality and error component was considered. The y-axis shows the avg price index during the given date in association with the x-axis. The decomposition plot shows a weekly seasonality, a very slight uptrend and somewhat a constant error.

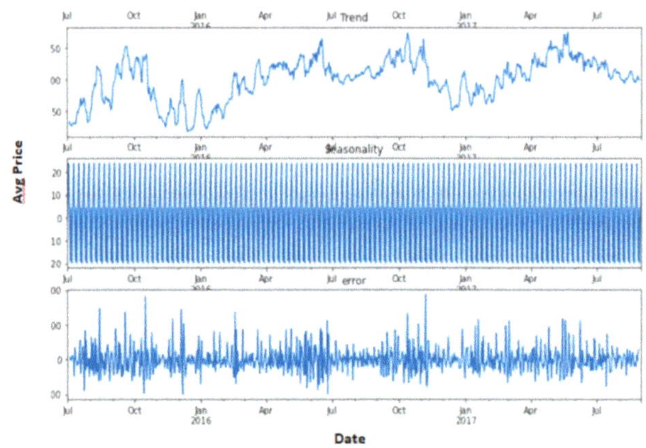

Fig. 1. Decomposition plot for Cereals and Millets

For Pulses: The Fig. 2 shows the decomposition plot for the Pulses dataset. The X-axis in the illustrated Fig. 2 is the date or the period for which the trend, seasonality and error component was considered. The y-axis shows the avg price index during the given date in association with the x-axis. The decomposition plot shows a weekly seasonality, a constant trend, and somewhat a constant error.

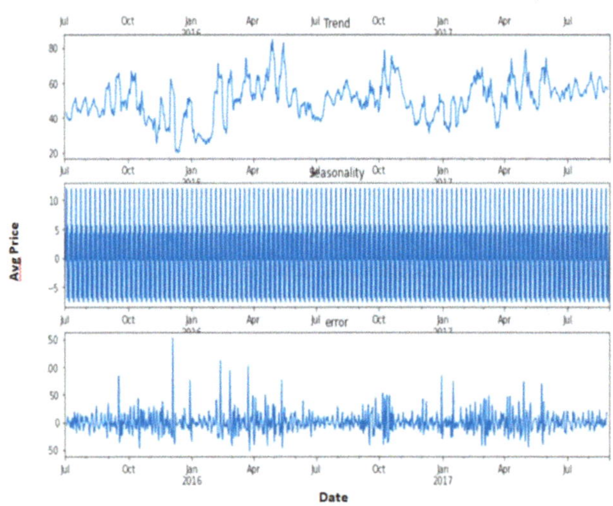

Fig. 2. Decomposition plot for Pulses

5.2 Results

Cereals and millet- For the purpose of testing, the final four months of the dataset are selected as the test sample. The three models, SARIMA, ETS, and FB Prophet, are applied to the test sample to calculate their respective RMSE and MAE values. Furthermore, the price index is forecasted using the developed models.

ETS Model: Holt Winter's seasonal method- To apply the Holt Winter's seasonal method of ETS model, the trend component for the Cereals and millet dataset was observed to have a linear pattern, which led to the adoption of an additive approach. Additionally, since the weekly seasonality component did not show any variation in magnitude over time, an additive approach was preferred for the seasonal component as well.

The chart displayed in Fig. 3 illustrates the ETS model's predictions compared to the actual values. However, the fit appears to be suboptimal since the forecasts do not account for the declining prices during the low season. Additionally, the seasonal component's amplitude appears to be inadequately aligned with the observed values.

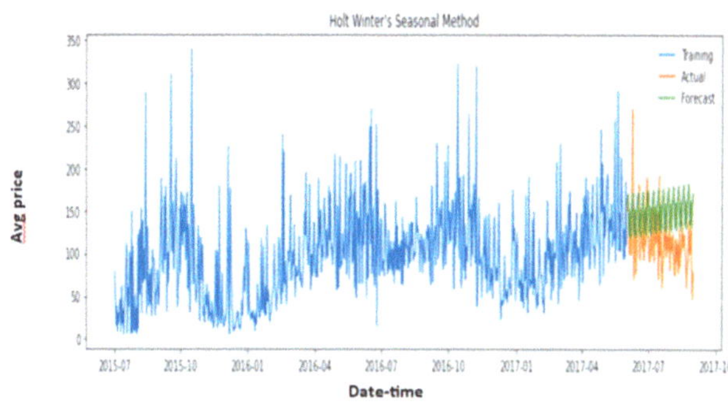

Fig. 3. Prediction of ETS model compared to Actual values

Seasonal Arima: SARIMA, which stands for Seasonal Autoregressive Integrated Moving Average, is a time series analysis technique that extends the capabilities of ARIMA to handle data with a seasonal component. This method introduces three extra parameters to determine the autoregressive (AR), differencing (I), and moving average (MA) of the seasonal component, along with a new parameter to specify the seasonal period.

Auto-correlation Function: The autocorrelation function (ACF) plot as illustrated in Fig. 4 below indicates a significant correlation up to the 30th lag, implying a persistent trend of increasing or decreasing values. To test for stationary, first differencing was applied to the series and the ACF after differencing is illustrated in Fig. 5.

The autocorrelation function (ACF) of the time series as illustrated in Fig. 5 indicates that there is only significant correlation at the first lag, which rapidly diminishes afterwards, indicating that the series has been successfully stationarized. The ACF also suggests that there is one autoregressive (AR) term for the non-seasonal component of the ARIMA model.

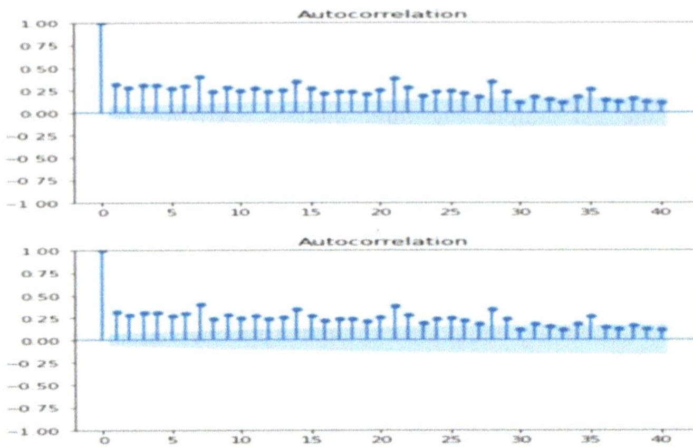

Fig. 4. Autocorrelation function before differencing

Furthermore, the ACF reveals a significant positive correlation at the seventh lag, indicating the need for one AR term to account for the seasonal component of the ARIMA model as well.

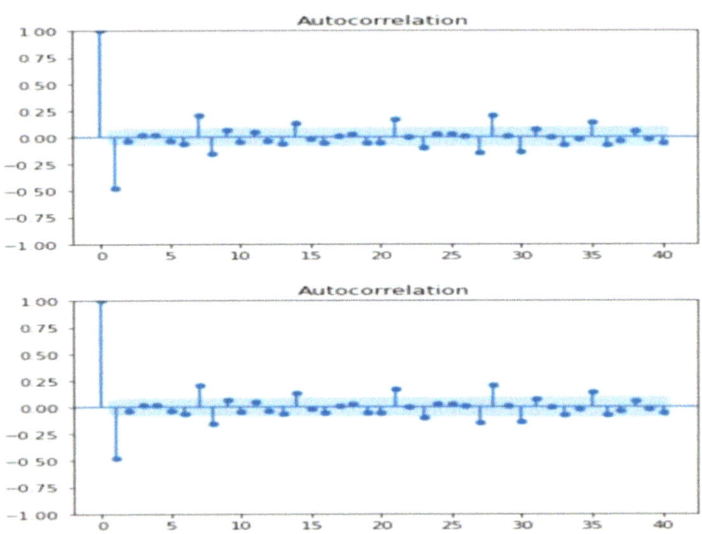

Fig. 5. Autocorrelation function after differencing

Partial Auto Correlation Function: The partial autocorrelation function (PACF) as illustrated in Fig. 6 supports the observation made from the autocorrelation function (ACF) plot, indicating the presence of a single autoregressive (AR) term in the non-seasonal component, as evident from the cutoff after the first lag. Based on this, the selected non-seasonal ARIMA model has a (1,1,0) order, while the seasonal ARIMA model has a (1,0,0) order, with a frequency of 7.

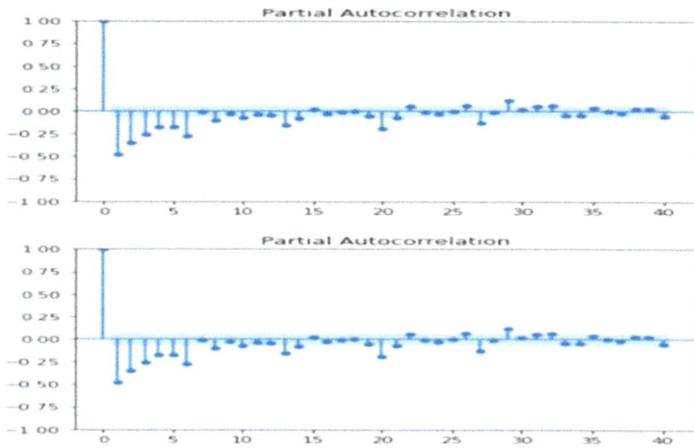

Fig. 6. Partial Autocorrelation function

FB Prophet: Prophet is a time series forecasting procedure that utilizes an additive model to predict future values. The model incorporates non-linear trends with yearly, weekly, and daily seasonal effects, as well as holiday effects. It is particularly effective for time series with pronounced seasonal patterns and abundant historical data. The Prophet algorithm can cope with missing data and changes in trend, and is typically resilient to anomalies in the data.

- Prophet decomposes time series data into trend, seasonality and holiday effect.
- Trend models the non-periodic changes in the time series data.
- Seasonality occurs as a result of recurring patterns, such as those that follow a daily, weekly, or yearly cycle.
- The holiday effect refers to the occurrence of irregular schedules either over a day or a span of days.
- Error terms are what are not explained by the model.

After conducting the analysis, it was observed that the Prophet model demonstrated better performance than ARIMA, with lower RMSE and MAE scores. Additionally, the forecast plot generated by the Prophet model indicated a close alignment with the actual values, further supporting its efficacy in predicting household food item demand (Fig. 7).

Comparing Models Performances: The model performance is compared based on the RMSE and MAE values of all the models. The RMSE is a scoring metric that quantifies the average error magnitude by squaring the differences between predicted and observed values and computing the average over the sample. The resulting value is then square-rooted to yield the final RMSE score. This scoring rule is commonly used in forecasting models and is expressed mathematically in both references provided. To evaluate the

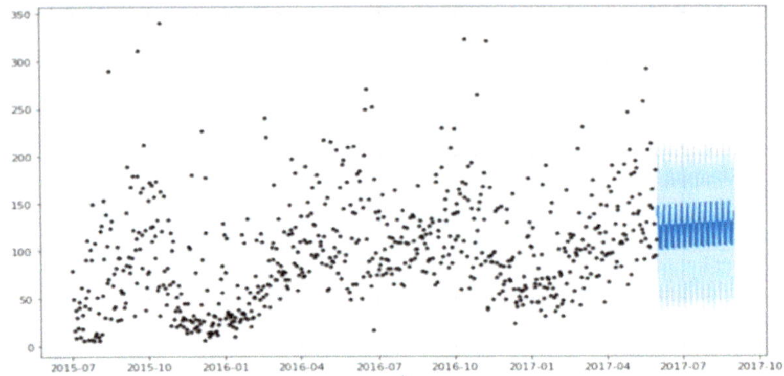

Fig. 7. Prediction of FB Prophet as compared to actual values

accuracy of a set of forecasts for continuous variables, the Mean Absolute Error (MAE) is used, which represents the average magnitude of errors without any consideration for their direction.

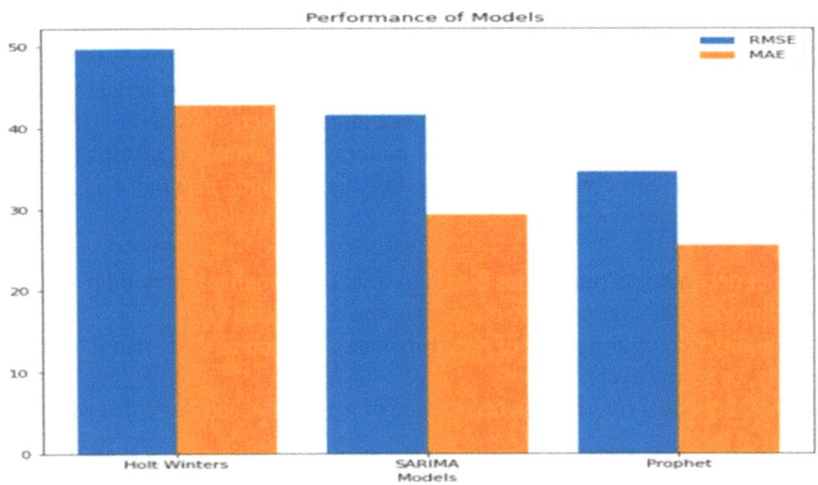

Fig. 8. Plot for comparing model performances

The plot illustrated in Fig. 8 shows the RMSE and MAE values of the three models that was attempted, it can be clearly seen that Prophet model is best suited as compared to the other two models that are used along with prophet model (Table 2).

Plotting the Forecasts with 95% Confidence Interval: To ensure a confidence level of 95%, the light blue intervals displayed in plot illustrated as Fig. 9 indicates the range within which 95% of the forecasts are expected to fall. Additionally, the RMSFE (calculated as the RMSE over the forecast residuals) can be used as an indicator of

Table 2. Quantitative analysis of the model's performance

Dataset	Model Used	Lag Value	p	d	q	P	D	Q	Frequency & seasonal period	RMSE	MAE
Cereals and millet	ETS (Holt winter's seasonal)	-	-	-	-	-	-	-	7	49.709	42.695
	SARIMA	40	1	1	0	1	0	0	7	41.507	29.241
	FB Prophet	-	-	-	-	-	-	-	-	34.588	25.441

the typical forecasting error magnitude. Multiplying the RMSFE by 1.96 (the standard deviation from the mean for a normal distribution) provides an interval within which the actual values are expected to lie 95% of the time.

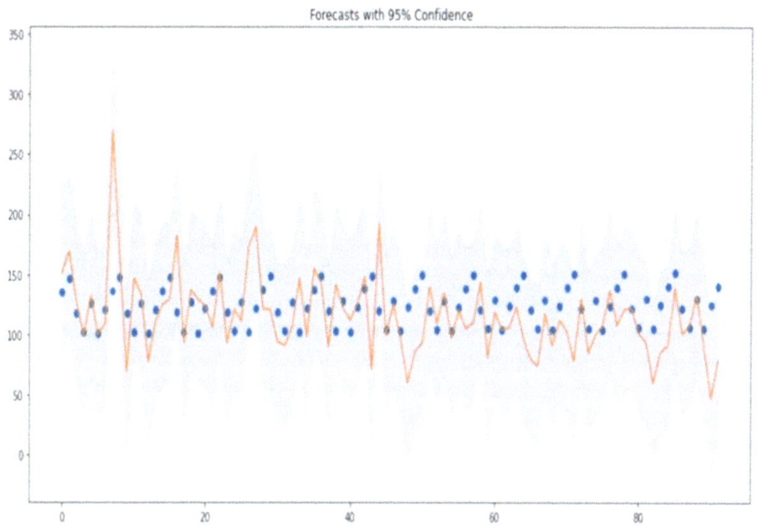

Fig. 9. Forecast for cereals and millet dataset

Pulses Dataset.

ETS Model: Holt Winter's seasonal method- Based on the decomposition plot of the Pulses time series, it appears that there is a negligible uptrend, which could be considered a constant trend. Moreover, there is a weekly seasonality pattern that does not exhibit any significant trend over time. As a result, an additive term for the seasonality would be appropriate for this time series analysis (Fig. 10).

The graph indicates a misalignment between the predicted values and the peaks and troughs observed in the actual data.

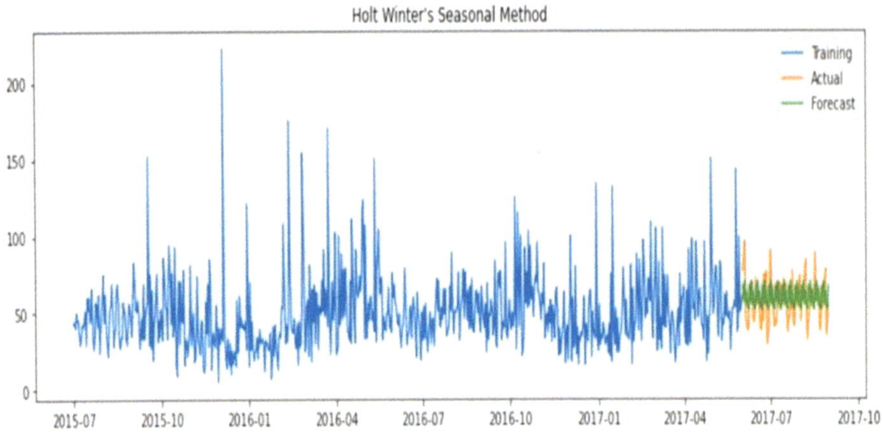

Fig. 10. Prediction of ETS model compared to actual values

Seasonal Arima: According to the autocorrelation plot, there is no significant correlation in most lags ACF, suggesting that the series is stationary. To confirm this observation, an ADF test was performed, and the p-value obtained was 0.0016, which is less than the significance level of 0.05. Therefore, the series is considered stationary, and there is no need to difference it (Fig. 11).

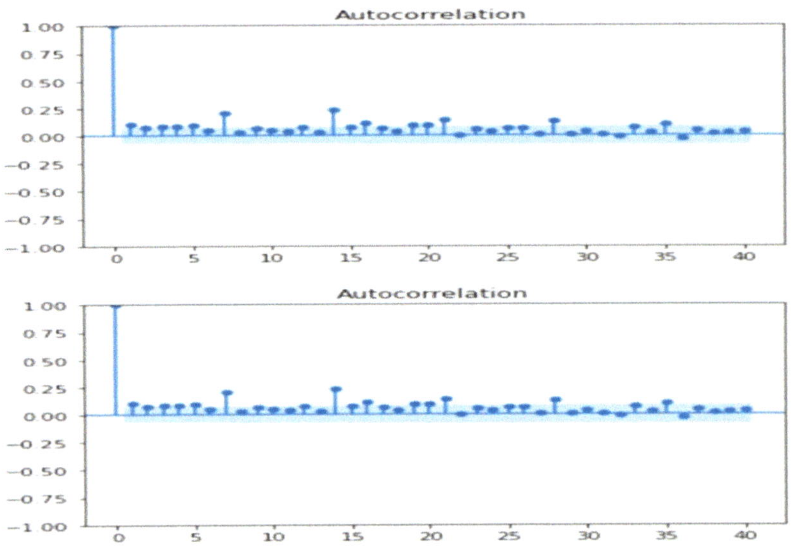

Fig. 11. Autocorrelation plot for pulses dataset

Based on the Autocorrelation Function (ACF) and Partial Autocorrelation Function (PACF), it has been determined that the appropriate choice for the non-seasonal component of ARIMA would be a single Autoregressive (AR) term with a positive first lag, while for the seasonal component, a single AR term would also be used, with a positive first seasonal spike. Therefore, the selected model specifications are Non-Seasonal ARIMA (1,0,0) and Seasonal ARIMA (1,0,0), with frequency of 7 (Fig. 12).

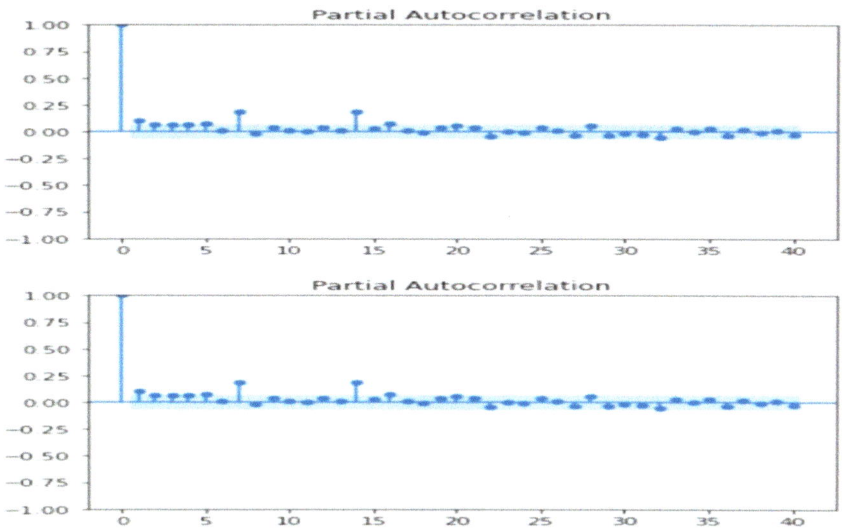

Fig. 12. Partial Autocorrelation plot for pulses dataset

The reason for the higher error in ARIMA is that it tends to closely align with the fluctuations in the data, sometimes even exceeding their magnitude. Despite this, ARIMA is still preferred over ETS because it effectively captures the overall movement of the data (Fig. 13).

FB Prophet: The results obtained by Prophet are better than those of ARIMA and only very slightly lower than ETS. After conducting the analysis, it was observed that the Prophet model demonstrated better performance than ARIMA, with lower RMSE and MAE scores. Additionally, the forecast plot generated by the Prophet model indicated a close alignment with the actual values, further supporting its efficacy in predicting household food item demand (Fig. 14).

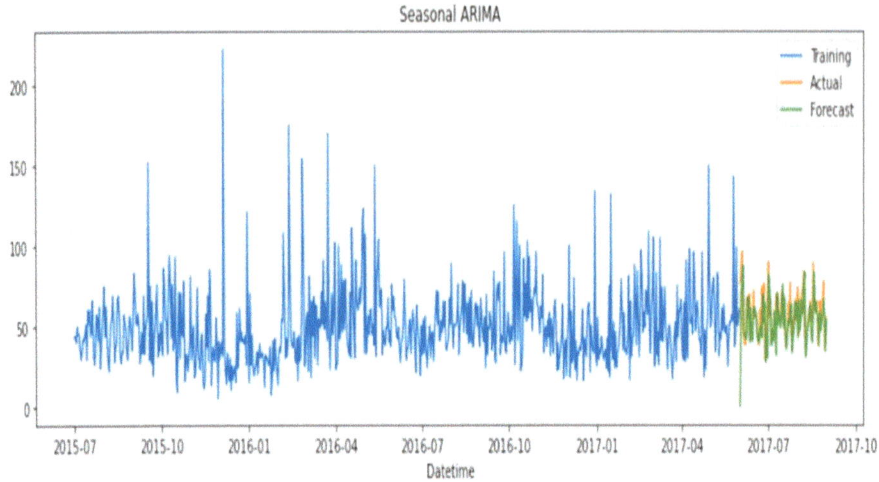

Fig. 13. Prediction of ETS model compared to actual value

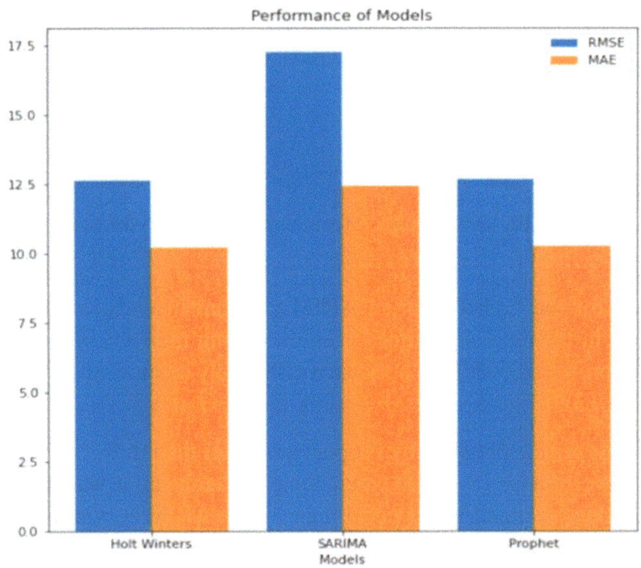

Fig. 14. Plot for comparing model performances for pulses dataset

Based on the displayed errors of the models, it is evident that both the Holt Winter's method and Prophet have comparable scores, whereas the ETS model exhibits superior performance (Table 3).

Plotting the forecasts with 95% confidence interval: To clarify, a confidence interval of 95% indicates that there is a high probability of 95% that the forecasted values will lie within the light blue intervals displayed on the plot (Fig. 15).

Table 3. Quantitative analysis of the model's performance

Dataset	Model Used	Lag Value	p	d	q	P	D	Q	Frequency & seasonal period	RMSE	MAE
Pulses	ETS (Holt winter's seasonal)	-	-	-	-	-	-	-	7	12.612	10.197
	SARIMA	40	1	0	0	1	0	0	7	17.257	12.405
	FB Prophet	-	-	-	-	-	-	-	-	12.589	10.104

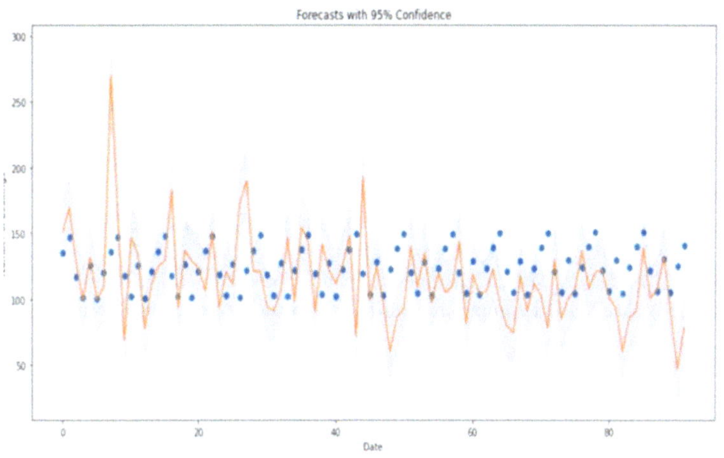

Fig. 15. Forecast for Pulses Dataset

6 Conclusion

Based on the analysis carried out using SARIMA, ETS, and FB Prophet, it can be concluded that food prediction for household food items such as cereals, millets, and pulses can be accurately predicted using time series models. These models can help households plan their food inventory better, reducing food waste and enabling them to make better purchasing decisions.

The study has also shown that the FB Prophet model outperformed the SARIMA and ETS models in terms of accuracy, with lower errors and better forecasting performance. Therefore, FB Prophet could be an effective tool for household food prediction, especially for households with limited resources for data analysis.

This project can have a significant impact on Indian households as it can help them plan their food consumption more effectively. With food prices constantly rising in India, the ability to accurately predict household food needs can help households save money and reduce food waste. Moreover, the prediction models can also aid farmers and food manufacturers in better planning their production and supply chains, leading to better food security in the country.

In conclusion, this study has demonstrated that food prediction using time series models is an effective way of managing household food inventory. The findings of this study can be useful for households, farmers, and food manufacturers in India and could lead to better food management, reduced waste, and improved food security.

References

1. Purohit, S.K., et al.: Time series forecasting of price of agricultural products using hybrid methods. Appl. Artif. Intel. **35**(15), 1388–1406 (2021)
2. Wang, Y., Ye, X., Huo, Y.: Prediction of household food retail prices based on ARIMA Model. 2011 International Conference on Multimedia Technology. IEEE (2011)
3. Sellam, V., Poovammal, E.: Prediction of crop yield using regression analysis. Indian J. Sci. Technol. **9**(38), 1–5 (2016)
4. Zhang, D., et al.: Prediction of soybean price in China using QR-RBF neural network model. Computers and Electronics in Agriculture **154**, 10–17 (2018)
5. Wamalwa, T.: An artificial neural network model for predicting maize prices in Kenya. Africa Journal of Physical Sciences (2020). ISSN: 2313-3317 4
6. Arunraj, N.S., Ahrens, D., Fernandes, M.: Application of SARIMAX model to forecast daily sales in food retail industry. Int. J. Operat. Res. Info. Sys. (IJORIS) **7**(2), 1–21 (2016)
7. Zahara, S., Ilmiddaviq, M.B.: Consumer price index prediction using Long Short Term Memory (LSTM) based cloud computing. Journal of Physics: Conference Series. Vol. 1456. No. 1. IOP Publishing (2020)
8. Sarangi, P.K., et al.: Machine Learning Approach for the Prediction of Consumer Food Price Index. 2021 9th International Conference on Reliability, Infocom Technologies and Optimization (Trends and Future Directions)(ICRITO). IEEE (2021)
9. Christensen, C., Wagner, T., Langhals, B.: Year-Independent Prediction of Food Insecurity Using Classical and Neural Network Machine Learning Methods. AI **2**(2), 244–260 (2021)
10. Oswari, T., et al.: Prediction analysis of food crop farmer index price during covid-19 pandemic using ARIMA and LSTM. J. Manage. Info. Decis. Sci. 25 (2022)
11. Deléglise, H., et al.: Food security prediction from heterogeneous data combining machine and deep learning methods. Expert Systems with Applications **190**, 116189 (2022)
12. Guo, Y., et al.: Agricultural Price Prediction Based on Combined Forecasting Model under Spatial-Temporal Influencing Factors. Sustainability **14**(17), 10483 (2022)
13. Veiga, D., Pereira, C., et al.: Demand forecasting in food retail: a comparison between the Holt-Winters and ARIMA models. WSEAS Trans. Bus. Econ. **11**(1), 608–614 (2014)
14. Choudhury, A., Jones, J.: Crop yield prediction using time series models. J. Econ. Econ. Educ. Res. **15**(3), 53–67 (2014)
15. Geweke, J., Porter-Hudak, S.: The estimation and application of long memory time series models. J. Time Ser. Anal. **4**(4), 221–238 (1983)

Efficient Mechanism of Text Identification and Retrieval

A. Thilagavathy(✉)

Department of Computer Science and Engineering, R.M.K. Engineering College, Kavaraipettai, Tamil Nadu, India
atv.cse@rmkec.ac.in

Abstract. Text recognition and extraction from videos suffers out of several issues. We introduce a hybrid method for edge detection in this study. We use the product of sobel and canny which proves to be effective when compared to the other edge detection mechanisms. Sobel, Canny Product (SCP) intensify the text pixels and helps in better understanding of the text in video. Once after detecting edges text classification is made that which eliminates the unwanted non text from the frame. This classification is made effective with the help of connected component method. Thus true text pixels are formed out of this classification technique. The resulting frame is then intersected with the Laplacian to obtain the text candidates. Bounding Box formation is the next stage in which text traversal is made. This method flip through the frame from top to bottom and forms a fringe over the text present in the frame. False positive elimination is the final stage, which aids in getting the true text pixels. According to the experimental results, the suggested system performs more stimulatingly than the current approaches.

Keywords: Sobel Canny Product (SCP) · Connected Component · Text Traversing · Text Candidates

1 Introduction

Understanding a situation involves a lot of text detection. Finding a collection of subregions in the image containing text and little to no non-text information is its goal. Over the past 20 years, a number of techniques have been put out to identify text that appears as signs, logos, and captions on pictures and videos. The developments in the fields of vision and image processing during their respective eras served as inspiration for the methods. The creation of strong and scalable machine learning algorithms also helped to improve performance and expand the applicability of these techniques to a wider range of images. The unprecedentedly high number of contestants in the ICDAR series demonstrates the growing importance and popularity of extracting and comprehending textual information embedded in natural landscapes in recent times. Text detection is a crucial step in the extraction and comprehension of textual data, as it is a requirement to the other procedures. The creation of features that allow text to be distinguished from backgrounds forms the basis of text detection. Effective features in deep learning-based

B. K. Mishra et al. (Eds.): ICT4GS 2024, ISEM 56, pp. 239–248, 2025.
https://doi.org/10.1007/978-3-031-94997-5_22

techniques are immediately learned from training data, whereas features in traditional approaches are manually developed to represent the attributes of scene text. One difficult visual recognition problem is identifying characters and text in scene photos. Similar to many other aspects of computer vision, the difficulties brought up by the intricacy of these pictures have been addressed by hand-crafted features and models that combine a variety of highly specialized previous knowledge elements. Character identification in scene photographs can potentially be significantly more complex than more traditional OCR challenges, where the characters are usually monotone on set backdrops. This is because there are numerous variations in background, lighting, texture, and font. Therefore, creating comprehensive systems for these situations necessitates creating representations that take into consideration each of these kinds of differences. It is true that a great deal of work has gone into developing these systems; the best ones integrate dozens of skillfully coupled features and processing stages. On the other hand, recent machine learning research has aimed to develop algorithms that can automatically learn higher level data representations for a variety of tasks. These kinds of solutions could be especially helpful in situations where specific features are required but difficult to construct by hand. These methods may also have the advantage of being easily able to produce huge quantities of features, which allow classification algorithms to operate at greater performance levels. We will utilize one such feature learning system in this work to investigate the potential utility of these methods for character recognition and scene text identification. Algorithms that use feature learning have demonstrated remarkable achievements in various domains, such as audio and visual identification. Sadly, one drawback of these systems is that they are frequently too computationally expensive, especially when applied to huge photos.

Digitization has led to exponential increase in image processing across all domains over time. Global demand for technology's application domains is quite high due to its advancement. Every element of life has seen a significant shift toward technology developments. Information was typically kept in books, on pages, printed materials, or through newspapers prior to the implementation of this modification. This type of digital storage allows for the reprocessing of the material for future use and creative interpretation. Clinical data and laboratory reports can be digitally recorded in the field of medical science with the aid of image processing models like text recognition. One of the key components of data extraction in image processing is text recognition, which has a variety of uses. For example, traffic signal cameras are installed and take pictures of passing cars' license plates; additional information about the driver can be obtained by using image text recognition software. One of the main uses is in the medical profession, where text recognition technology can be used to make doctor's prescriptions readable. For the aim of diagnosis, text extraction from medical records can be accomplished with ease, improving the assessment's quality. Text recognition systems are also used for information collecting and tracking, among other purposes. Using this method, we may extract a variety of information from texts and movies, some of which may contain the data we needed. Consequently, text extraction and recognition must be carried out. Traffic cameras are one of the main uses for text recognition technology. It may be stated that deep learning architecture is useful for text recognition when applied to several applications. Multiple approaches are used in the recognition, identification, and prediction of text. Text recognition techniques can be carried out by using both the text's characters and their sequence. Text recognition relies on the regularity or irregularity of image frames. Text that is not uniformly written, such as text that is written vertically or that may be both curved and horizontal, is referred to as irregular text. Character identification, text prediction, and text recognition are all combined into a single model in a text recognition architecture. The model's nature is bidirectional, and the decoders that are typically used are likewise bidirectional. The proposed network is also trained using the suggested text recognition architectures. Text recognition methods are used and evaluated in large-scale trials on a fixed dataset.

2 Related Work

In the year before, a number of experiments were carried out on scene text recognition methods. However, despite being widely employed in many real-world applications, the arbitrary-orientation scenario is rarely given much attention in most text semantics studies. The utilization of digital video is spreading more and more. Digital video technology and personal media archives are commonplace in many homes nowadays. The need for image and video data in applications has also increased as a result of the substantial advancements in processing technologies. More and more video content is being made available online, which emphasizes the need for effective algorithms that can retrieve this content according to its content. Video text enhances visual comprehension. In order to improve the viewing experience and provide a more straightforward description of semantics, most broadcasting videos typically employ additional overlay text. For instance, news video headlines provide a summary of the stories, and subtitles in a documentary play aid in the material comprehension of the audience. Text seen in the videos may generally be divided into two categories: overlay text and scene text. While the caption or overlay text has been placed outside to the film, the scene text appears naturally in the video. Effective text retrieval remains a difficult challenge despite the decades-long development of numerous techniques. This is due to factors such as unrestricted colors, sizes, letter alignments, color bleeding, poorer resolution, object distance, etc.

In order to address the issue of poor contrast, edge-based techniques are suggested. Utilizing statistical characteristics that were taken out of the Sobel edge maps for each direction, Liu et al. [5] Sort pixels employing K-means into clusters that contain text and those that don't. This approach does not work well with complicated backgrounds, However, it is not effective when the text has minimal contrast or tiny font dimensions. A hybrid method was presented by Pan et al. [8, 9] to locate and identify text in photos of nature scenes. In features like text component segmentation, region-based information is highly helpful, and CRF Classifier performs better than local classifiers in classifying text and non-text. These make it successful in unconstrained scene text localization. Even if it performs admirably in some situations, it nevertheless falters in others. There is still need to increase speed. Text localization and text recognition ought to work together. Support vector machines (SVM) are used by Lee et al. [10] to categorize each pixel as text or non-text. The feature vector of the center pixel for each M_M window is made up of 4M _ 3 neighboring gray scale values. Texture-based techniques can be trained on a variety of databases because they typically employ classifiers like SVM and neural networks. Nevertheless, a sizable samples, both textual and non-textual, in the training set is needed for these classifiers, and it can be particularly challenging to guarantee that the examples without text are typical.

Employing the local zone of the wavelet reduction responses characteristics that are identified by a neural network classifier, Li et al. [7] suggested an approach for text detection in videos. Next, text regions are projected back to the original image map at each pyramid layer. By integrating multiple levels Laplacian of Gaussian (LOG) edge detection, Chen et al. [6] presented a text detection system using hierarchical signs. Ultimately, layout analysis is performed using a Gaussian mixture model (GMM). Canny edge characteristics and a morphological close operation were employed by Chen et al. [4] to identify potential text blocks. In order to achieve efficient text detection with minimal false positives, Wong and Chen in [1] combined edge and slope characteristics; nevertheless, this approach was predicated on the assumption that the text remained horizontal. The aforementioned techniques frequently provide good recall rates, but they also frequently result in false positives as, like text, Additionally, the background could have a noticeable slope or edge [2, 3]. In [11], Cai and Lyu also employed edge and color features for low contrast text detection. To control contrast, custom thresholds were established using the masking procedure. Mariano and Kasturi presented the idea of using color features-based clustering to recognize caption texts in videos [12]. This technique functions best when words and characters in text lines have the same color. A technique for extracting multi-oriented special effects text is proposed by Crandall et al. [13], however it is only applicable to visual text with a few instructions. The related works [6, 14], which concentrate especially on text orientation, deal with scene text in camera-based photographs. These techniques are effective in images taken on cameras, but less effective in video images since they need highly resolved text in order to fully extract each character's distinct shape [15].

3 Proposed Method

The proposed methodology is described under four divisions. In section A, Sobel and canny product has been used to enhance the text details. In the next section classification of text and non text is made with the help of connected component segmentation. In the third section text-growing technique discussed based on the idea of closest neighbors. The last section involves the false positive elimination. The step involved in the proposed method is shown in Fig. 1.

3.1 Text Enhancement

It has been observed that there are frequently numerous discontinuities between text and background regions as well as between background and text in text regions. Consequently, this attribute provides a clear signal that text is present. In this section, we attempt to take use of this trait by combining the outcomes as follows and utilizing Sobel and Canny. To acquire finely improved features of text pixels, we employ a 3×3 mask.

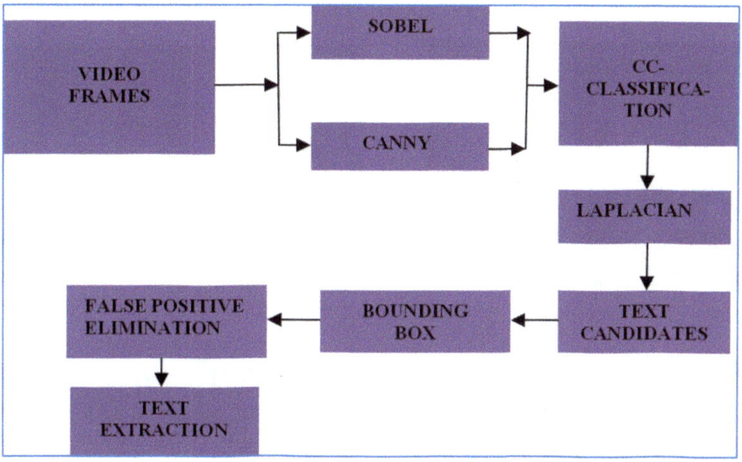

Fig. 1. Block Diagram for the proposed method

Canny is an edge detector. When applied alone on a frame it is not much effective in obtaining the enhancement and hence the product of Sobel and canny has been proposed.

It reduces noise in relatively flat areas while preserving information at text pixels with high and low contrast. Furthermore, we do tests utilizing solely Sobel and SCP for text detection to prove that SCP is better than Laplacian alone using quantitative measurements. The experimental portion will include a presentation of the experiment's findings.

3.2 Text Classification

Our suggestion is to show text lines in a distinctive way by using CCs. In addition, we suggest using skeletonization to divide CCs into distinct text lines. CCs come in two varieties: simple and complex. Simple CCs, like the ones at the bottom of Fig. 5b, are made up of either a single text string or a false positive. A complicated CC, on the other hand, consists of several text strings that are linked to one another as well as to false positives that occur in the background (Fig. 2).

Connected Component Classification process helps to better classify the text and non text. As a result, only the text pixels present in the frame will be shown and the non text including the logos will be hidden or removed from the frame. Thus effective classification of text and non text is seen using this method. This is shown in Fig. 3.

In the Fig. 3 only the text pixels are seen. The non text like the folder icons present in the original image frame has been removed from the frame.

3.3 Text Traversing

Moving through the textual pixels found by the classification process in the text direction for the purpose of fixing a closed bounding box is the primary challenge in text detection. This is because navigating solely text pixels becomes more engaging and tough due to the video's intricate background.

Fig. 2. SCP Process. (a).Input. (b), Sobel. (c),canny. (d).SCP

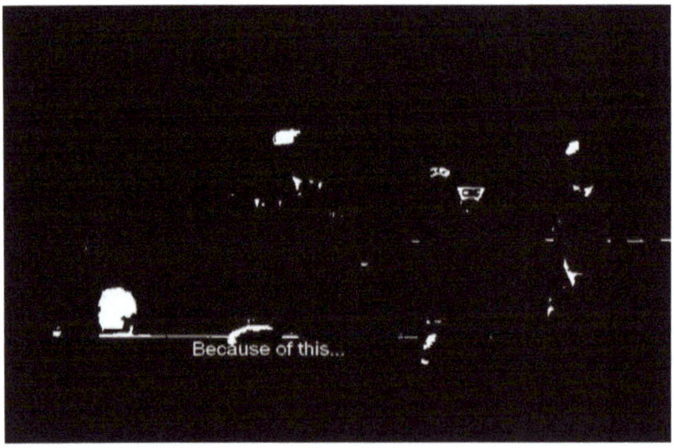

Fig. 3. Text classification

Text obtained from the previous method called the text classification will be intersected with the Laplacian to obtain text candidates. Then it is let as an input to the next method where text traversing is done. Text traversing is done with the help of text growing method in which bounding boxes are formed.

This method's foundation is the image's text lines, which consistently display words and characters at regular intervals in a single direction. The process creates bounding boxes and uses them to scan a text candidate image. Line by line, the boundary boxes extend gradually from the top of the frame to the bottom. The distance between characters, words, and lines is used to empirically establish when a text line ends. Figure 4 describes the boundary growing process. It shows how the text traversing takes place Word by Word.

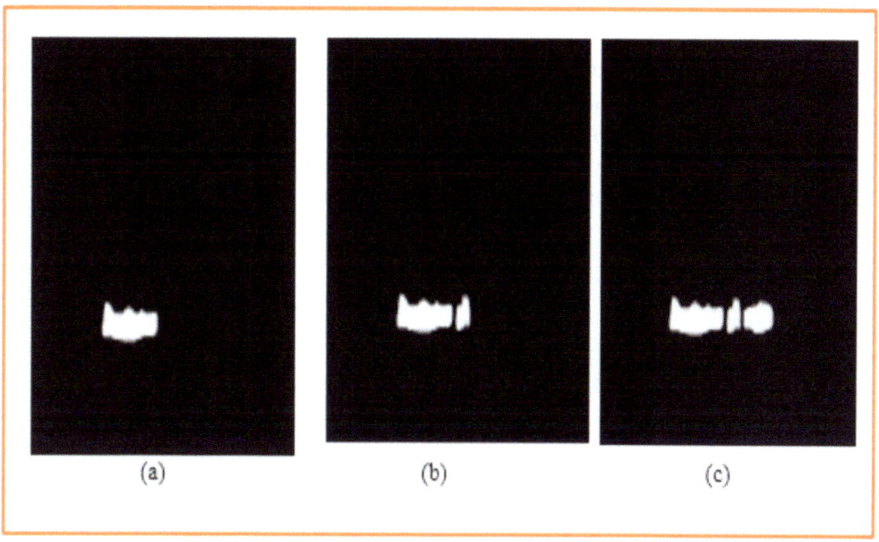

Fig. 4. Text Growing Method, (a), TG for first Word. (b), TG for second word (c) TG for last word.

The suggested approach might result in more false positives. False positive removal is known to be tough and demanding. Since geometrical features of text blocks are frequently used in the research on false positive removal, we use them for this purpose in this paper as well.

4 Experimental Results

We have taken into consideration a range of datasets, such as photographs of sports news, little contrast, diverse fonts and sizes, dissimilar orientations to demonstrate the effectiveness and robustness of the suggested method regard to measurements and text lines with multiple orientations recognition. We build our own dataset for evaluation since there isn't a common database for identifying text in videos.

Three approaches that are currently in use have been used for comparison such as edge based, color based, gradient based. And then compared these results with the proposed approach. It shows that the text detection is more effective in the proposed when compared to the existing methods.

The unchanging constant criteria for text pixel categorization and the requirement that text lines contain characters of the same color are the primary causes of the three approaches' subpar performance. The recall and precisions are calculated with this experiment. They are calculated as follows,

$$\text{Precision Rate} = \frac{\textit{Correctly detected words}}{\textit{Correctly detected words} + \textit{False positives}} x100\% \quad (1)$$

$$\text{Recall Rate} = \frac{\textit{Correctly detected words}}{\textit{Correctly detected words} + \textit{False negatives}} x100\% \quad (2)$$

The precision rate accounts for false positives, or areas of the image that are not text but were mistakenly identified as text by the algorithm. The recall rate accounts for false negatives, or text words in the picture that the computer was unable to identify. As a result, recall rates and precision are helpful metrics to assess how well an algorithm finds the right text regions and removes non-text regions.

Table 1. Experimental results.

Methods	Recall	Precision	F-measure
Proposed (SCP)	0.86	0.84	0.85
Edge-based	0.58	0.68	0.63
Color-based	0.55	0.45	0.50
Gradient-based	0.66	0.83	0.74

To show that SCP gives better results than Sobel, we had conducted experiments on 845 video images. At first, we worked with the Laplacian alone and then on SCP. The results are reported in Table 1. Figure 5 depicts the proposed system's recall. Precision and f-measure against existing edge detection methods.

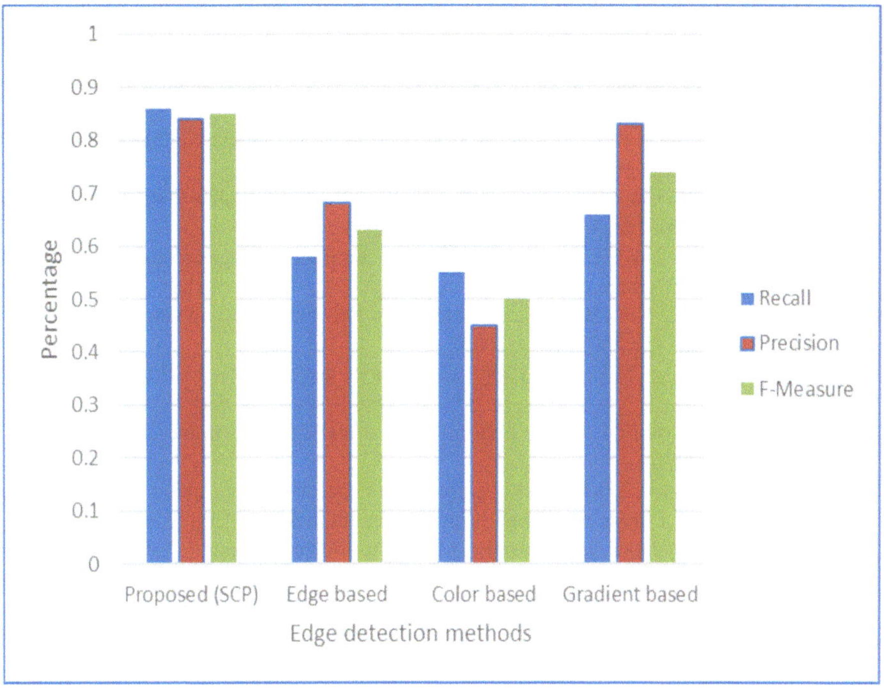

Fig. 5. SCP method vs existing methods

5 Conclusion

In our paper we had proposed a new method called the Sobel and canny product that seems to perform well. Connected component classifier is used that which helps out in classifying text from the non-text. It works better than the local classifiers. Text traversing has been performed in which the boundary growing is involved. Then the original text extraction takes place. The suggested approach performs better than the current ones and provides good recall, precision and F-Measure. In future we had planned to extend our work by including the extraction of scrambled or deformed text.

References

1. Wong, E.K., Chen, M.: A New robust algorithm for video text extraction. Pattern Recognit. **36**, 1397–1406 (2003)
2. Lyu, M.R., Song, J., Cai, M.: A Comprehensive method for multilingual video text detection, localization, and extraction. IEEE Trans. Circuits Syst. Video Technol. **15**(2), 243–255 (2005)
3. Phan, T.Q., ShivaKumara, P., Tan, C.L.: A Laplacian method for video text detection. In Proc. Icdar, pp. 66–70 (2009)
4. Chen, D., Odobez, J.M., Thiran, J.P.: A Localization/verification scheme for finding text in images and video frame based on contrast independent features and machine learning. Signal Process. Image Commun. **19**(3), 205–217 (2004)

5. Liu, C., Wang, C., Dai, R.: Text detection in images based on unsupervised classification of edge-based features. Proc. ICDAR, pp. 610–614 (2005)
6. Chen, X., Yang, J., Zhang, J., Waibel, A.: Automatic detection and recognition of signs from natural scenes. IEEE Trans Image Process. **13**(1), 87–99 (2004)
7. Li, H., Doremann, D., Kia, O.: Automatic text detection and tracking in digital video. IEEE Trans. Image Process. **9**(1), 147–156 (2000)
8. Pan, Y.F., Hou, X., Liu, C.L.: A hybrid approach to detect and localize texts in natural scene images. IEEE Trans. Image Process. **20**(3), 800–813 (2011)
9. Pan, Y.F., Hou, X., Liu, C.L.: Text localization in natural scene images on conditionalrandom field. In: Proc. ICDAR pp. 6–10 (2009)
10. Lee, C.W., Jung, K., Kim, H.J.: Automatic text detection and removal in video scenes. Pattern Recognit. Lett. **24**(15), 2607–2623 (2003)
11. Cai, M., Song, J., Lyu, M.R.: A new approach for video text detection. In: Proc, ICIP, pp. 117–120 (2002)
12. Mariano, V.Y., Kasturi: Locating uniform-colored text in video frames. In: Proc, ICPR, pp. 539–542 (2000)
13. Crandall, D., Antani, S., Kasturi, R.: Extraction of special effects caption text events from digital video. Int'l J. Document Analysis and Recognition **5**(2/3), 138–157 (2003)
14. Roy, P.P., Pal, U., Llados, J., Delalandre, M.: Multi-oriented and multi-sized touching character segmentation using dynamic programming. Proc, Int'l Conf. Document Analysis and Recognition, pp. 11–15 (2009)
15. Tang, X., Gao, X., Liu, J., Zhang, H.: A spatial-temporal approach for video caption detection and recognition. IEEE Trans. Neural Netw. **13**(4), 961–971 (2002)

Enhancing Security Through Dual Layer Techniques for Encrypting Text Messages Within Images Using LSB Image Steganography and AES Encryption Algorithms

Harshal V. Patil[1(✉)] and Vaibhav P. Sonaje[2]

[1] Sandip University, Nashik, India
harshal4patil@gmail.com
[2] Department of Computer Science and Application, SOCSE, Sandip University, Nashik, India
vaibhav.sonaje@sandipuniversity.edu.in

Abstract. Ensuring data security has become a serious concern as it has evolved into a multifaceted problem. Over the past few decades, many security measures have been developed to address this issue. However, cryptography and steganography are the most commonly used tools to ensure the security of sensitive data. Cryptography involves encoding secret messages, while steganography involves hiding messages within digital media and images. While both methods have their advantages, cryptography is more suited for secure communication, while steganography is better for hiding sensitive information.

In our paper, we propose a two-tier encryption approach to ensure data security. We use the AES algorithm to generate a top-secret message in the first layer, and the LSB image steganography method to encrypt the information in the second layer. By combining these two encryption methods, we ensure the confidentiality, integrity, and availability of the data. We perform comprehensive testing and analysis on our proposed methodology to validate its efficiency, effectiveness, and reliability.

Keywords: Cryptography · Steganography · Cover-Image · Stego-Image · Plain-Text · Cipher Text · AES · LSB

1 Introduction

Information is a crucial resource for all computer and internet users, and each user aims to maintain their records with the highest level of security. Information security refers to the process of safeguarding information from unauthorized access. Two primary methods used to maintain information security are cryptography and steganography. Cryptography is a technique that converts readable text into unreadable text, often referred to as ciphertext, while steganography conceals secret data within a cover object to protect it from unauthorized access [1].

Some commonly used terms in the field of information security include plain text, cipher text, cryptanalysis, encryption, and decryption.

B. K. Mishra et al. (Eds.): ICT4GS 2024, ISEM 56, pp. 249–258, 2025.
https://doi.org/10.1007/978-3-031-94997-5_23

The use of cryptography and steganography for secure communication involves integrating both techniques to enhance the security of communication. The project's aim is to develop a cross-platform tool that effectively conceals a message within a digital image file in a secure and sturdy manner [2]. This is demonstrated in Fig. 1, which illustrates the combined concept of cryptography and steganography.

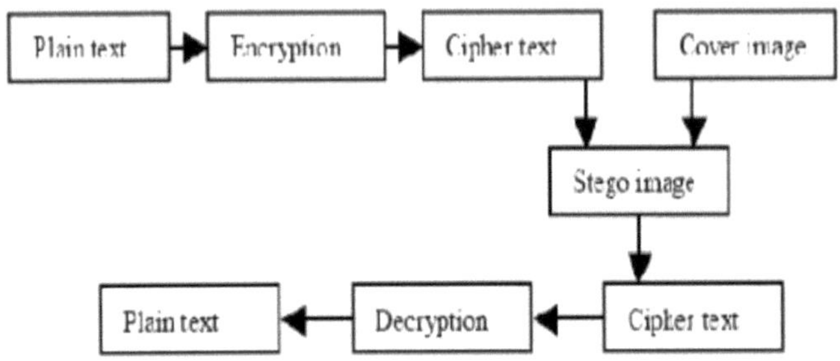

Fig. 1. Combination of Cryptography and Steganography

1.1 LSB Image Steganography

Strategography is a process that involves the creation and use of various mediums to convey information to the target audience. This system often involves the use of tactics, strategies, and secret messages to achieve a specific purpose. The strategy graphy system includes various mediums such as audio, video, prototypes, and protocols. By implementing strategies, we can create digital files that contain data that is essential for the target audience. Therefore, embedding messages in digital files is possible and effective [3]. To prepare a strategic message, three elements are necessary: covert medium, secret message, and strategic content [4].

1.2 Advance Encryption Standard

The AES is a widely used and committee-approved block cipher that is commonly employed. Specifically, AES operates on 128-bit blocks with a key length of 128 bits. AES provides superior security; numerical strength, resistance to known plaintext attacks, collision-free encryption, and is suitable for real-time applications and hardware and software applications across multiple categories [5]. In contrast, DES and 3DES encryption algorithms are less secure, as they possess vulnerabilities, whereas AES offers superior strength with no known weaknesses [5].

2 Previous Work

This section will examine some of the proposals made in the information hiding field.

It is crucial to note that while both cryptography and steganography aim to maintain the confidentiality of information, relying on one method alone may not be enough to safeguard against advanced and automated attacks. To address this, researchers have recommended combining these two approaches to improve the security of sensitive data transmitted over public networks [6, 7].

S.M. M. Karim and colleagues presented a method that ensures strong data security using a LSB-based approach with a secret key to conceal sensitive data, which is dispersed throughout the LSBs of the image [8].

The King and colleagues proposed a scheme based on the human visual system (HVS) for JPEG, which can enable perceptual coding of RGB components for all types of images on various aircraft [9].

According to the author's [10] research, they have improved the algorithm of Image Stagnography using the LSB model. However, the model has been prohibited from revealing confidential information, reducing the color imbalance, and enhancing the quality.

Dr. Young and his colleagues proposed an innovative LSB-based approach to improve the quality of strategy and to enhance the scalability of the algorithm to achieve high performance and low computational complexity. The Pixellated-Ladder-Based Algorithm is integrated with this approach to achieve these goals [11].

The author [12] employs vektor quantization to conceal secret information, resulting in increased efficiency and a smaller size.

Nouf A. Al-Otaibi et al. [13] A two-tier security method was developed to protect personal information that is sensitive and requires privacy. The system comprises cryptographic and steganographic layers. The steganographic layer utilizes the Least Significant Bit (LSB) algorithm to encode information. This system is based on a fundamental platform and has been enhanced through experimentation to improve its capacity to protect information.

In a paper published in [14], an image-steganography technique based on LSB was presented. This technique used a common bit pattern to conceal data by modifying the least significant bits of pixels based on the message and pattern bits. However, this approach suffered from a low hidden capacity.

During the study conducted with Kamali et al. [15], the improved AES image encryption algorithm (MAES) provided enhanced security for the prototype data. The Shifted Transform Function's application through the MAES algorithm resulted in improved encryption outcomes and more diverse statistical results.

Zhou et al. employed [16] a significant effort was made to refine, where a modified LSB steganography technique was employed to encode control messages using the RSA encryption algorithm. To ensure secrecy of the information, the encrypted messages were then transmitted through secure channels. The sender was confirmed to be genuine by the receiver by observing the proper color and shape of the Channel emblem on the received message. The control information was extracted and decoded from the encrypted message by the receiver using the same channel's emblem's red and blue colors.

A study conducted by Sridevi and colleagues proposed a technique for secure data transmission through the use of the LSB technique for information drive in the cover-image, followed by encryption of the stego-image using the AES algorithm [17]. However, this method has a limitation in that the encrypted image showed to the receiver does not be similar to the original image due to encryption being performed after the data embedding process.

Song et al. [18] The use of steganography and cryptography to create a secure compression protocol is discussed. The LSB-M machine and stream cipher functions were used to process data and ensure high security. Despite this, the idea of grey scale primitivity remains unrealized.

Narayan and Prasad [19] used steganography and cryptography together to secure data transmission. In the first method, the S-DES algorithm was used to encrypt each byte of the secret message using the S-boxes and the resulting image was masked with the stego image in the LSBs using the XOR encoding technique. In the another method, the top-secret message was encoded using the S-DES procedure and implanted in the cover-image using a description.

In the middle of creating an encrypted three-tier protection scheme, Usha and others employed [20] Playfair cipher and AES algorithm to secure the encrypted message. The process of converting the encrypted information into a secure format was carried out using the LSB method, and the encrypted image was stored in the cloud. The encryption method ensured the data's security and confidentiality, as seen by the ciphertext coverage in the scatter plot.

In [21], Joshi and Yadav proposed an image steganography and cryptography technique for grayscale images in the spatial domain. The authors initially encrypted the confidential message using the cipher-text to improve data security. The encrypted message was then inserted in the cover-image using the LSB-S procedure. While the method demonstrated a satisfactory level of data-hiding capacity, it was limited to grayscale images only.

Pavar and Gavandi [22] have employed strategic and cryptographic tactics to integrate mobile (M)-commerce and e-commerce applications using the Steganography and Cryptography techniques. The encypted information was processed using a 128-bit AES algorithm, and the encrypted data was stored using LSB Stategraphy for better data hiding. This approach made the research more challenging and exciting.

3 Proposed Work

3.1 Proposed Methodology:

The importance of data security is paramount for any organization. As such, we are presenting our work, titled "Dual Layer Security of Encrypting Text Messages within Images Using LSB Image Steganography Method and AES Encryption Algorithm." Our study aims to offer a double layer of security for sensitive text messages.

Our method for enhancing the security of data utilized both cryptography and steganography. By combining these two techniques, we were able to offer a higher level of security. Our approach is comprised of two stages, the first of which employs the AES encryption algorithm, while the second stage uses the LSB image steganography method [23, 24].

Phase – I Embedding Process using Cryptography and Steganography

The first phase of the process involves using cryptography and steganography to embed the message. Both methods involve encrypting the message using the Advanced Encryption Standard (AES) algorithm, with the plain text being converted to cipher text through AES. The ciphertext is then concealed within a cover image using a steganography algorithm, which can be chosen based on its suitability and performance. The selection criteria for the stegano method may include factors such as encryption and decryption times, as well as measures such as MSE, PSNR, and SNR. The final output is the stego image that contains the encrypted message [25, 26] (Fig. 2).

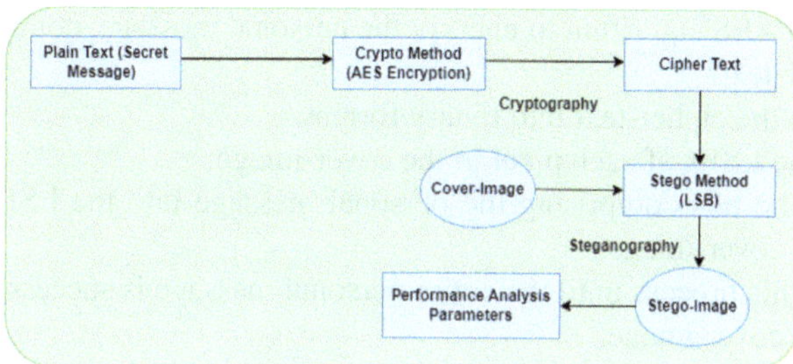

Fig. 2. Crypto-Stego Encryption and Embedding Model

Phase – II Extracting Process

When the steganography process is carried out, the order is reversed compared to when the embedding procedure is. In order to access the concealed cipher text, the stego image is fed into the algorithm specified during the embedding process. Following this, the corresponding decoding method subjects the cipher text to the AES decryption technique, which ultimately uncovers the confidential or plain text [25, 26] (Fig. 3).

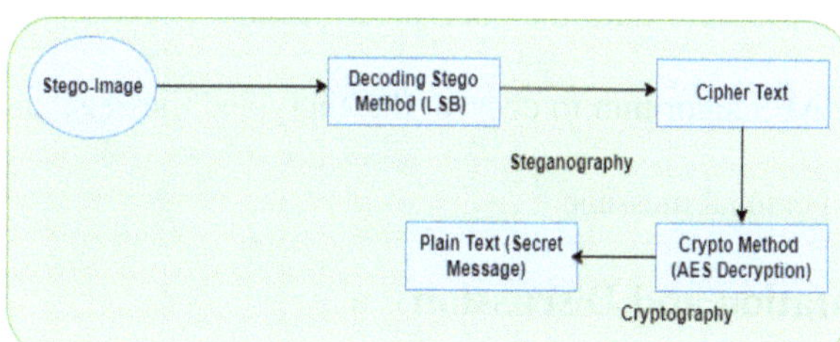

Fig. 3. Crypto-Stego Extraction and Decryption Model

3.2 Proposed Crypto-Stego Algorithm

Embedding Algorithm

Input: A concealed message and a cover-image
Output: A stego-image
Steps:

1. Analyze the cover-image and the personal message intended to be embedded within it.
2. Utilize the AES algorithm to encrypt the personal message, converting plain-text into cipher-text.
3. Transform the cipher-text into binary format.
4. Identify the LSBs of each pixel in the cover-image.
5. Integrate the bits comprising the personal message into the LSBs of the pixels within the cover-image.
6. Reiterate this process until the entire personal message is successfully embedded within the cover-image.
7. The resulting steganographic image, or stego-image, is then generated.

Retrieval Algorithm

Input: Stego-image
Output: A confidential message
Steps:

1. Choose the steganographic image or stego-image.
2. Extract the LSBs of each pixel in the stego-image.
3. Recover and retrieve the LSBs of each pixel in the stego-image.
4. Repeat the procedure until the encrypted message is extracted from the stego-image.
5. Utilize the AES algorithm to decrypt the encrypted message and convert it into plain text.
6. Obtain the personal message.

4 Implementation and Discussion

Our system employs Crypto-Stego algorithms that are grounded in the proposed algorithm. We have successfully implemented the Crypto-Stego system using Visual Studio 2010, which integrates AES and LSB. Our principal objective is to rigorously scrutinize the Crypto-Stego system and assess various situations to increase its relevance in both academic and industrial research domains. Moreover, the implementation details of the proposed algorithm for text messages and text files are depicted in Figs. 4 and 5 below.

4.1 Encryption and Embedding Procedure for Text Message

The accompanying illustrations in Fig. 4 depict the graphical representation of the dual layer data security system's implementation interface. The system consists of four functional components, namely Select Cover-Image, Encrypt Original Text Message, Hide Encrypted Text into Image, and Save Stego-Image. The Select Cover-Image feature utilizes the original image and encrypts the Original Message using the AES encryption algorithm with a password or key to generate Cipher-Text. The Cipher-Text is then inserted into the Cover-Image using the LSB technique, resulting in a Stego-Image. Finally, the Stego-Image is saved. Although the Cover-Image and Stego-Image appear visually identical, the latter contains the concealed Cipher-Text.

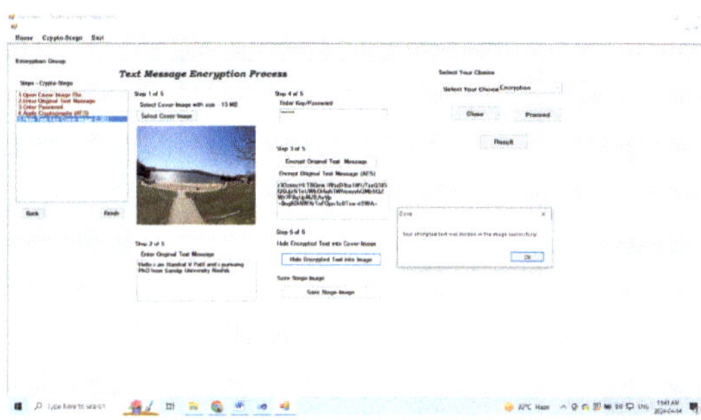

Fig. 4. Text Message Encryption and Embedding Procedure

4.2 Extracting and Decryption Procedure for Text Message

Three buttons are accessible for completing the process: Select Stego-Image, Get Encrypted Text from Stego-Image, and Decrypt Original Text Message. These buttons are illustrated in Fig. 5, where the input supplied to this layer consists of Stego-Image and password/encryption key. The encrypted Text Message is generated using the LSB technique, and it can be decrypted into the Original Message by clicking the Decrypt Original Text Message button, which employs the AES method.

Fig. 5. Text Message Extracting and Decryption Procedure

4.3 Comparison Cover-Image and Stego-Image

This part presents all significant images, including the cover-image and stego-image respectively. A close examination reveals no noticeable disparity among the stego-image and cover-image. However, the dimensions of the stego-image are smaller than those of the cover-image. The use of steganography to conceal data within the cover-image is discreet, with modifications made at the pixel level that are almost undetectable, thereby ensuring a high level of security (Table 1).

Table 1. Displays the efficiency parameter of the proposed method for text messages.

Cover-Image	Original Message	Cipher Message	Stego-Image
	Hello i am Harshal V Patil and i pursuing PhD from Sandip Universit y Nashik	r3Ozxec HLTBQew1 RhaDlha1 WUTxxQ3I GQ6UjzNT eUWkZ50 G5RiB2ZI2J H/ORZue MUj/ipvR2 G1PH/wPa qyaxQO5Y rsBxraRGN W8iO1AV Hsw=	

5 Conclusion and Future Work

Our study has revealed a method that provides twofold security for confidential information. By integrating steganography and cryptography, we aim to accomplish the intended objectives. Our system, built using Visual Studio 2010 IDE, is designed to minimize the likelihood of security breaches during the transmission of sensitive data over a network. The proposed system is user-friendly, making it accessible to anyone with basic computer expertise. For image steganography, we have utilized the LSB algorithm, which is a simple yet powerful method. For the encryption and decryption layer, we have employed the AES algorithm, which is a flawless encryption technique.

In future endeavors, we plan to enhance the steganography level by refining the current LSB algorithm. Furthermore, we intend to investigate other steganography methods to optimize our results. We also plan to evaluate our system with various symmetric and asymmetric key encryption algorithms to determine their efficacy.

References

1. Parah, S.A., Sheikh, J.A., Akhoon, J.A., Loan, N.A., Bhat, G.M.: Information hiding in edges: a high capacity information hiding technique using hybrid edge detection. Multimedia Tools and Applications **77**(1), 185–207 (2018)
2. Saraireh, S., Al-Saraireh, J.A.A.F.E.R., Al-Sbou, Y.A.Z.E.E.D., Saraireh, M.: A hybrid text-image security technique. J. Theoret. Appl. Info. Technol. **96**(9) (2018)
3. Chandramouli, R., Memon, N.: Analysis of LSB based Image Steganography Techniques. Proceeding of IEEE International Conference on Image Processing, October 7–10, Thessaloniki, Greece (2001)
4. Raja, K.B., Chowdary, C.R., Venugopal, K.R., Patnaik, L.M.: A secure image steganography using LSB, DCT and compression techniques on raw images. Proceeding of 3rd IEEE International conference on Intelligent Sensing and Information Processing (ICISIP), December. Bangalore, India (2005)
5. Singh, S., Maini, R.: Comparison of Data Encryption Algorithms. Int. J. Comp. Sci. Commun. **2**(1) (2011)
6. Zhou, X., Gong, W., Fu, W., Jin, L.: An improved method for LSB based color image steganography combined with cryptography. In: Proc. IEEE/ACIS 15th Int. Conf. Comput. Inf. Sci., Jun, pp. 1–4 (2016)
7. Reddy, M.I.S., Kumar, A.P.S.: 'Secured data transmission using wavelet based steganography and cryptography by using AES algorithm. Proc. Comput. Sci. **85**, 62–69 (2016)
8. Karim, S.M.M., Rahman, M.S., Hossain, M.I.: A New Approach for LSB Based Image Steganography using Secret Key. Proceedings of 14th IEEE International Conference on Computer and Information Technology, December 22–24. Dhaka, Bangladesh (2011)
9. Qing, X., Jianquan, X., Yunhua, X.: A high capacity information hiding algorithm in color image. Proceedings of 2nd IEEE International Conference on E-Business and Information System Security, May 22–23. Wuhan, China (2010)
10. Gupta, S., Gujral, G., Aggarwal, N.: Enhanced Least Significant Bit Algorithm for Image Steganography. Int. J. Computat. Eng. Manage. **15**(4) (2012)
11. Yang, H., Sun, X., Sun, G.: A high-capacity image data hiding scheme using adaptive LSB substitution. J. Radio Eng. **18**(4) (2009)
12. Sachdeva, S., Kumar, A.: Colour image steganography based on modified quantization table. Proceedings of IEEE 2nd International Conference on Advanced Computing & Communication Technologies, January 7–8. Rohtak, India (2012)

13. Al-Otaibi, N.A., Gutub, A.A.: 2-Leyer Security System for Hiding Sensitive Text Data on Personal Computers. Lecture Notes on Information Theory **2**(2) (2014)
14. Channalli, S., Jadhav, A.: Steganography an Art of Hiding Data. Int. J. Comp. Sci. Eng. (IJCSE) **1**(3) (2009)
15. Kamali, S.H., Shakerian, R., Hedayati, M., Rahmani, M.: A new modified version of Advanced Encryption Standard based Algorithm for Image Encryption. proceeding of IEEE International Conference on Electronics and Information Engineering (ICEIE), August 1–3. Kyoto, Japan (2010)
16. Zhou, X., Gong, W., Fu, W., Jin, L.: 'An improved method for LSB based color image steganography combined with cryptography. In: Proc. IEEE/ACIS 15th Int. Conf. Comput. Inf. Sci., Jun., pp. 1–4 (2016)
17. Sridevi, R., Paruchuri, V., Rao, K.S.: Image steganography combined with cryptography. Int. J. Comput. Technol. **9**(1), 976–984 (2010)
18. Song, S., Zhang, J., Liao, X., Du, J., Wen, Q.: 'A novel secure communication protocol combining steganography and cryptography.' In: Proc. Eng. **15**, 2767–2772 (2011)
19. Narayana, S., Prasad, G.: Two new approaches for secured image steganography using cryptographic techniques and type conversions. Signal Image Process. Int. J. **1**(2), 60–73 (2010)
20. Usha, S., Kumar, G.A.S., Boopathybagan, K.: 'A secure triple level encryption method using cryptography and steganography.' in Proc. Int. Conf. Comput. Sci. Netw. Technol., Dec., pp. 1017–1020 (2011)
21. Joshi, K., Yadav, R.: 'A new LSB-S image steganography method blend with cryptography for secret communication. In: Proc. 3rd Int. Conf. Image Inf. Process., Dec., pp. 86–90 (2016)
22. Pawar, P.Y., Gawande, S.H.: 'M-commerce security using random LSB steganography and cryptography.' Int. J. Mach. Learn. Comput. **2**(4), 427 (2012)
23. Borole, V.Y., Nimbhore, S.S., Kawthekar, S.S.: Image processing techniques for brain tumor detection: a review. Int. J. Emerg. Trends & Technol. Comp. Sci. (I JETTCS) ISSN 2278-6856 Volume 4, Issue 5(2), September – October, 28–32 (2015)
24. Gore, D.B., Kawthekar, S.S., Borole, V.Y.: Ear recognition techniques for biometrics using digital image processing: A review. In Int. Conf. on Recent Advances in Computer Science, Engineering and Technology, pp. 22–26 (2016)
25. Patil, H.V., Sonaje, V.P., Barhate, B.H., Borole, V.Y.: Performance analysis of various media for information security using Aes and Rsa cryptography algorithm. J. Nonlinear Analysis and Optimizat. Theory and Appl. **15**(2(i)), 63–69 (2024). ISSN: 1906-9685
26. Patil, H.V., Sonaje, V.P.: A Study of The Performance of Various Media for Information Security Via LSB Steganography Method for Text Messaging. Electrical Systems Volume-20-No-9s ISSN: 1112-5209, pp. 1159–1165 (2024)

RUL Prediction of Electric Battery with Error Indices Using Machine Learning Techniques

C. L. Sravanthi(✉) and J. N. Chandra Sekhar

Department of Electrical and Electronics Engineering, Sri Venkateswara University College of Engineering, Sri Venkateswara University, Tirupati, India
sravanthi.cl@gmail.com, chandrasekhar.jn@svuniversity.edu.in

Abstract. Remaining use life (RUL) prediction is becoming more and more popular as a way to allay worries about security and dependability of lithium-ion batteries in electric cars. Even though battery operational characteristics are thoroughly recorded, the current prediction methodologies for evaluating battery performance are inadequate. To address the issues of inadequate local feature learning and insufficient capacity to handle large datasets, a data-driven prediction method of machine learning algorithms is proposed. The intention of this paper is to use specific machine learning techniques to improve prediction resilience and accuracy. Selected machine learning techniques are evaluated for accuracy prediction using real life battery cycle data set from the Hawaii National Energy Institute (HNEI). For each of the ML methods, performance error indices are computed, including Mean Square Error (MSE), Root Mean Square Error (RMSE), Mean Absolute Error (MAE) and R-Squared. Relevant inferences are then exhibited, highlighting the prospective of battery RUL prediction near most accurate values. The findings demonstrate that, in comparison to other approaches, the random forest method is more resilient and considerably increases RUL prediction accuracy while reducing prediction error.

Keywords: RUL prediction · Lithium-ion battery · Machine learning algorithms · Performance error indices

1 Introduction

Lithium-ion battery technology has become extensively used in various modern industries, such as automotive, home appliances, communications, and aerospace. A key component for the anticipated integration of electric vehicles into the contemporary transportation sector is the energy storage system (ESS). LIBs are extensively utilized in EV's because of their high energy density, superior cycle pursuance, efficient charging capabilities, and extended lifespan [1]. When using lithium batteries, several irreversible chemical reactions might happen as a result of internal defects. This would reduce the battery's capacity and safety performance. In real-world scenarios, battery deterioration can result in mishaps like electric car batteries dying. For this reason, real-time battery state monitoring requires a Li-ion battery management system (BMS) [2]. A

B. K. Mishra et al. (Eds.): ICT4GS 2024, ISEM 56, pp. 259–272, 2025.
https://doi.org/10.1007/978-3-031-94997-5_24

high-performing BMS can enhance customer satisfaction and better sustain battery performance. The dependability and security of the battery when it is operating may be guaranteed by an accurate battery life forecast. Battery RUL is a popular subject for BMS studies.

A battery management system concentrates on three essential elements: state of charge, state of health and remaining usable life in order to guarantee the security of Lithium-ion batteries [3]. Maintaining SOH and RUL of battery on a regular basis helps reduce the frequency of repairs. Due to their crucial function in battery system functioning, SOH and RUL estimates have been the subject of much research. The battery's remaining usable life (RUL) is an important factor that has impact on how user feels. Since result of prediction can usually be used as a basis for calculating the approximate condition of the battery both now and in the future, an accurate RUL forecast is essential for improving equipment safety and reducing maintenance costs [4].

Many academics have been researching and delving deeply into the topic of predicting RUL of LIB in recently. Model-based and data-driven approaches are two broad groups into which these RUL prediction techniques may be divided [5]. Prediction accuracy can be improved by using the model-based RUL prediction strategy when external conditions are sufficiently consistent. However, in certain situations, it can be challenging to create an accurate mechanistic model, and the model will malfunction as soon as a new kind of battery is installed. Consequently, the use of such technologies is limited by the challenge of gathering adequate a priori information in the real sector. Since the emergence of neural networks and the advent of the digital era, many scholars and researchers have developed a great interest in data-driven methodologies. Instead of relying too heavily on physical failure models and related expert knowledge, data-driven approaches model the system based on empirical knowledge or even directly on the data itself, and then use machine learning, artificial intelligence, and other algorithms to extract the failure characteristics of the data to achieve RUL prediction [6]. As machine learning continues to advance, an increasing number of new models are being applied for RUL prediction. In the big data age, ML is a valuable technology used automatically create analytical models that are extensively applied in many different sectors. Because of this, data-driven approaches are crucial for forecasting the long-term condition of batteries.

This article is structured as follows. Section 2 presents recommended algorithms, structure, and general methodology. The performance evaluation metrics are covered in Sect. 3. Experiments and data analysis are shown in Sect. 4 and in Sect. 5 conclusion is provided.

2 Methodology

2.1 RUL Definition

Under certain charge and discharge circumstances, number of cycles required for a battery's maximum usable capacity to degrade a predetermined failure threshold is known as battery's remaining useful life as seen Fig. 1, amount of time that equipment performance drops to the failure threshold for the first time or that it approaches its end of life. It indicates the amount of time that has passed till the equipment's operational life is over. The EOL of battery is the moment at which its properties, as defined by time and number of charge-discharge cycles, reach replacement threshold. Accurate RUL estimation makes predictive maintenance of the equipment possible.

$$RUL = N_{EOL} - N \tag{1}$$

Here, N_{EOL}: maximum number of battery cycles
N: quantity of battery cycles.

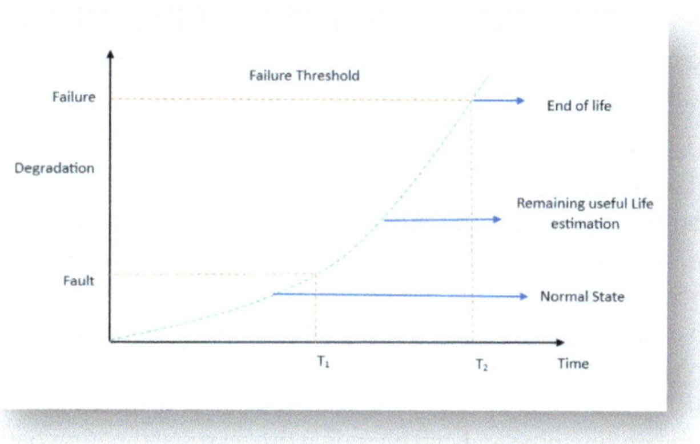

Fig. 1. Illustration of RUL Estimation

2.2 RUL Prediction Methods Classification

Currently, a number of methods have been established for battery RUL estimation. LIB RUL prediction models have been condensed and divided into many groups. There are

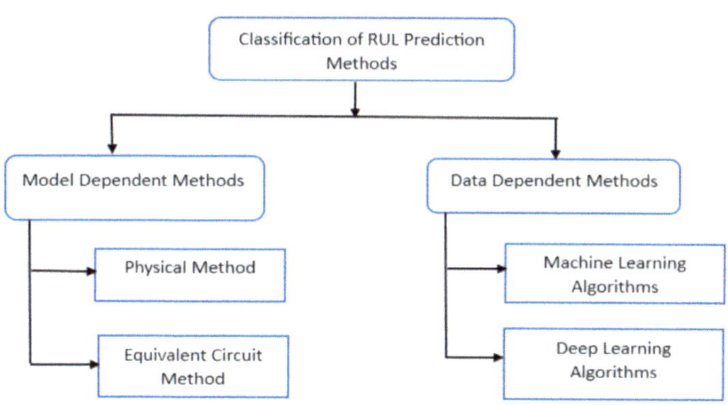

Fig. 2. RUL classification Methods

three types of estimation of RUL models: model dependent, data dependent and hybrid. An extensive evaluation of the prediction methodologies illustrated in Fig. 2. Model-based approach builds deterioration and practical model for illustrating decreasing path of related system using modelling based on physics. A variety of empirical and algebraic equations are employed in the construction of model-based systems. Conversely, data-driven models derive the degradation model for various applications using machine learning (ML) techniques and historical data.

A thorough evaluation of many model dependent, data dependent and ensembled techniques for LIB RUL estimation has been assisted. The study considers a wide range of aspects, such as research gaps, benefits, downsides, contributions, techniques, attributes, and classifications [7]. The findings indicate that model-based techniques provide a comprehensive examination of internal battery ageing, whereas data-driven approaches may be utilised with previous battery data for RUL prediction.

Model Dependent Methods

Model dependent techniques comprise scientific models, an algebraic collection of empirical equations, and related parameters. They rely heavily on empirical data and experimentation. Regression models or physics models may be used to build the degrading performance of battery using model-based methodologies, and findings can be extrapolated to estimate battery performance [8]. The physics model is based on electrochemical and physical processes occur inside the battery. This model for RUL prediction is developed using the concepts of electrode porosity and reaction kinetics. When it comes to comprehending physical and chemical processes estimating battery conditions hence it is more helpful. Thus, using circuit analysis techniques to create a mathematical model, the battery deterioration mechanism may be explained [9]. Furthermore, the application of ECM leads to the simulation of several critical parameters under varied operating situations, including power, energy, and heat generation. It has been discovered that when the SEI film layer grows, the battery's internal resistance would gradually rise.

Data Dependent Methods

Using historical details, a data-driven algorithm forecasts pattern of battery degradation. Without using physics-based methods, training data used to determine the weight parameters in a data-driven mathematical framework. Data-driven models are already gaining popularity throughout the world because of their adaptability and versatility. The categories apply to the data-driven method categorization of RUL prediction techniques are ML and DL algorithms. The machine learning approach fits variable degradation model to the data by utilizing its ability to do nonlinear fitting. Furthermore, link between the degradation criterion, i.e., relevant health indicators (HI) and battery RUL, is constructed using a mapping connection [10]. A number of machine learning (ML) models are frequently used for RUL estimation of batteries, including Naïve Bayes, Support Vector Machine and Gaussian Process regression. In [11] different ML algorithms are used for early identification of battery life cycle and experiments are performed to estimate the error indices and accuracy of the models [12, 13]. Many studies have used deep learning (DL) approaches for machine health monitoring; however, only a small number of these studies have used DL to the prediction of RUL. This [14] study provides a brief introduction to RUL prediction. Growing interest in RUL prediction has been seen in relation to DL approaches, which points to a bright future for its use in manufacturing.

2.3 Related Work

The growing number of electric vehicles has resulted in a significant problem for the infrastructure, electrical system, and charging station requirements. As a result, a lot of study has gone into battery life prediction. A typical LIB in EVs is a time-varying, dynamic electrochemical system that exhibits nonlinear behavior and complex internal mechanics. The LIB's life and performance steadily decline with charge and discharge cycles. Degradation of batteries can occur for a variety of causes, such as temperature fluctuations, mechanical stress, chemical reactions, and changes in physical processes. Predicting the battery's remaining lifespan also becomes a very difficult process as a result of deterioration. Still, in order to guarantee dependable performance of the battery management system, this is necessary. Health indicators (HI) may be used to display the capabilities and performance of LIBs. Battery characteristics like internal resistance, capacity, and voltage can be employed as HIs.

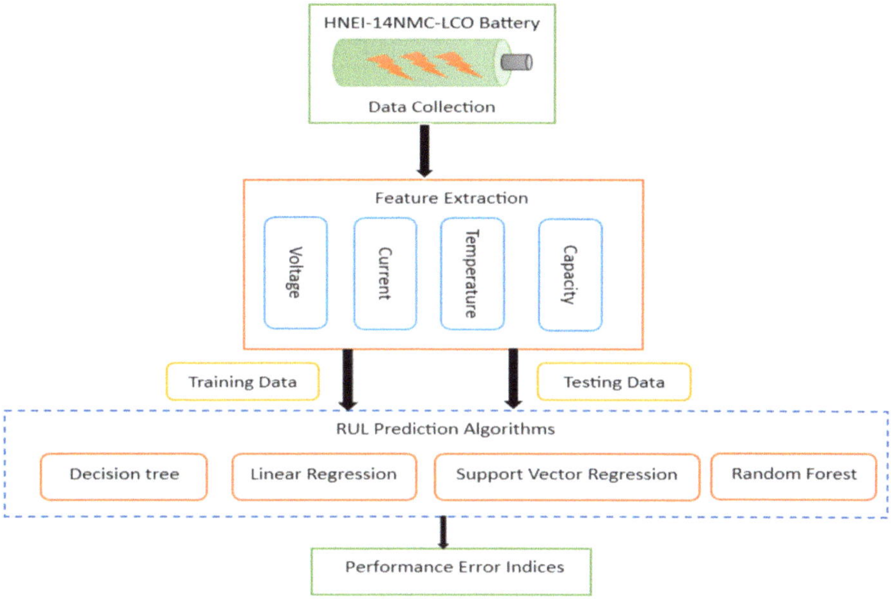

Fig. 3. Framework of Prediction of RUL using ML algorithms

These characteristics allow for the calculation of RUL forecasts as well as failure threshold predictions (when battery capacity drops to 80% of original value). Machine earning techniques primarily use cycle counts, internal impedance, current, voltage, temperature, and charge/discharge capacity as training data for predicting RULs. Since cycle life is known in the dataset, characteristics that best indicate battery performance are chosen first using the supervised learning framework. Figure 3 demonstrates the primary machine learning techniques. This article will give a summary of the current advancements in ML-based RUL prediction, starting with these basic methods.

Numerous academics have looked at RUL behavior prediction algorithms. The most popular algorithms from these investigations are covered in this section below:

2.4 Proposed Algorithms

Linear Regression (LR)

It is most fundamental and often used machine learning methods. It's a statistical technique for forecast analysis. "Linear regression" is the process of demonstrating a linear connection between one or more independent (t) variables and a dependent (S) variable. Since linear regression displays a linear connection, it may be used to identify how the value of dependent variable varies in response to value of the independent variable.

A LR can be mathematically represented as:

$$\mathrm{S} = \boldsymbol{a}_0 + \boldsymbol{a}_1\boldsymbol{b} + \epsilon \tag{2}$$

Here,

S = Dependent parameter
B = Independent parameter
a0 = line intercept
a1 = regression coefficient.
ε = random error.

In suggested battery, a linear regression is used to predict RUL. A method used for regularising a linear regression model is called Lasso Regression (L1 Regularisation); it involves adding a penalty term to target function of LR to stop overfitting. Following is lasso regression function.

$$\mathrm{L}(\theta) = \frac{1}{2n}\sum\nolimits_{i=1}^{n}\left(\overline{Y_i} - Y_i\right) + \lambda\sum\nolimits_{j=1}^{m}|\theta_L| \tag{3}$$

The initial term, least squares loss, represents squared difference between the predicted and actual values. The second term, L1 regularising, penalises regression coefficient θ $_{\mathrm{L}}$ total absolute values. [15] If result is continuous variable, linear regression procedures are used. Instead, when output is divided in groups, grading algorithms are applied.

Decision Tree (DT)

In machine learning, decision trees are a typical supervised learning technique used to model and predict outcomes based on input data. Recursively dividing the data according to the values of several characteristics is the first step in creating a decision tree. The method finds the best characteristic to divide data at each internal node based on variables like information gain or Gini impurity. Finding the characteristic for the root node at each level in the Decision Tree is a significant task. We call this procedure "attribute selection." The entropy varies as we divide the training examples into smaller groups using a decision tree node. This shift in entropy is measured as information gain. For Suppose set of instances 'i', attribute 'a', i_v is subset of i, 'v' represents an individual value that attribute a can take and Values (a) is set of all possible values of a, then

$$\text{Gain}(\text{i, a}) = \text{Entropy}(\text{i}) - \sum\nolimits_{v}^{a} \left| \frac{i_v}{i} \right| \cdot \text{Entropy}(\boldsymbol{i_v}) \quad (4)$$

A terminal tree with leaves representing ultimate prediction and branches represents observations is produced by the DT classifier [16]. While the procedure is ongoing, the dataset is divided into many subsets, which are then used to build the DT. For both classification and regression, the well-known classifier DT can be used. Thus, DT may be used with both category and numerical data. Because DTs perform well in situations when the dataset contains missing values, they are helpful in a variety of applications.

Random Forest (RF)

This is an effective machine learning algorithm. In the training phase, it generates a lot of Decision Trees. To measure a random subset of qualities in each division, each tree is built using a random subset of the data set. The randomization makes each tree more varied, which reduces possibility of overfitting and improves prediction performance overall. To generate predictions, approach averages each tree's output (for regression tasks) or votes (for classification tasks). The insights from several trees are utilised in this cooperative decision-making process to yield accurate and reliable outcomes. Considering that random forests can provide accurate forecasts, minimise overfitting, and handle complicated data.

There are several similarities between random forests and decision trees. The most fundamental difference is that RF produces more DTs when the set of rules is applied in the educational setting. The three main advantages of RF are its high predicted accuracy, ease of usage, and ability to solve the overfitting issue. When employing random forests that rely solely on categorization data, one must comprehend the meaning of the widely used Gini index.

$$\text{Gini}(\text{S}) = 1 - \sum\nolimits_{i=1}^{c} \boldsymbol{P}_i^2 \quad (5)$$

Calculating remaining usable life of batteries ahead of time is essential to guaranteeing a consistent energy supply and making the most use of that energy [17] predict RUL for BHMs, choose to use a random forest model to forecast RUL. This model may be used to collect, preprocess, and classify data in order to forecast RUL.

Support Vector Regression (SVR)

By evaluating the risk ratio, Support Vector Regression (SVR) seeks to determine the ideal margin values independent of error rates. Initially, Support Vector Machines were created to recognize patterns, or Support Vectors—small subsets of training samples that serve as decision boundaries. It works pretty well in nonlinear interactions between variables and may be successfully used to extremely nonlinear systems. The final formulation of the SVM method in terms of the support vectors, which reduces the massive training data to a much smaller subspace of SVs, is its primary benefit for classification and regression [18]. Furthermore, no computationally demanding mathematical procedures are needed for the formulation. By taking use of these SVM benefits, the suggested technique produces a computationally effective RUL estimate algorithm. Using the SVR technique, the SVM may be modified for a regression situation where the goal is to identify an optimum function.

$$\mathrm{F(X)} = (\mathrm{w}, \mathrm{X}) + \mathrm{b} \tag{6}$$

SVR function maximizes flatness while guaranteeing that the largest deviation f(x) from any training data is less than a user-defined quantity 'e'.

$$\phi(w, \xi) = \frac{1}{2}w^2 + C\sum\nolimits_i \xi_i \tag{7}$$

The resultant optimized target function is

$$\mathrm{F(x)} = \sum_{i=1}^{l} \sigma_i(x_i, x) + b \tag{8}$$

The final formulation of the SVM method in terms of the support vectors, which reduces the massive training data to a much smaller subspace of SVs, is its primary benefit for classification and regression. Furthermore, no computationally demanding mathematical procedures are needed for the formulation. By taking use of these SVM benefits, the suggested technique produces a computationally effective RUL estimate algorithm.

3 Performance Error Indices

Analysing the effectiveness of classification and regression algorithms, particularly in engineering and scientific practice, is essential to establish appropriate assessment measures for AI/ML models. This includes the use of measures for performance fitness and error to ensure accurate and relevant evaluations [19].

Most regression learning techniques forecast a target value using independent variables. Various performance metrics and error matrices, such as those mentioned below, are commonly used in engineering studies. [ERR] The majority of the reviewed papers rely on established measures like MAE, MSE, RMSE, and R2 as primary indicators for assessing prediction error.

***Mean Absolute Error (MAE)*:** It calculates absolute variation between what was expected and what was observed.

$$\mathrm{MAE} = \frac{1}{N}\sum \left| Y - \overline{\overline{Y}} \right| \tag{9}$$

Mean Square Error (MSE): It estimates average squared difference between values that the model predicts and the actual value that it provides.

$$\mathrm{MSE} = \frac{1}{N}\sum \left| Y - \overline{\overline{Y}} \right|^2 \tag{10}$$

Root Mean squared Error (RMSE): It is MSE square root.

$$\text{RMSE} = \sqrt{\frac{1}{N}\sum\left|Y - \overline{\overline{Y}}\right|^2} \tag{11}$$

R-Square Error(R^2): The R-Squared (Coefficient of Determination) indicates how well the predicted values match the original values.

$$\text{R}^2 = 1 - \frac{MSE(Model)}{MSE(Baseline)} \tag{12}$$

4 Results and Analysis

In Fig. 3, we described the procedures for implementing our investigation. We initially provide some background information on the battery dataset in the subsections that follow. We will next go over the implementation outcomes.

4.1 Data Set Description

To estimate the RUL prediction methods, researchers have used publicly accessible battery datasets. 14 NMC-LCO 18650 batteries with 1.5 Ch nominal capacity and over 1000 cycles at 25 °C with a C/2 CC-CV charge rate and 1.5 C discharge rate were studied by Hawaii Natural Energy Institute. Features that highlight behaviour of voltage and current during each cycle are constructed using the original dataset [20]. These characteristics allow calculation of the batteries' life. The 14 batteries' summaries are included in the dataset.

Dataset characteristics

- Sequence Index: Total cycles;
- D1: Discharge Time (s);
- D2: Time at 4.15 V (s);
- D3: Constant Current Time (s);
- D4: Decrement 3.6–3.4 V (s);
- D5: Maximum Voltage Discharge (V);
- D6: Minimum Voltage Charge (V);
- D7: Charging Time (s);
- Execution time (s);
- Final prediction.

Table 1. Estimation of Performance error indices for RUL prediction using ML Algorithms

ML ALGORITHMS	MSE	RMSE	MAE	R-Squared
Support Vector Regression	35.734	8.903	8.903	0.9994
Linear Regression	49.795	7.056	4.477	0.9995
Random Forest	**16.116**	**3.575**	**2.080**	**0.9999**
Decision Tree	25.084	5.008	2.380	0.9997

4.2 RUL Estimation Results

Different algorithms used for RUL estimation using performance error indices have been provided in above Section were used in our trials. Depending on the approach used, the various performance error indices are estimated and given in Table 1 below.

Table displays the projected RUL findings along with the mean errors. Figure 4 compares the performance of SVR, LR, RF, and DT. It can be shown that the random forest approach yields the lowest MSE measurement value (16.116), compared to the other methods. As a result, the learning prediction approach is more applicable and robust (Fig. 4).

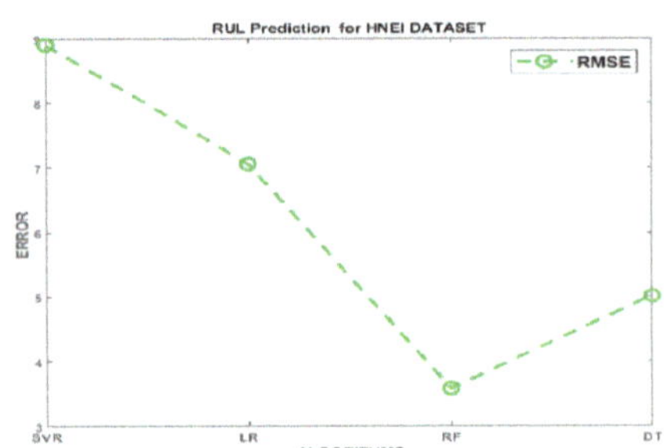

Fig. 4. RMSE for ML Algorithms

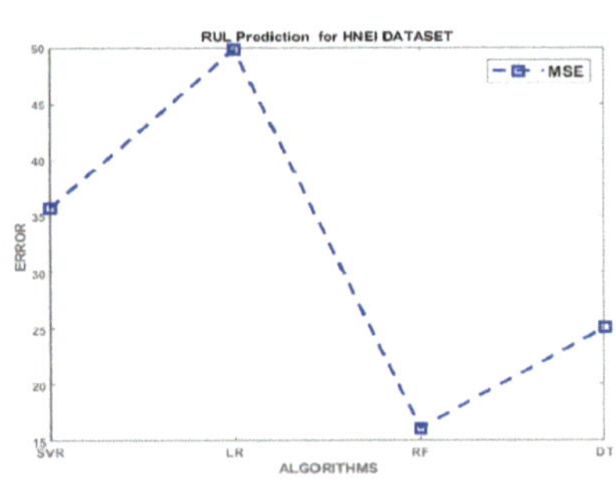

Fig. 5. MSE for ML Algorithms

Figure 5 shows the battery RUL prediction training and prediction outcomes based on RMSE measurement. Furthermore, it is clear that random forest performs best prediction in terms of RMSE. Which further confirms the benefit excellence for each of the five machine learning algorithms.

Figure 6 displays the MAE forecast of battery RUL. This establishes that random forest regression model has strong prediction capability determining a battery's RUL and validates four distinct machine learning techniques for prediction. This further confirms that the model is a reliable predictor.

The algebraic indicator R-Squared is predicted for SVR, LR, RF, and DT, shown in Fig. 7. The performance indicator shows proportion of uncertainty that can be explained by an absolute parameter. These performance error indices show overall best accession and which strategy yields great efficiency.

From the related research work a comparative table of different machine learning algorithms used for prediction of battery RUL along with different performance error indices are indicated in Table 2 below.

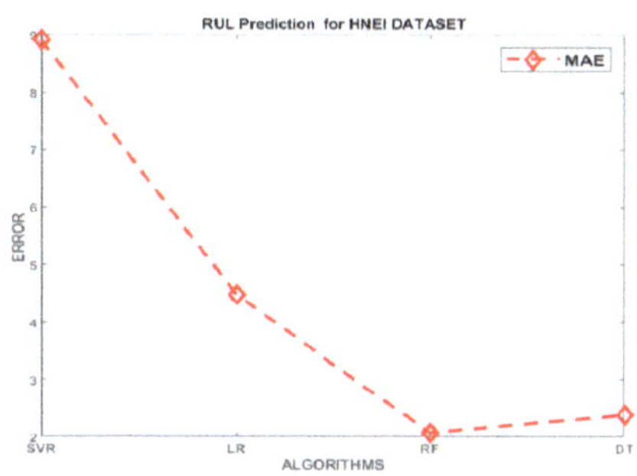

Fig. 6. MAE for ML Algorithms

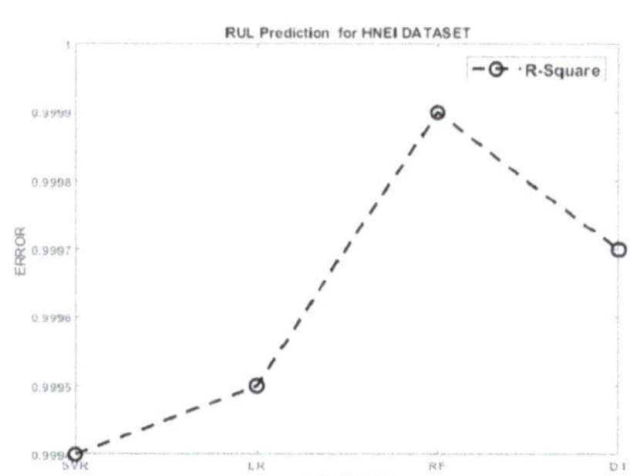

Fig. 7. R-Square for ML Algorithms

This study's primary contribution is that it locates the optimized version and accurately and successfully captures the properties of the RUL model prediction. Figure 8 displays sketch of the four performance error indices employing each four machine learning techniques.

Heat Map

To illustrate the relationships between different factors, these coefficients are presented as a heatmap. This visualization helps in identifying patterns that are beneficial for

Table 2. Comparison of Performance error indices for RUL prediction with different models

Reference	Model	MSE	RMSE	MAE	R-Square
[11]	SVM	46.55	-	18.32	-
	RF	19.69	-	11.12	-
[15]	LR	54.543	7.385	4.644	0.9994
	DT	26.848	5.816	2.296	0.9997
[16]	DT	72.13	84.67	-	-
	SVM	76.25	79.32	-	-
[17]	LR	226.50	148.527	-	-
	RF	516.33	22.72	-	-
	DT	133.71	36.56	-	-

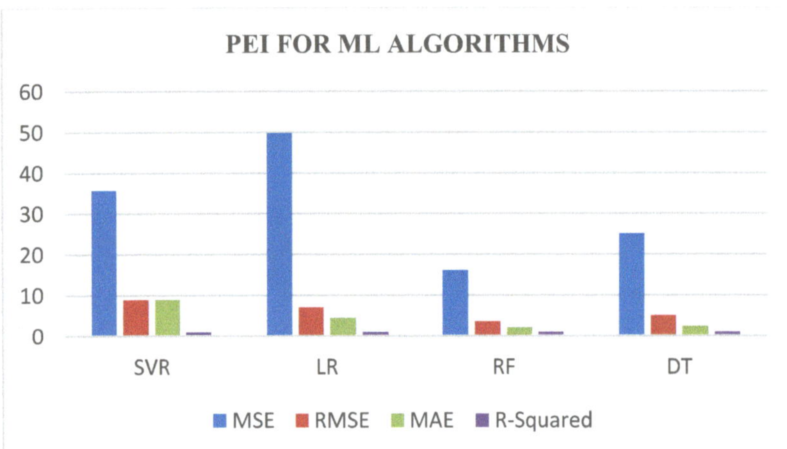

Fig. 8. Bar graph representation of PEI for different ML algorithms

developing machine learning models. The correlation matrix is converted into a color-coded representation using the heatmap. The correlation matrix displays the degree of relationship between the variables. Based on the heatmap in Fig. 9, the following inferences can be made:

- Time Constant Current and Charging Time show a high positive link.
- Using 0.78 coefficient, RUL and Maximum Voltage Discharge exhibit a robust positive association.
- Discharge Time also show a positive correlation, coefficient of 0.78.
- Minimum Voltage Discharge has robust negative connection.

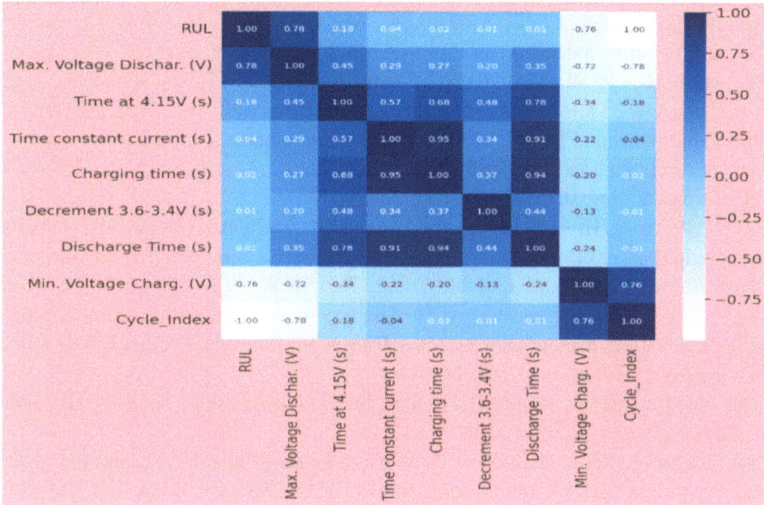

Fig. 9. Heat map representing the original dataset's characteristics

5 Conclusion

This research presents a data-driven machine learning technique for predicting the RUL of lithium-ion batteries. This work studied RUL prediction for the real-life battery dataset from Hawaii Natural Energy Institute using a range of ML methods, such as SVR, DT, RF and LR. The error metrics for battery dataset were then determined, including R-Squared, MAE, RMSE and MSE. When the four methods were compared with different performance error indices, they all displayed a notable variance in relevance. The findings indicate that the random forest method was more suitable for accurate RUL prediction and had better optimised evaluation criteria as compared to other methods. The study results covered in this article may improve scientists' ability to predict battery RUL accuracy, leading to an overall improvement in battery efficiency. Subsequent research endeavours may focus on the utilisation of Deep and Hybrid Learning techniques to enhance the precision of forecasts.

References

1. Hasib, S.A., et al.: A comprehensive review of available battery datasets, RUL prediction approaches, and advanced battery management. IEEE Access **9**, 86166–86193 (2021). https://doi.org/10.1109/ACCESS.2021.3089032
2. Chen, Y., Kang, Y., Zhao, Y., et al.: A review of lithium-ion battery safety concerns: the issues, strategies, and testing standards. J. Energy Chem. **59**, 83–99 (2021). https://doi.org/10.1016/j.jechem.2020.10.017
3. Chandra Sekhar, J.N., Hitesh Chandra, S.: Batter1y management systems for electric vehicles using lithium-ion batteries. Int. J. Sci. Res. **11**(12), 1255–1261 (2022)
4. Zhao, J., Tian, L. Cheng, L., Zhang, Y., Zhu, C.: Review on RUL Prediction Methods for Lithium-ion Battery. In: 2022 IEEE/IAS (I&CPS Asia). https://doi.org/10.1109/ICPSAsia55496.2022.9949753

5. Elmahallawy, M., Elfouly, T., Alouani, A., Massoud, A.M.: A comprehensive review of lithium-ion batteries modeling, and state of health and remaining useful lifetime prediction. IEEE Access **10**, 119040–119070 (2022). https://doi.org/10.1109/ACCESS.2022.3221137
6. Ben Ali, J., Azizi, C., Saidi, L., Bechhoefer, E., Benbouzid, M.: Reliable state of health condition monitoring of Li-ion batteries based on incremental support vector regression with parameters optimization. Proc. Instit. Mech. Eng., Part I: J. Syst. Control Eng. **237**(4), 717–727 (2020). https://doi.org/10.1177/0959651820950849
7. Ansari, S., Ayob, A., Hossain Lipu, M.S., Hussain, A., Mohamad Hanif, M.S.: Remaining useful life prediction for lithium-ion battery storage system: a comprehensive review of methods, key factors, issues and future outlook. Energy Reports **8**, 12153–12185 (2022)
8. Wang, Y., et al.: A comprehensive review of battery modeling and state estimation approaches for advanced battery management systems. Renew. Sustain. Energy Rev. **131**, 110015 (2020). https://doi.org/10.1016/j.rser.2020.110015
9. Hu, X., Xu, L., Lin, X., Pecht, M.: Battery lifetime prognostics. Joule **4**, 310–346 (2020). https://doi.org/10.1016/j.joule.2019.11.018
10. Ge, M.F., Liu, Y., Jiang, X., Liu, J.: A review on state of health estimations and remaining useful life prognostics of lithium-ion batteries. J. Int. Meas. Confed. **174**, 109057 (2021). https://doi.org/10.1016/j.measurement.2021.109057
11. Dokgoz, M., Yashan, Y.: A comparison of machine learning algorithms on lithium-ion battery cycle life prediction. In: 2021 6th ICCSE(UBMK). https://doi.org/10.1109/UBMK52708.2021.9558946
12. Reddy, V., et al.: Data Augmentation based Cross-Lingual Multi-Speaker TTS using DL with Sentiment Analysis. In: ACM Transactions on Asian and Low-Resource Language Information Processing 23 Oct 2023
13. Kumar Raja, J., et al.: Deep learning based multilingual speech synthesis using multi feature fusion methods. In: ACM Transactions on Asian and Low-Resource Language Information Processing (2023)
14. Wang, Y., Zhao, Y., Addepalli, S.: Remaining useful life prediction using deep learning approaches: a review. Procedia Manuf. **49**, 81–88 (2020)
15. Chandra Sekhar, J.N., Domathoti, B., Santibanez Gonzalez, E.D.R.: Prediction of battery remaining useful life using machine learning algorithms. Sustainability **15**(21), 15283 (2023). https://doi.org/10.3390/su152115283
16. Harippriya, S., Esakki Vigneswaran, E., Jayanthy, S.: Battery management system to estimate battery aging using deep learning andmachine learning algorithms. In: International Conference on Electronic Circuits and Signalling Technologies, vol. 2325, p. 012004 (2022)
17. Ha, V.T.: Experimental study on remaining useful life prediction of lithium-ion batteries based on three regressions models for electric vehicle applications. Appl. Sci. **13**, 7660 (2023)
18. Madan, A.K., Kaushik, S., Chaturvedi, T., Srivastava, V.: Machine learning predictive models for lithium-ion battery life expectancy. IJRTI **8**(5) (2023)
19. Naser, M.Z., Alavi, Amir H.: Error metrics and performance fitness indicators for artificial intelligence and machine learning in engineering and sciences. Archit. Struct. Construct. **3**(4), 499–517 (2021). https://doi.org/10.1007/s44150-021-00015-8
20. GitHub. How the Dataset Was Built. https://github.com/ignavinuales/Battery_RUL_Prediction. Accessed on 1 March 2022

Optimising Academic Performance: Leveraging Custom GPT and Prompt Engineering for Effective Study Companionship

Alan K. Sijo and Rhythm Choudhary(✉)

Maharaja Surajmal Institute, New Delhi, India
{alan01814902021,rhythmchoudhary}@msijanakpuri.com

Abstract. In today's digital age, the integration of artificial intelligence (AI) technologies into educational settings has become increasingly prevalent. This study explore the utilisation of custom Generative Pre-trained Transformers (GPT) and prompt engineering techniques to enhance the role of ChatGPT as a study companion. Through a combination of qualitative and quantitative analysis, this study investigates the reliability and efficiency of personalised GPT models tailored to educational texts and preferences of each student and the strategic use of prompts to guide learning interactions. By harnessing these innovative approaches, students can potentially improve study efficiency and comprehension enhancing the whole studying experience. The findings shed light on the benefits and challenges of integrating AI-driven study companions into educational practices and provide insights into future directions for leveraging AI technologies in learning environments.

Keywords: GPT · ChatGPT · AI · Prompt Engineering · Custom GPT · LLM

1 Introduction

Over the past few years, great leaps have been made in the field of artificial intelligence. It has positioned itself at the forefront of many fields from medicine to education and even in areas of cooking. The ability of computers to mirror the human brain to be "artificially intelligent" has reached new levels specifically over the recent years. Primarily, this entails obtaining information in the form of outside facts, learning via instruction, and applying that knowledge to accomplish the intended goals [1].

This rapid development has seen more impact in areas where the need for automation and creativity was very high. Artificial Intelligence has been very useful in serving these unique areas where the difference of each individual needs to be respected and taken into account to produce the best result. Education is one such field the impact of artificial intelligence has been of paramount importance [2, 3]. The successful application of AI-powered chatbots to improve student engagement and assist pedagogical methods in educational settings has also been highlighted by recent studies [4]. We have seen not only a technological but also a cultural shift where people have identified the benefits of AI and have actively tried to use it whenever possible.

B. K. Mishra et al. (Eds.): ICT4GS 2024, ISEM 56, pp. 273–287, 2025.
https://doi.org/10.1007/978-3-031-94997-5_25

The integration of artificial intelligence in education has yielded notable results, including enhanced educational process efficiency, global learning promotion, personalised learning, more intelligent content creation, and optimised educational management in terms of effectiveness and efficiency [5]. Given all these advances it is also necessary to keep in check the challenges and ethical questions the implementation of AI brings. Students have embraced these new and effective AI technologies, with Chat-GPT being the most popular, but issues like data privacy, equitable access to education, and the role of instructors need to be carefully considered [6].

A particular kind of GPT language model called Chat GPT (Generative Pre-trained Transformer) has been trained to produce text in response to inputs in natural language. It can be used in many different applications, such as chatbots, virtual assistants, and language translation systems, and is intended to mimic human-like discourse [7, 8]. Because Chat GPT is built on sophisticated machine learning algorithms and has been pre-trained on extensive text datasets, it can produce incredibly complex responses when users input [9].

There are several ways in which GPT language model is being used in education

- Teaching Assistance: Chatbots and virtual tutors that support teachers and students in their efficient learning can be developed using GPT.
- Writing assistance: GPT is being used as both a tool and an analyzer to both help write content and analyse the written content.
- Personalised Learning: With GPT, user-specific learning experiences may be created. This is very useful for students. Each student has different preferences and priorities and models like chat GPT allows personalised customisation both over prompts and under the hood customisations allowing to fully cater to the individuality of the student.

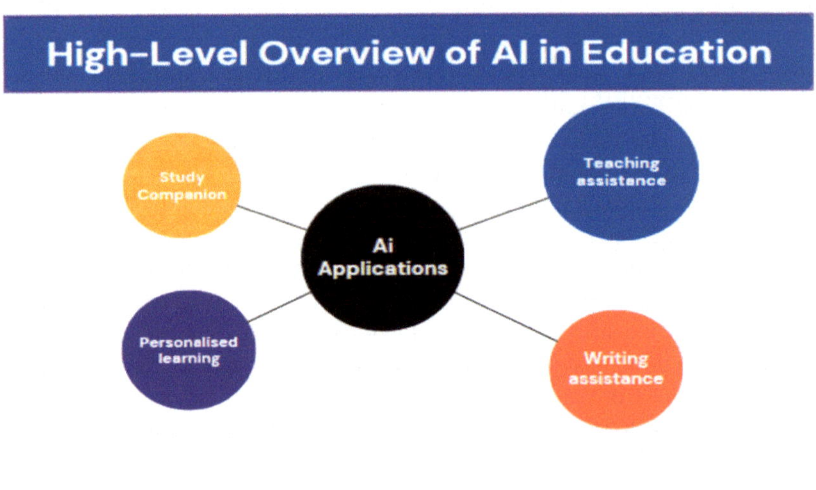

Fig. 1. Applications and impacts of AI in education

As illustrated in Fig. 1, the applications and impacts of AI in education. It covers teaching assistance, writing support, and personalized learning plans. It highlights the benefits like efficiency and personalization, and challenges such as data privacy and equitable access. The diagram provides a comprehensive view of how AI is transforming the educational landscape.

Personalised learning has seen a huge demand due to its efficiency and flexibility which has not only attracted students but also huge edtechs and startups. Everyone has been eager to involve AI to help cater to the needs of each student without deploying extra resources. Consequently, the primary goal of this study is to examine how businesses and students have used chatGPT—a particular type of artificial intelligence—as a study aid or companion, and to determine how best to maximise its effectiveness [10].

The following inquiries are addressed in this context:

1. How is the usage of chatGPT as a study aid doing right now?
2. What are the major challenges to using chat GPT and what are the solutions to these challenges?
3. What are custom GPTs and how can it be used as a better alternative to the generic chatGPT?
4. What is prompt engineering and how can it be used to tweak chat GPT and get better results?
5. What are the ways to merge these services and techniques to find an optimised Ai based study companion solution?

2 The Current Use of ChaptGPT

2.1 ChatGPT As Study Tool

The infusion of technology to academics and studying is not something new. We have seen it time again, that the evolving tech has made its way into classrooms and how curriculums itself are designed. Be it the introduction of smart teaching mediums like smart boards or interactive applications software, the tools available to the assistance of teachers has always been a boon. But the popularity of chatGPT has started a trend where this movement in technologies is being captured by students first rather than teachers. The use of ChatGPT started with it being a smart hack to do your homework/assignments [11, 12].

In order to scan vast volumes of text input and identify linguistic patterns, GPT use deep neural networks, which enable it to produce cohesive and grammatically sound phrases. With over 175 billion parameters and extensive training on internet text data, GPT-3, the most recent and sophisticated version of GPT, is one of the most potent language models on the market right now [13].

This allowed students to use it flexibly for any subject or purpose regardless of the domain or situation. This quick and easy access data and its simple interpretation combined with the human chat-like feel of ChatGPT made it the most sought after source of custom content generation. This has revolutionised the way the students used the internet as they used to rely on extensive google searches and manual interpretation of data for specified results. ChatGPT helps in avoiding all these steps altogether by providing a one stop platform for all your needs. The extensive database of ChatGPT combined with the text generation capabilities has resulted in its extensive usage by students and learners [14].

2.2 Impact of ChatGPT

The fact that ChatGPT has just recently been used in educational settings and is still in its early phases of development means that there isn't much overall scientific research on the topic. Twelve papers in all published in 2023, were deemed pertinent and gathered for a systematic evaluation. This is due to the fact that ChatGPT's use in educational settings is still relatively new—it was only introduced at the end of 2022—so research on the impact on education has not been done in sufficient length, and meaningful data has not been gathered until the first few months of the following year [15].

3 Custom GPT

Custom GPTs, or Custom Generative Pre-trained Transformers, represent a tailored approach to natural language processing (NLP) tasks by fine-tuning or retraining the base GPT model to suit specific applications, domains, or tasks. Introduced as an extension of OpenAI's groundbreaking GPT architecture, custom GPTs offer enhanced capabilities and efficiency in generating and understanding text within targeted contexts [16, 17].

At their core, custom GPTs leverage the power of transfer learning, a technique where a model trained on a large, diverse dataset is further refined on a smaller, more specialised dataset. This fine-tuning process allows the model to adapt its learned representations to better align with the nuances and intricacies of the desired task or domain.

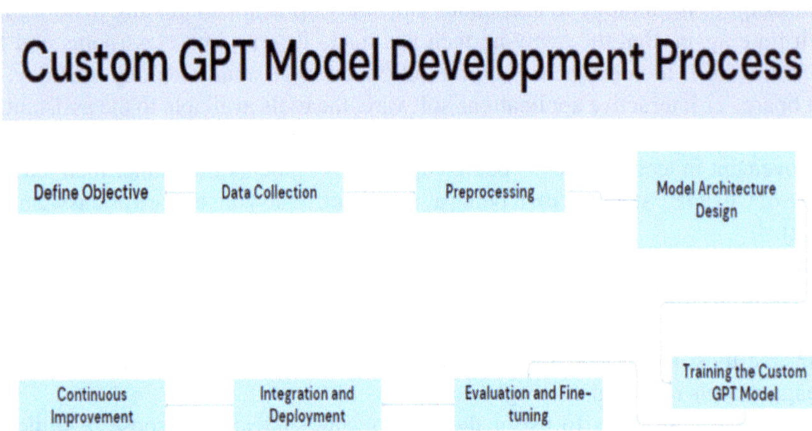

Fig. 2. Custom GPT Model Development Process

The introduction of custom GPTs has opened up a myriad of possibilities across various fields and industries. For instance, in healthcare, custom GPTs can be trained on medical literature and patient records to assist with diagnosis, treatment recommendations, and patient-doctor interactions. In customer service, these models can be fine-tuned on support ticket data to provide more accurate and personalised responses to inquiries [18]. Additionally, custom GPTs find applications in content generation, sentiment analysis, language translation, and many other NLP tasks.

Figure 2 indicates the steps to create a custom GPT model. It starts with defining objectives and collecting data. The data is then pre-processed and the model architecture is designed. After training the model, it's evaluated and fine-tuned. The final steps involve integrating and deploying the model, followed by continuous improvement to keep it updated.

3.1 Building Custom GPTs

1. Dataset Collection: Gathering a relevant dataset that captures the language patterns and characteristics specific to the target domain or task.
2. Model Fine-tuning: Using the pre-trained GPT model as a starting point, the model is further trained on the specialised dataset, adjusting its parameters to better suit the desired objectives.
3. Evaluation and Iteration" Assessing the performance of the fine-tuned model through validation datasets and iterative refinement to optimise its effectiveness and accuracy.
4. Deployment and Integration: Incorporating the custom GPT into applications, systems, or workflows where its enhanced language capabilities can be leveraged to solve real-world problems.

Custom GPTs can have a huge impact on how Ai is used for personalised learning. The flexibility of building your very own customised chat bot which is tailored to your need is basically everything a student could ask in an AI based study tool.

3.2 Differences in Custom GPT and ChatGPT

Both general GPT and custom GPT models have their own advantages and disadvantages, depending on the use case and the goal. Here are some of the main pros and cons of each type of model:

Pros and Cons of General GPT:

- Versatile and flexible: A general GPT model can generate texts on a wide range of topics and tasks and adapt to different inputs and contexts. A general GPT model can also generate texts in different languages, styles, and tones, depending on the data it was trained on and the parameters you set.
- Powerful and advanced: A general GPT model is trained on a large and diverse corpus of texts, which allows it to learn general linguistic patterns and knowledge, as well as capture the diversity and complexity of natural language. A general GPT model can also generate coherent and fluent texts on almost any topic and task, with high quality and accuracy.
- Accessible and easy to use: A general GPT model is available to anyone through various platforms and tools, such as Playground, Codex, and Custom GPTs. A general

GPT model can also be used without any coding or technical skills required, as it can be interacted with through natural language inputs and prompts.
- Generic and broad: A general GPT model may not perform as well on specific or niche topics or tasks, as it may not have enough relevant or accurate data or knowledge to generate texts that are appropriate and satisfactory for the desired purpose. A general GPT model may also generate texts that are inconsistent or irrelevant, depending on the input and the context.
- Expensive and resource-intensive: A general GPT model requires a lot of computational power and resources to train and run, as it has a large and complex architecture with billions of parameters. A general GPT model may also be costly to use, depending on the platform and the tool that you use to access it.
- Unreliable and unpredictable: A general GPT model may generate texts that are inaccurate, incomplete, or misleading, as it may not have enough factual or logical information or reasoning to generate texts that are correct and reliable. A general GPT model may also generate texts that are harmful, offensive, or unethical, as it may not have enough moral or social awareness or responsibility to generate texts that are respectful and appropriate [18].

Pros and Cons of Custom GPT

- Focused and specialised: A custom GPT model can generate texts that are relevant and accurate for a specific task or domain, as it is guided by additional data and instructions that are provided by the user. A custom GPT model can also generate texts that are consistent and satisfactory, depending on the data and instructions that are provided to it.
- Cheap and resource-efficient: A custom GPT model does not require a lot of computational power and resources to create and run, as it is a smaller and simpler version of ChatGPT that can be tuned to excel at a specific task or domain. A custom GPT model is also free to use, as it is a feature of ChatGPT that is available to anyone.
- Reliable and predictable: A custom GPT model can generate texts that are correct and reliable, as it is based on factual and logical information and reasoning that are provided by the user. A custom GPT model can also generate texts that are respectful and appropriate, as it is based on moral and social awareness and responsibility that are provided by the user.
- Limited and narrow: A custom GPT model may not perform as well outside its trained domain or task, as it may not have enough general or diverse data or knowledge to generate texts that are suitable and acceptable for different purposes. A custom GPT model may also generate texts that are redundant or boring, depending on the data and instructions that are provided to it.
- Complex and difficult to create: A custom GPT model requires a lot of data and instructions to create and optimise, as it needs to be tailored to a specific task or domain. A custom GPT model may also require some coding or technical skills to create and modify, depending on the complexity and the specificity of the task or domain.

- Risky and insecure: A custom GPT model may expose your data to a global audience, potentially compromising your privacy and security, as it may contain sensitive or personal information that you provide to it. A custom GPT model may also be misused or abused by others, as it may be shared or accessed by anyone.

Since custom GPTs are much new there is not much data available on the precise impact and results of using custom GPT over chatGPT. But the ability to train the model to your specific subject/course unlocks the ability to make a custom AI bot that is personal to you. This greatly reduces the errors and inconsistency introduced by generic models like chat GPT by streamlining the entire process to revolve around the student.

3.3 Leveraging Custom GPTs for Personalised Learning

Creating a custom AI study bot for students using custom GPT models involves several steps, from designing the model architecture to training it on relevant data. Let's break down the process in detail:

1. Define the Objective: The first step is to clearly define the objective of the AI study bot. What should it help students with? For example, it could assist with homework, provide explanations for difficult concepts, quiz students on their knowledge, or offer personalised study plans.
2. Data Collection: Gather a diverse set of data relevant to the study bot's objective. This can include textbooks, lecture notes, past exam papers, online resources, and student queries. The more varied and comprehensive the dataset, the better the AI model can learn from it.
3. Preprocessing: Clean and preprocess the data to remove noise, standardise formatting, and enhance the quality of the dataset. This may involve tasks such as text normalisation, tokenization, and data augmentation.
4. Model Architecture Design: Design a custom GPT model architecture tailored to the study bot's requirements. This involves determining the size of the model (number of layers, hidden units), the input/output format, and any additional features or modules needed for specific functionalities (e.g., question-answering, knowledge retrieval).
5. Training the Custom GPT Model: Train the custom GPT model on the preprocessed dataset using techniques such as transfer learning. Start by fine-tuning a pre-trained GPT model on the study-related data to adapt it to the specific domain and task. This step requires significant computational resources and may take several days or weeks depending on the size of the dataset and complexity of the model.
6. Evaluation and Fine-tuning: Evaluate the performance of the trained model using metrics such as accuracy, perplexity, and user feedback. Fine-tune the model further based on the evaluation results to improve its effectiveness and address any shortcomings.
7. Integration and Deployment: Integrate the trained model into a user-friendly interface or platform that students can access easily. This can be a web application, mobile app, chatbot interface, or integration with existing learning management systems. Ensure that the deployment infrastructure is scalable, reliable, and secure.
8. Continuous Improvement: Monitor the performance of the AI study bot in real-world usage and collect feedback from users. Use this feedback to iteratively improve the model by retraining it on updated data, fine-tuning parameters, or adding new features and functionalities.

4 Prompt Engineering

Artificial intelligence (AI) systems that use generative reasoning are designed to generate specific outcomes based on how well prompts are provided. Prompt building generative artificial intelligence models enable them to comprehend and respond more effectively to a wide range of queries, from the most simple to the most complicated. In essence, positive impulses produce positive results [19]. Iteratively developing diverse fast engineering techniques is fundamental to generative AI's ability to efficiently learn from a range of input data sources and adapt to minimise biases, confusion, and provide more accurate results.

To bridge the gap between meaningless AI-generated answers and raw queries, prompt creation must be done carefully. Engineers can greatly optimise the quality and relevance of outputs to solve for both the specific and the general by fine-tuning effective prompts. By reducing the need for post-generation editing and human review, this procedure eventually saves time and effort in reaching the desired results [17].

4.1 How Does Prompt Engineering Work?

As simple as it may sound, prompt engineering is as simple as uniquely specifying to the generative AI what you want and how you want it. It uses transformational architecture, which enables neural networks to handle enormous volumes of data and comprehend the complexity of language. Rapid engineering is turning out to be essential for maximising the potential of the foundation models that drive generative artificial intelligence. Large language models (LLMs) known as foundation models are constructed using transformer architecture and contain all the data required by a generative artificial intelligence system [20].

Text-to-image generative AI systems like DALL-E and Midjourney use stable diffusion, a model that excels at generating images from text descriptions, in conjunction with an LLM. Effective rapid engineering combines technical know-how with a deep understanding of natural language, vocabulary, and context to get the greatest results with the fewest adjustments [21].

Top-k sampling, tokenization, and model parameter modification are among prompting strategies that guarantee AI models produce useful answers [22].

4.2 Benefits and Necessity of Prompt Engineering

The capacity to produce optimal outputs with little post-generation work is the main advantage. Because of the randomness or immensity of the subject, generative AI can occasionally be unclear and imprecise. Prompt engineering allows to minimise this gap without employing any surplus resources. This also helps in getting instant results as we aren't changing anything fundamentally to the model itself.

Composing prompts for Google Bard is not the same as creating prompts for Open AI's GPT-3 or GPT-4. Bard can be told to incorporate more recent information into its results because it can obtain information via Google Search. However, since it was ChatGPT's main purpose upon construction, it is a superior tool for reading and summarising content. AI models are guided to produce more accurate, individualised, and relevant responses by well-written prompts [23].

4.3 LLM and their Constraints when Using Prompts

Natural language processing has advanced significantly thanks in large part to Large Language Models (LLMs), especially those built on the Transformer architecture. These models have impressive language abilities since they have been pre-trained on large datasets to predict future tokens. Nevertheless, despite their sophistication, LLMs are limited by innate features that have an impact on their usefulness [24].

- Transient State: As LLMs are designed without permanent memory or state, they require extra software or systems to handle and retain context.
- Probabilistic Nature: LLMs' stochastic nature causes responses to vary, even in response to the same cues, making application consistency difficult. This implies that, even while using the same prompt, you can receive somewhat different replies each time.
- Content Fabrication: Often called "hallucination," LLMs have the ability to produce content that appears credible yet is factually false.
- Outdated Information: LLMs are limited to historical knowledge by reliance on pre-training data, which prevents real-time awareness or changes.
- Domain Specificity: Despite being naturally generalist, LLMs frequently need domain-specific information to perform well in specialised jobs.
- Resource Intensity: The large size of LLMs results in high computational and financial expenses, which affect accessibility and scalability.

4.4 How to Prompt Efficiently

A clear purpose describes the precise information or action you anticipate from the model, whereas a clear context describes the backdrop or circumstance to which the prompt relates. These two components increase the likelihood that the LLM will provide an accurate and helpful result by reducing the range of possible outcomes depending on your input [10].

Figure 3 outlines the process of refining prompts for language models. It begins with defining the task, creating an initial prompt, and testing the output. Based on the results, the prompt is refined and validated. Once the quality is ensured, the prompt is implemented into the system.

Let us look at a simple example of how two different prompts on the same topic can generate very different results.

Prompt: "Can you provide a brief overview of the causes of World War II?".

Optimised Prompt: "Could you outline the key political, economic, and social factors that contributed to the outbreak of World War II, emphasising the role of treaties and alliances in Europe during the 1930s?".

- Contextual Relevance: The prompt provides context by specifying the subject (World War II) and the aspect the student wants to understand (causes). It also adds specificity by focusing on political, economic, and social factors, as well as the role of treaties and alliances, which directs ChatGPT's response toward a comprehensive analysis of the topic.

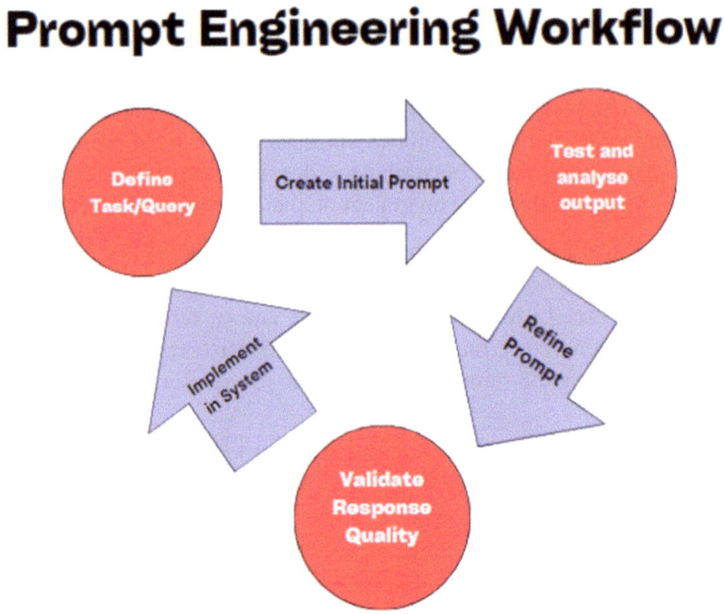

Fig. 3. Prompt Engineering Workflow for language model

- Clarity and Specificity: Unlike the previous prompt, which was more general, this prompt is clear and specific about what information the student is seeking. It narrows down the scope and guides ChatGPT to provide a targeted and informative response.
- Depth of Inquiry: By asking for an analysis of specific aspects such as treaties and alliances, the prompt encourages a deeper exploration of the topic, which can lead to a more insightful and detailed response from ChatGPT. This enhances the student's understanding and engagement with the subject matter.
- Feedback Incorporation: If the previous response from ChatGPT was lacking in certain areas or contained inaccuracies, this optimised prompt allows the student to refine their request based on previous interactions, thus optimising the learning experience.
- Language Optimization: The language used in the prompt is clear, concise, and academic in tone, aligning with the expectations of a research-oriented inquiry. This ensures effective communication with ChatGPT and facilitates better comprehension of the student's needs.

5 Methodology

The main goal was to comprehend the existing use of ChatGPT as a study tool by educational institutions and students. A comprehensive review was conducted on twelve relevant papers published in 2023 to gain an understanding of the rising trends and effects of ChatGPT in educational contexts. The methodology comprises a comprehensive evaluation of the application and efficacy of artificial intelligence-based solutions in many educational settings, analysing their advantages and disadvantages.

5.1 Understanding ChatGPT

To investigate the utilization and influence of ChatGPT in educational settings, the research delves into numerous studies and surveys that highlight its application. This section explores ChatGPT's role in enhancing student learning, supporting research activities, and providing immediate assistance with coursework. The methodology includes a thorough review of literature documenting the experiences of students and educators with ChatGPT, emphasizing both advantages and drawbacks.

5.2 Exploring Custom GPTs

The development of Custom GPTs suited to certain educational demands was also investigated in the study. Although ChatGPT is flexible, custom GPTs were shown to be more useful in educational settings since they could provide more focused and domain-specific answers (Fig. 2). The adaptability and efficiency of customised GPT models in meeting certain educational needs are the main areas of analysis. This section of the research examines studies that demonstrate how GPT models may be developed and tailored for certain subjects or learning environments.

5.3 The Role of Prompt Engineering

Prompt engineering is one crucial component that has been found to be essential to maximizing ChatGPT's effectiveness as a study partner. By enhancing prompts to provide clear context and focused queries, users might instruct ChatGPT to produce more insightful and accurate responses tailored to educational requirements (Fig. 3). The methods and approaches used in prompt engineering to maximize interactions with AI tools such as GPT are examined in the section Prompt Engineering. A thorough analysis of the literature on rapid design, including recommended techniques and typical roadblocks, is part of the research. Research examining how well different kinds of prompts elicit desired answers from AI models is examined to determine important variables impacting performance.

6 Results

The results of this study highlight the effectiveness of leveraging custom Generative Pre-trained Transformers (GPT) and prompt engineering techniques in enhancing academic performance.

Currently Using ChatGPT as a Studying Tool
The results of the analysis showed that students primarily utilise ChatGPT as a study tool for a variety of academic duties, such as help with homework, concept clarification, and content creation. Its rapid generation of pertinent data and explanations across various topics and areas is the reason for its widespread use (Fig. 1).

ChatGPT's Effect on Education
Numerous beneficial effects, including enhanced student engagement, personalised

learning experiences, and greater learning efficiency, were noted by the impact evaluation. Pupils valued ChatGPT's immediate availability and thoroughness of material, which successfully complemented conventional teaching strategies.

Although students gain a great deal from using AI-based technologies, their effect affects more than just a student's academic achievement; it also affects the larger educational environment. These tools, which assist teachers with curriculum preparation, individualised instruction, and administrative duties, are being incorporated into teaching methods more and more. The effects of AI on educators and different educational processes show how AI technology is changing the nature of contemporary education. Below are the list of AI-based tools [25]:

- Virtual Mentor

Professors and lecturers frequently utilise virtual mentors as apps to share notes, assignments, assessments, and quizzes. Students can submit assignments and ask questions on these platforms, which expedites the evaluation process and improves communication between teachers and students.

- Smart Content

Intelligent content technology facilitates more effective sharing and searching of programmable digital books and material information. Digital libraries frequently employ this technology to help with the organised and efficient classification of learning materials.

- Automatic Assessment

Platforms for automated assessments make the process of creating and administering exams and quizzes more efficient. These technologies evaluate and fix questions automatically, saving teachers' and tutors' manual labour and boosting productivity.

- Personalized Learning

Personalised learning platforms serve as students' personal assistants by providing individualised educational experiences. By adjusting to each user's needs and preferred method of learning, these tools enhance the standard and usefulness of education.

- Intelligent Tutoring System (ITS)

Adaptive teaching is offered via intelligent tutoring systems, which can change based on the skills of the pupils. These tools boost educational outcomes by personalising the learning process and providing individualised instruction and support.

In summary, the integration of AI-based tools in education has significantly transformed both teaching and learning processes. The survey results indicate a widespread adoption of AI tools among students across various fields of study, with a notable impact on their academic performance and efficiency. Tools like ChatGPT, DeepL, and Google Assistant have become essential in aiding research, enhancing comprehension, and providing personalized learning experiences. The diverse applications of AI, from virtual mentors to intelligent tutoring systems, have not only facilitated student learning but also streamlined administrative tasks for educators. As AI technology continues to evolve, its role in education is expected to expand, offering even more innovative solutions to enhance the learning experience and educational outcomes.

7 Conclusion

The introduction of AI has been revolutionary to almost all domains of life, especially in fields like technology, healthcare and education where the need to cater to different and unique requests has been ever growing. AI has helped us revolutionise the way work is done in any industry. The thought process to tackle any problem nowadays is incomplete without automation or inclusion of AI.

As discussed above, Ai has also played a pivotal role in how both teachers and students now view education. AI which may have been introduced into this field as a "quick hack" or "helper" is now undeniably a crucial part not only for students but teachers as well. The involvement of AI in the form of GPTs has paved a new way in education. However, there are drawbacks and restrictions associated with this, such as worries about preconception, data privacy, and cheating. Also the use of any GPT model or all tasks is not very effective as they are not trained to a specific domain. This begs the question of how we can make GPT models suitable and accurate to the field of education and hence to each student. The need for custom GPTs and LLMs are therefore necessary to create tailored experience. However time and resources are required to develop and test these models before they are implemented and used.

The optimisation of AI or specifically chatGPT under the current circumstances can only come from using the current leading models like GPT-4, custom GPT and the efficient use of prompts. This can greatly help students contextualise and filter what they want to study and how they want to study. Ultimately, the use of AI as a personal study tool or assistant will depend on how students and teachers are able to leverage the available resources to create optimised teaching and assisting models. We can realise the potential of this document to revolutionise education and enable the upcoming generation of learners by implementing the strategies and tactics covered in it.

References

1. Montenegro-Rueda, M., Fernández-Cerero, J., Fernández-Batanero, J.M., López-Meneses, E.: Impact of the implementation of ChatGPT in education: a systematic review. Computers **12**, 153 (2023). https://doi.org/10.3390/computers12080153
2. Renz, A., Vladova, G.: Reinvigorating the discourse on human-centered artificial intelligence in educational technologies. Technol. Innov. Manage. Rev. **11**(5), 5–16 (2021). https://doi.org/10.22215/timreview/1438
3. Božić, V., Poola, I.: Chat GPT and education (2023). https://doi.org/10.13140/RG.2.2.18837.40168
4. Bekeš, E.R., Galzina V.: Exploring the pedagogical use of AI-powered chatbots educational perceptions and practices. IN: 2023 46th MIPRO ICT and Electronics Convention (MIPRO), pp. 636–641. Opatija, Croatia (2023). https://doi.org/10.23919/MIPRO57284.2023.10159734

5. Nazir, A., Wang, Z.: A Comprehensive Survey of ChatGPT: Advancements, Applications, Prospects, and Challenges. Meta Radiol. **1**(2), 100022 (2023). https://doi.org/10.1016/j.metrad.2023.100022. Epub 2023 Oct 7. PMID: 37901715; PMCID: PMC10611551
6. GPTStore.ai. GPTStore. https://gptstore.ai/
7. OpenAI. Introducing GPTs. https://openai.com/blog/introducing-gpts
8. Dempere, J., Modugu, K., Allam, H., Ramasamy, L.: The impact of ChatGPT on higher education. Front. Educ. **8**, 1–13 (2023). https://doi.org/10.3389/feduc.2023.1206936
9. Wang, T., Zhu, Q.: ChatGPT – technical research model, capability analysis, and application prospects. In: 2024 IEEE 7th Advanced Information Technology, Electronic and Automation Control Conference (IAEAC), pp. 787–796. Chongqing, China (2024). https://doi.org/10.1109/IAEAC59436.2024.10503756.
10. Menick, J., Trebacz, M., Mikulik, V., Aslanides, J.: Teaching language models to support answers with verified quotes (2022). https://arxiv.org/abs/2203.11147
11. Kuzman, T., Ljubešić, N.: Automatic genre identification: a survey. Lang. Resou. Evaluat. **59**(1), 537–570 (2023)
12. Hebrem Joander, P., Hezron Belix, P., Hency Juliet, A.: Revolutionizing education: harnessing the power of AI for unparalleled student learning advancements. current trends inform. Technol. **14**(01), 18–25 (2024). https://journals.stmjournals.com/ctit/article=2024/view=138289
13. Javaid, M., Haleem, A., Singh, R.P., Khan, S., Khan, I.H.: Unlocking the opportunities through ChatGPT Tool towards ameliorating the education system. BenchCouncil Trans. Benchmarks, Stand. Evaluat. **3**(2), 100115 (2023)
14. John Rush. AllGPTs. https://allgpts.co/
15. Liu, Y., et al.: Summary of ChatGPT-related research and perspective towards the future of large language models. Meta-Radiol. **1**(2), 100017 (2023). https://doi.org/10.1016/j.metrad.2023.100017
16. Cooper, K.: OpenAI GPT-3: Everything You Need To Know (2021)
17. Lo, P.: An Autoethnographic Reflection of Prompting a Custom GPT Based on Oneself. 1–9 (2024). https://doi.org/10.1145/3613905.3651096
18. Clarisó, R., Cabot, J.: Model-driven prompt engineering. In: 2023 ACM/IEEE 26th International Conference on Model Driven Engineering Languages and Systems (MODELS), pp. 47–54. Västerås, Sweden (2023). https://doi.org/10.1109/MODELS58315.2023.00020
19. Giray, L.: Prompt engineering with ChatGPT: a guide for academic writers. Ann. Biomed. Eng. **51**, 3 (2023). https://doi.org/10.1007/s10439-023-03272-4
20. Park, D., An, G.T., Kamyod, C., Kim, C.G.: A study on performance improvement of prompt engineering for generative ai with a large language model. J. Web Eng. **22**(8), 1187–1206 (2023)
21. Allam, H., Dempere, J., Akre, V., Parakash, D., Mazher, N., Ahamed, J.: Artificial intelligence in education: an argument of chat-GPT use in education. In: 2023 9th International Conference on Information Technology Trends (ITT), pp. 151–156. Dubai, United Arab Emirates (2023). https://doi.org/10.1109/ITT59889.2023.10184267
22. Spasić, J., Janković, D.S.: Using ChatGPT standard prompt engineering techniques in lesson preparation: role, instructions and seed-word prompts. In: 2023 58th International Scientific Conference on Information, Communication and Energy Systems and Technologies (ICEST), pp. 47–50. Nis, Serbia (2023). https://doi.org/10.1109/ICEST58410.2023.10187269
23. Lo, L.: The CLEAR path: a framework for enhancing information literacy through prompt engineering. J. Acad. Librariansh. **49**, 102720 (2023). https://doi.org/10.1016/j.acalib.2023.102720

24. Wang, Z.Q., Ayibuen, B.: Beyond ChatGPT: Possible value and traps of GPT-4 application in future university teaching (2023). https://doi.org/10.13927/j.cnki.yuan.20230712.002
25. von Garrel, J., Mayer, J.: Artificial Intelligence in studies—use of ChatGPT and AI-based tools among students in Germany. Humanit. Soc. Sci. Commun. **10**, 799 (2023). https://doi.org/10.1057/s41599-023-02304-7

Improving the Accuracy and Comprehensibility of XAI Explanations

Janvi, Rhythm Choudhary(✉), and Manoj Kumar

Maharaja Surajmal Institute, New Delhi, India
rhythmchoudhary@msijanakpuri.com

Abstract. The increasing integration of artificial intelligence (AI) into various domains necessitates an understanding of its decision-making processes to ensure reliability and ethical use. This paper addresses the "black box" problem, where the opacity of AI models impedes trust and accountability.

Objective: To enhance the transparency and comprehensibility of AI systems, particularly through explainable artificial intelligence (XAI).

Methodology: It includes a detailed examination of current XAI techniques such as Local Interpretable Model-Agnostic Explanations (LIME), Partial Dependency Plots (PDP), and decision trees, alongside an exploration of their limitations and potential improvements.

Findings: It highlights that while these methods contribute significantly to model interpretability, they often fall short in accuracy and user comprehension. This research proposes hybrid approaches combining multiple XAI techniques to address these gaps.

Conclusion: The study emphasizes the need for on-going development in XAI to foster responsible AI usage and paves the way for future advancements that can better balance transparency, privacy, and ethical considerations. This study underscores the critical role of XAI in enhancing the trustworthiness of AI, particularly in sensitive applications like healthcare, and outlines future research directions to overcome existing challenges. By addressing the limitations of current XAI methods and proposing improvements, this paper aims to promote responsible AI development, ultimately unlocking the full potential of AI for societal benefit.

Keywords: Explainable machine learning · PDP · LIME · Decision tree · XAI · Dimensionality reduction

1 Introduction

1.1 Background and Motivation

The rapid integration of artificial intelligence (AI) into various aspects of our lives, from healthcare and finance to travel and entertainment, has brought undeniable benefits and transformative power. However, the main obstacle to widespread adoption and societal impact is the lack of transparency in many AI paradigms, often referred to as the "black

B. K. Mishra et al. (Eds.): ICT4GS 2024, ISEM 56, pp. 288–305, 2025.
https://doi.org/10.1007/978-3-031-94997-5_26

box" problem. This phenomenon impedes reliability and accountability, limits statistical power, and can perpetuate bias and discrimination. For example, denying a loan application without explicitly explaining the reason, the applicant may perceive bias or injustice, leading to a loss of trust in the system. Similarly, healthcare professionals may be reluctant to accept AI findings or treatment recommendations that they do not fully understand, potentially jeopardizing patient outcomes [1, 2].

In response to these challenges, the field of Explainable Artificial Intelligence (XAI) has emerged as a critical research area. XAI aims to bridge the gap between AI models and human understanding by providing explanations for model decisions.

Several XAI techniques have been developed to achieve this goal, ranging from feature importance analysis to model visualization, counterfactual interpretation, and local interpretive model-agnostic methods such as LIME (Local Interpretable Model-Agnostic Explanations) [3].

Despite significant progress in XAI research, several limitations remain, particularly in terms of the accuracy and understandability of explanations. Existing methods often struggle to provide precise descriptions of the model's decision-making process in a manner easily understood by humans, especially non-experts. Addressing these limitations is crucial for maximizing the benefits of XAI and enabling responsible AI development.

This paper aims to address these challenges by:

- Investigating the limitations of current XAI techniques.
- Proposing solutions to overcome these limitations.
- Advancing the field of XAI to unlock its full potential.
- Comparing various XAI techniques with respect to a single problem.

In the context of explainable AI (XAI), interpretability is crucial for understanding, trust, and accountability. "Black box" models, despite their power, lack transparency in their decision-making processes. In contrast, interpretable models provide clear, understandable explanations for their predictions, enhancing trust.

Interpretable models enable feature analysis as shown in Fig. 1 to show how different features influence learning, offer transparency by explaining what the model learns, and visualize decision-making processes. This is particularly vital in high-stakes domains like healthcare, where understanding the rationale behind predictions can aid in informed decision-making.

1.2 The Root of the Problem

The "black box" trouble arises from the complicated internal workings of many AI fashions, mainly those primarily based on deep studying techniques. These fashions research complex relationships from huge quantities of facts, often resulting in tricky selection-making methods which might be tough to comprehend for human beings [5]. This loss of transparency hinders our potential to:

Validate the version's reasoning: We can't verify whether the model is making selections based on relevant elements or whether or not biases are found in its choice-making manner.

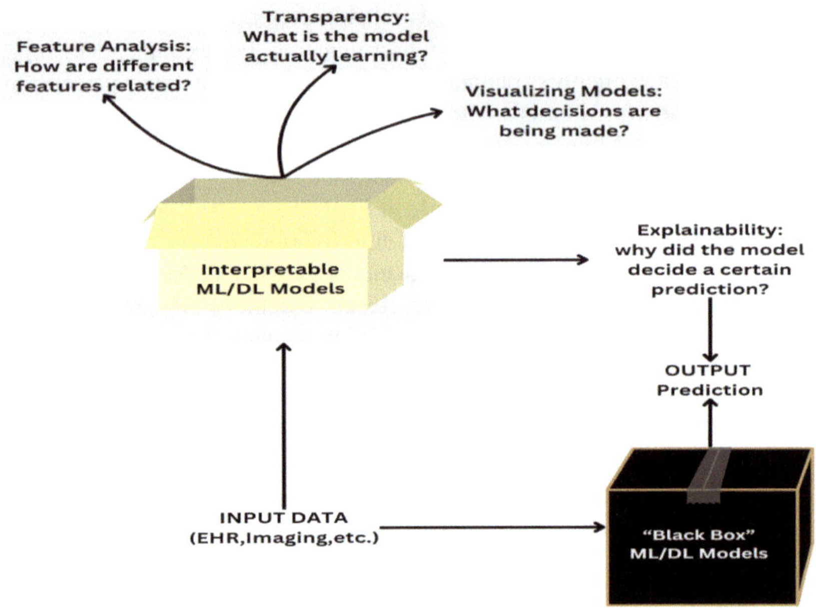

Fig. 1. Black box AI models versus interpretable and explainable AI models [4]

Identify and mitigate bias: If we can't understand why the model makes sure choices, it will become difficult to discover and deal with ability biases which can cause discriminatory or unfair effects.

Build accept as true with and accountability: Without know-how the model's inner workings, users may be hesitant to believe its decisions, specifically whilst those selections have massive consequences.

Ensure accountability of AI development: It will become difficult to broaden AI structures which are honest, responsible, and obvious while we lack knowledge of the way they perform.

1.3 The Negative Implications

The "black box" hassle can have various poor implications for the development and deployment of AI-

- Reduced acceptance as true with and adoption: Users may be reluctant to accept as true with AI systems that they cannot apprehend, hindering the adoption of AI in numerous domain names.
- Perpetuation of bias: Biased records can lead to biased AI fashions, and without transparency, these biases can be tough to discover and mitigate.
- Lack of duty: When AI choices are opaque, it will become tough to hold developers and users accountable for capability poor consequences.
- Hindered studies and development: Difficulty in understanding how fashions' paintings can impede similarly research and improvement efforts inside the field of AI.

Langer et al. [6] highlighted how a lack of transparency in AI models can make it difficult to identify and address biases that may be present in the training data, leading to discriminatory or unfair outcomes. Also, Lundberg, S. M. et al. [7] discussed the need for explainable AI techniques to advance research and development in the field of AI by allowing researchers to better understand how models work and improve their performance. The importance of explainability for holding developers and users accountable for the consequences of AI decisions was proposed via a framework for understanding explainability in AI by Miller, T [8]. At last, Amodei, D. et al. [9] argued that a lack of explainability can lead to a lack of trust and hinder the adoption of AI in various domains.

1.4 Addressing the Challenge

Researchers are actively exploring numerous techniques to address the "black box" problem and make AI models more transparent and interpretable. These techniques include-

- Model-agnostic techniques: These methods can be applied to any model and provide insights into its decision-making process. One prominent example is LIME (Local Interpretable Model-Agnostic Explanations) [10]. LIME works by approximating the model locally around a specific prediction, allowing for easier interpretation of the factors influencing that particular decision.
- Feature importance evaluation: This method identifies the features within the input data that have the most significant effect on the model's decision. Techniques like SHAP (SHapley Additive exPlanations) [11] can be used to assign importance scores to different features, helping humans understand which aspects of the data hold the most weight in the model's decision-making.
- Counterfactual causes: This technique generates hypothetical situations that explain why the model made a particular choice. It explores what changes to the input data would have caused the model to reach a different conclusion. While Pearl's [12] focused on broader causal inference concepts, it provides a foundation for understanding how counterfactuals can be applied to explain model decisions.
- Model visualization: This method creates visual representations of the model to help humans understand how it works. Techniques like decision trees or rule-based models are inherently interpretable, but for complex models, other visualization methods like attention maps can be used to highlight which parts of the input data receive the most focus from the model. Ashish Vaswani et al. [13], introduced the concept in the context of neural machine translation.

While these techniques offer promising solutions, further research and development are needed to ensure that AI models are not just powerful but also transparent, and trustworthy, and contribute to a positive and inclusive future.

Figure 2 shows the emphasis and scope of research of XAI. The counterfactual explanation techniques in XAI libraries are used to identify the counterfactual instances, as well as the change in features used to attain the positive class.

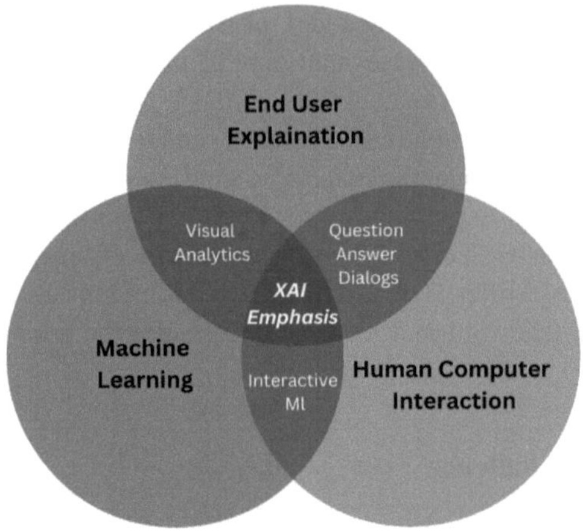

Fig. 2. Emphasis and Scope of Explainable AI models [14]

2 Literature Review

The realm of artificial intelligence (AI) and machine learning (ML) is rapidly evolving, and interpretability has emerged as a critical factor in this landscape. As data scientists, researchers, and engineers, they strive not only for models with high accuracy and efficiency but also for transparency and trustworthiness. This literature review delves into the world of interpretable machine learning, bridging the gap between intricate, often opaque models and the need for clarity in their decision-making processes. Authors exploration encompasses various methods, ranging from fundamental concepts to sophisticated techniques applied to complex models like neural networks.

The field of healthcare presents a compelling case for interpretable AI (XAI). Angati Kalyan KUMAR et al. [15] critically examined existing research on XAI in healthcare, emphasizing the importance of transparent and comprehensible AI models for users, especially medical professionals. Their study evaluated state-of-the-art XAI techniques such as LIME, SHAP, PDP, and decision trees. While these methods enhance trust in AI models, the challenge lies in providing user-friendly explanations for both doctors and patients.

The field of XAI often caters to users with strong technical backgrounds, leaving non-experts struggling to grasp XAI methods. Amodei D et al. [9] introduced "x-[plAIn]," a novel approach that aims to democratize XAI. This custom Large Language Model (LLM), developed using ChatGPT Builder, generates clear, concise summaries of various XAI methods. The key feature is its adaptability – x-[plAIn] tailor explanations for different audiences, including business professionals and academics. Results from use-case studies demonstrate its effectiveness in providing easy-to-understand, audience-specific explanations, regardless of the XAI method employed. This adaptability bridges the gap between complex AI technologies and practical applications, paving the way for a more accessible XAI landscape.

Moving beyond healthcare, Rajjliwal et al. [16] explored the use of decision trees, a type of inherently interpretable ML model, in the medical field. Decision trees represent a series of questions about the data, with each question leading to a branch in the tree and ultimately culminating in a prediction. This inherent simplicity allows for interpretability because the path through the tree reveals the sequence of features that led to the final outcome. Feature importance in a decision tree context refers to the influence each symptom or test result has on the final diagnosis. A feature with a higher importance score signifies that it plays a more critical role in differentiating between diagnoses. This interpretability is particularly valuable in medical settings, where understanding the reasoning behind a diagnosis is crucial for both patients and healthcare professionals.

Building upon the interpretability of decision trees, Lundberg et al. [7] introduced the concept of Partial Dependence Plots (PDP) as a tool for visualizing feature importance in random forests, a popular ensemble method that combines multiple decision trees. Random forests are known for their high accuracy, but their inherent complexity makes them less interpretable than individual decision trees. PDP plots depict the average prediction of the model for a range of values for a single feature, while averaging out the effects of all other features. By analyzing these plots, we can gain insights into how a specific feature influences the model's predictions. This approach allows us to interpret the complex behaviour of random forests by breaking them down into the simpler, interpretable components of individual decision trees.

While decision trees and random forests offer some level of inherent interpretability, they may not be suitable for all machine learning tasks. Miller, T. [8] explored two advanced techniques, LIME (Local Interpretable Model-Agnostic Explanations) and SHAP (SHapley Additive exPlanations), for explaining individual predictions made by complex models. LIME works by approximating a complex model locally around a specific prediction using a simpler, interpretable model, such as a linear regression model. This local explanation highlights the features that were most influential for that particular prediction. SHAP, on the other hand, leverages game theory to distribute the prediction credit among all features based on their contribution. This approach provides a more nuanced understanding of feature importance by considering how each feature interacts with others in influencing the final prediction. Miller, T. [8] compared LIME and SHAP, highlighting their strengths and weaknesses. LIME is generally faster to compute explanations, making it suitable for real-time applications. However, SHAP can provide more accurate explanations, particularly for complex models with feature interactions.

Beyond individual techniques, research is ongoing to integrate XAI practices seamlessly into the software development process. Guleria P et al. [17] conducted a systematic literature review focusing on XAI methods that align with different stages of software development. Their work highlighted the importance of understanding "what" features influence model outcomes and "how" they contribute to predictions. By incorporating XAI techniques throughout development, we can enhance the overall transparency and interpretability of real-world AI applications. However, challenges remain, such as balancing computational complexity with achieving a desired level of interpretability.

Rachael Hagan et al. [18] proposed a novel XAI method called specific-input LIME that addresses this challenge. This interpretable approach combines the strengths of both feature importance and partial dependence plots (PDPs). By identifying which features a deep learning model focuses on for tabular data and how these features impact the model's predictions, specific-input LIME achieves a balance between global and local interpretations. This method offers valuable insights for practitioners building interpretable deep learning models.

While individual techniques like LIME and SHAP offer valuable insights, integrating XAI practices throughout the entire software development process is crucial. Brożek et al. [5] presented XAIR, a comprehensive meta-review of promising XAI methods and tools. Unlike existing reviews, XAIR aligns its results with the software development process, encompassing requirement analysis, design, implementation, evaluation, and deployment. By mapping XAI techniques to these steps, the study facilitates real-world AI applications with built-in explainability. It ensures transparency from the very beginning of the development cycle.

Lundberg et al. [11] conducted an extensive survey on XAI methods, exploring tools and applications across diverse domains. The survey assessed current techniques, addresses challenges, and emphasizes the growing importance of XAI in Generative AI, where models like Transformers, GPT-4, and DALL-E are gaining prominence. By categorizing and analyzing existing literature, this comprehensive review provides valuable insights into achieving transparency and interpretability across various fields. It serves as a roadmap for researchers and practitioners navigating the ever-evolving XAI landscape.

Langer et al. [6] guided on interpretability in machine learning empowers practitioners by dissecting various models and methods. This resource offers practical insights applicable to research and projects. Whether the goal is to enhance model transparency or comply with regulations, Adams' guide equips practitioners with a valuable toolkit for implementing XAI effectively.

Pearl et al. [12] demonstrated the practical benefits of XAI by enhancing a hotel review management model. Specifically, they leverage LIME and SHAP to provide both local and global explanations. LIME highlights features influencing individual predictions (e.g., identifying specific keywords in reviews that impact sentiment analysis). SHAP offers a global perspective on feature importance, revealing which aspects of reviews generally have the most significant influence on the model's overall sentiment classification. By integrating XAI, the model becomes more transparent, enabling stakeholders to validate decisions, identify potential biases in reviews, and make informed choices based on reliable insights. This study exemplifies how XAI not only enhances model performance but also fosters user trust and understanding.

As Mridha et al. [19] emphasized, XAI is crucial for building trust and understanding in AI models impacting medical decisions. They provided a comprehensive overview of XAI concepts and challenges, highlighting the need for interpretability in healthcare applications. P. A. Moreno-Sanchez et al. [20] further reinforce this necessity in the context of cardiovascular disease (CVD) prediction. They explored various XAI techniques like feature importance visualization, SHAP values, and partial dependence plots, demonstrating their effectiveness in explaining CVD risk predictions.

The importance of XAI in medical diagnosis is echoed by Bharati et al. [21]. Furthermore, Gautam et al. and Angati Kalyan KUMAR et al. [15, 22] specifically underscores the need for interpretable machine learning in CVD risk prediction. Gupta, S. et al. [23] offered a broader survey on XAI for disease prediction, which can be conceptually adapted to CVD classification data, particularly when narrowed down using relevant keywords.

Interpretable machine learning is not just a technical consideration; it's a critical step towards fostering responsible AI development and deployment, particularly in fields like healthcare. By exploring the research papers presented in this review, we gain valuable insights into various methods and techniques for making complex machine learning models more explainable.

The reviewed literature highlights the importance of striking a balance between model accuracy and interpretability. While complex models often achieve high accuracy, their inherent opacity can hinder trust and understanding. XAI techniques like LIME, SHAP, and PDP offer valuable tools for explaining both individual predictions and overall feature importance, allowing us to gain insights into how models arrive at their decisions.

The research presented in this review underscores the critical role of XAI in responsible AI development. As we continue to explore and develop new techniques, interpretability will remain a central theme in ensuring the ethical and trustworthy integration of AI into various aspects of our lives. By prioritizing explainability, we can pave the way for a future where AI solutions are not only powerful but also transparent and understandable for all stakeholders.

3 Understanding the Need for XAI: Specific Reasons and Implications

The growing integration of Artificial Intelligence (AI) into various facets of daily life necessitates a deeper understanding of its decision-making processes. Explainable Artificial Intelligence (XAI) addresses this need by offering transparency and insights into how AI models derive their conclusions. XAI enables users to make informed decisions, detect potential biases, and hold AI developers accountable. Several key aspects underscore the importance of XAI in the responsible development and deployment of AI:

3.1 Trust and Accountability

Opaque Decision Making: Many AI models, specifically deep learning architectures, perform as "black boxes," making it difficult to understand the reasoning and logic behind their choices. This lack of transparency hinders trust, especially when the decisions have significant consequences, including in healthcare, finance, and criminal justice [5, 24].

Lack of Explainability: Without clear reasons for AI decisions, it becomes hard to hold developers and users accountable for potential biases, mistakes, or discriminatory consequences. Explainable AI (XAI) helps bridge this gap by providing insights into the model's reasoning, facilitating responsible and accountable AI development [9].

3.2 Identifying and Mitigating Bias

Here's the section on identifying and mitigating bias with references for XAI techniques-

- Data Bias: AI models trained on biased data can perpetuate and amplify existing societal biases, leading to unfair and discriminatory outcomes. XAI techniques, along with counterfactual explanations and feature importance analysis, can help identify data biases in the model and the data it was trained on.
- Counterfactual explanations: This technique, as mentioned earlier [25], can be used to identify data biases by generating hypothetical scenarios that would have led the model to a different, unbiased decision. By analyzing the changes needed in the data for a different outcome, data biases can be revealed.
- Feature importance analysis: Techniques like SHAP (SHapley Additive exPlanations) [11] can help identify features in the data that have a disproportionate influence on the model's decisions. By examining which features hold the most weight, data biases can be surfaced, particularly if sensitive features (e.g., race, gender) show unexpected importance.

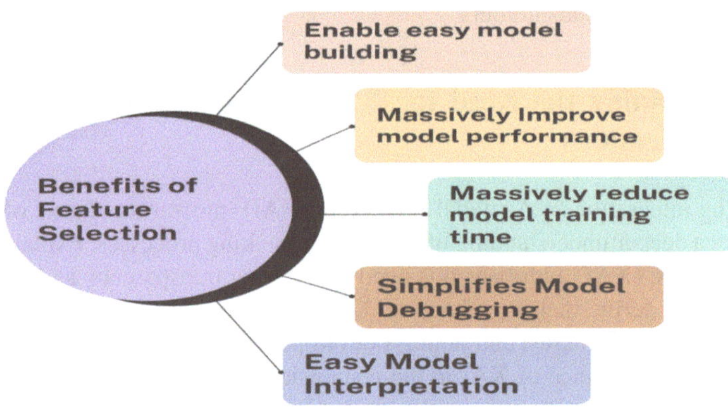

Fig. 3. Benefits of the Feature Selection Technique.

Feature selection is a process used in machine learning to select the most relevant features from a dataset. This can improve the performance of a machine learning model by reducing the amount of data that needs to be processed, and by making the model easier to interpret and has other benefits as shown in Fig. 3.

Algorithmic Bias: Even with unbiased data, algorithmic design choices can result in accidentally biased outcomes. XAI techniques can assist developers in becoming aware of and mitigating these algorithmic biases, ensuring that AI models are fair and equitable.

Here, specific XAI techniques for algorithmic bias detection are a developing area of research. However, general XAI methods like model visualization can be helpful. For instance, attention maps. It can reveal if the model is focusing on irrelevant parts of the data during decision-making, potentially indicating an algorithmic bias.[13].

3.3 Auditing and Debugging

Model Debugging: Understanding how AI fashions arrive at their selections is crucial for debugging and troubleshooting issues. XAI provides valuable insights into the version's internal workings, allowing builders to pick out and fasten insects that might result in incorrect or harmful consequences.

Model Auditing: XAI facilitates the auditing of AI fashions, ensuring that they are functioning as meant and aren't being used for malicious purposes. This is particularly essential in excessive-chance domain names where AI choices have a sizable societal impact [8].

3.4 User Engagement and Satisfaction

Understanding Predictions: Providing motives for AI predictions can improve consumer delight and reputation. By know-how how the model arrived at its selection, customers can feel greater assured in its accuracy and reliability [9, 26].

Improving User Experience: XAI can beautify the person revel in by presenting interactive visualizations and reasons, allowing customers to explore the model's behavior and benefit a deeper know-how of its capabilities and obstacles [27, 28].

3.5 Scientific Progress and Innovation

Model Improvement: XAI techniques may be used to advantage insights into the strengths and weaknesses of AI fashions, informing in addition studies and improvement efforts.

Theoretical Understanding: By expertise how AI fashions paintings, researchers can broaden new theories and algorithms that are extra transparent and interpretable, leading to advances within the area of AI.

3.6 Past Surveys

Model Improvement: According to a systematic meta-survey on XAI challenges and future opportunities [29], explainability is needed in many areas of AI like AI planning, explainable recommendations, explainable agency, and explainable embodied agents.

Theoretical Understanding: S. Chen et al. proposed a new framework, based on distinct images and vocabulary, in their work [30] to use a multi-modal explanation component. This model provides better reasoning and interpretability.

The Table 1 below summarizes the need for XAI (Explainable Artificial Intelligence) according to recent surveys (2023). It highlights how XAI can address various challenges in AI, including improving model generalizability, ensuring privacy while explaining decisions, and incorporating human preferences into explanation design, etc. Finally, the table highlights some more prospects of why we need XAI other than the ones talked above.

Table 1. The need of explanations by various Surveys done on XAI

Need for XAI	Description	Reference
Improved Model Generalizability	XAI techniques can help identify factors limiting a model's generalizability to unseen data, improving its performance in real-world applications	Guidotti et al., 2023. "A Survey of Methods for Explainable Artificial Intelligence" (https://arxiv.org/pdf/2101.09429) [31]
Privacy-Preserving XAI	Research is ongoing on developing XAI methods that protect sensitive data while still providing explanations for model decisions	Liu et al., 2023. "A Survey of Privacy-Preserving Explainable Artificial Intelligence" (https://arxiv.org/abs/2204.08237) [32]
Human Preferences in XAI Design	Understanding human preferences for explanation formats and levels of detail helps design effective XAI methods for specific user groups	Miller et al., 2023. "Explainable AI: A User-Centric Perspective" (https://arxiv.org/abs/2204.03209) [33]
Regulatory Compliance	As AI becomes more prevalent, regulations are being developed to ensure fairness and accountability. XAI can help ensure models comply by providing explanations for their decisions	Arrieta et al., 2023. "Explainable Artificial Intelligence (XAI): A Review" (https://arxiv.org/abs/2006.00093) [34] (This survey explores XAI applications in human-AI collaboration)
XAI for Countering Adversarial Attacks	XAI techniques can help identify weaknesses in models that might be exploited by adversarial attacks, improving their robustness	Xu et al., 2023. "Adversarial Attacks and Defenses in Explainable AI: A Survey" (https://arxiv.org/abs/2204.03232) [35]
XAI for Explainable Machine Learning Pipelines	As machine learning pipelines become more complex, XAI becomes crucial for understanding not just the final model, but the entire decision-making process	Xu et al., 2023. "Adversarial Attacks and Defenses in Explainable AI: A Survey" [35]

4 Comparative Analysis of XAI Techniques Based on Input Parameter

Let's delve deeper into each of these XAI techniques and the input parameters they utilize:

4.1 Model-Agnostic Methods

Input Parameters: These methods generally take the input data and the corresponding predictions made by the AI model as parameters. They don't rely on the internal architecture or parameters of the model itself.

Explanation Process: Model-agnostic strategies purpose to provide insights into the decision-making method of any system studying model without requiring getting right of entry to its internal shape. They generate motives by perturbing input facts and observing the model's response, permitting users to recognize how the version behaves with no need for specific understanding of its internal workings.

Example: Local Interpretable Model-Agnostic Explanations (LIME) [10] is a popular model-agnostic technique. LIME perturbs input data points round the example of interest and trains a neighborhood interpretable version to approximate the conduct of the black-box model round that instance [36].

4.2 Feature Importance Analysis

Input Parameters: This technique directly examines the input features of the data used to train the model.

Explanation Process: Feature importance evaluation identifies which enter functions contribute the maximum to the version's decision-making manner. By analyzing the weights or significance scores assigned to each characteristic, users can recognize which capabilities the model considers maximum relevant whilst making predictions.

Example: Techniques like permutation significance or SHAP (Shapley Additive causes) [11, 37] analyzed the impact of character capabilities on version predictions by using systematically altering function values and staring at changes in model outputs.

4.3 Counterfactual Explanations

Input Parameters: Counterfactual causes generally require the authentic enter data factor for which evidence is sought, alongside the prediction made by way of the model for that data point.

Explanation Process: This technique generates hypothetical times which might be much like the authentic data point however with altered feature values to illustrate how modifications in input functions would affect the version's selection. By providing those counterfactual situations, customers benefit insights into why the version made a selected prediction.

Example: Counterfactual examples can be generated via perturbing input characteristic values at the same time as ensuring that the modified example remains near the unique records point inside the feature space [25, 38].

4.4 Model Visualization

Input Parameters: Model visualization techniques may require access to the model's architecture, parameters, or intermediate representations, depending on the specific visualization method employed.

Explanation Process: Model visualization creates visual representations of the AI version, its inner layers, decision boundaries, and characteristic interactions. These visualizations assist users apprehend the version's conduct with the aid of imparting intuitive insights into how it tactics enter data and makes predictions.

Example: Techniques like Partial Dependency Plots (PDPs) are graphical gear used to visualize the marginal effect of a function at the predicted final results of a device learning version even as preserving different functions steady [39]. They help to recognize the connection among a particular input feature and the version's predictions even as controlling for the effect of different functions [40].

These techniques offer diverse approaches to explaining AI model decisions, each leveraging different input parameters to provide insights into model behavior. The Table 2 below illustrates various XAI techniques, showcasing how different input parameters and model predictions are explained using example data.

5 Addressing the Limitations of XAI Techniques

Here's an overview of the limitations of each XAI technique and potential solutions:

5.1 Model-Agnostic Methods (LIME)

Limited Accuracy: LIME can generate inaccurate explanations, especially for complex models.

Incomplete Explanations: LIME only provides local explanations for individual predictions, not global insights into the model's behavior.

Oversimplification: LIME may simplify complex relationships within the model, leading to misleading explanations.

Solutions:

Develop hybrid approaches incorporating knowledge-based techniques to improve accuracy and explainability [41].

Implement ensemble LIME models for more robust explanations across different data points.

Explore explainable model-agnostic methods (EMAM) that combine LIME with other XAI techniques.

5.2 Feature Importance Analysis

Limited Context: Feature importance analysis doesn't explain how features interact or influence the model's decision.

Static and Oversimplification: This technique provides static global explanations that don't consider the influence of specific input values.

Model-Dependent: The relevance of features is specific to the model and may not generalize to other models.

Solutions:

Incorporate feature interaction analysis techniques to understand how features influence each other.

Table 2. Various XAI techniques, input parameters and model predictions using example data

XAI Technique	Input Parameters	Model Prediction	Example Data	Model Prediction of Example Data
Model-Agnostic Methods	Input Data: {Age, Income, Education, Marital Status}	Predicted Probability	Input Data: {Age: 35, Income: $50,000, Education: Bachelor's, Marital Status: Married}	Model Prediction: 0.75
Feature Importance Analysis	Input Data: {Age, Income, Education, Marital Status}	Predicted Outcome	Input Data: {Age: 35, Income: $50,000, Education: Bachelor's, Marital Status: Married}	Model Prediction: Approved Loan
Counterfactual Explanations	Original Input Data: {Age, Income, Education, Marital Status} Modified Input Data (Counterfactual): {Age, Income, Education, Marital Status}	Predicted Outcome	Original Input Data: {Age: 35, Income: $50,000, Education: Bachelor's, Marital Status: Married} Modified Input Data: {Age: 40, Income: $60,000, Education: Master's, Marital Status: Single}	Model Prediction: Denied Loan
Model Visualization	Model Architecture: {Layers, Neurons} Input Data Representation: Feature Embeddings	Model Decision Boundary	Model Architecture: {Layers: [Input, Hidden 1, Hidden 2, Output], Neurons: [1, 10, 15, 20]} Input Data Representation: [0.75, -0.22, 0.43, 1.12]	Model Decision Boundary: Linear Separation between Classes A and B

Develop dynamic feature importance methods that account for variations in input data.

Employ model-agnostic feature analysis methods that are not specific to any particular model architecture.

5.3 Counterfactual Explanations

Limited Counterfactual Generation: Generating relevant counterfactuals that are close to the original data point can be challenging.

Ethical Concerns: Counterfactual explanations may reveal sensitive information or lead to discrimination if not used responsibly.

Computational Cost: Computing counterfactuals for complex models can be computationally expensive and slow.

Solutions:

Explore advanced counterfactual generation techniques based on causal inference methods.

Develop ethical guidelines for using counterfactuals to avoid bias and discrimination.

Investigate efficient algorithms and optimization techniques to reduce computational costs.

5.4 Model Visualization (PDP)

Limited Accuracy: Partial Dependency Plots (PDP) may not fully capture complex model behaviors, especially in nonlinear or interactive models, potentially reducing accuracy.

Incomplete Explanations: PDPs offer partial insights into the effects of individual features on predictions, omitting the full scope of model behavior.

Oversimplification: PDPs may oversimplify feature-prediction relationships, leading to potentially misleading interpretations.

Solutions:

Develop hybrid approaches that combine Partial Dependency Plots with other XAI techniques, such as Shapley values or individual conditional expectation (ICE) plots, to improve the accuracy and interpretability of explanations.

5.5 Ethical Considerations and Bias

Explaining Bias: XAI strategies can also amplify or maybe introduce biases into the reasons they generate.

Transparency and Privacy: Balancing transparency with records privateness and security may be hard, specifically when touchy data is worried.

Solutions:

Develop truthful XAI techniques which might be strong to bias and provide impartial factors.[42].

Implement privateness-retaining XAI techniques that guard sensitive records and user privacy.

Establish ethical suggestions and guidelines for XAI development and deployment.

By actively addressing these boundaries, we can pave the manner for more correct, understandable, and ethical XAI strategies that make contributions to the responsible and useful use of AI in society.

6 Conclusion and Future Scope

The "black box" problem has emerged as a significant challenge hindering the widespread adoption and ethical implementation of AI. As discussed throughout this paper, opaque decision-making in AI models raises concerns about trust, fairness, accountability, and hinders both research and development. Explainable AI (XAI) techniques offer a promising solution to address these critical issues.

This paper has explored the various ramifications of the "black box" problem in AI. We examined how the lack of transparency in model decision-making can lead to reduced user acceptance, perpetuate biases, hinder accountability, and impede advancements in the field. Subsequently, we delved into several XAI techniques that hold promise for mitigating these challenges. Techniques like model-agnostic methods, feature importance analysis, counterfactual explanations, and model visualization offer valuable insights into the inner workings of AI models.

- While current XAI techniques offer significant advancements, there's still room for improvement. Future research should focus on:
- Enhancing Explainability: Developing more comprehensive and user-friendly XAI methods that cater to diverse audiences with varying technical backgrounds.
- Addressing Algorithmic Bias: Refining XAI techniques to better detect and mitigate biases embedded within algorithms themselves.
- Human-Centered Design: Integrating human-centered design principles into XAI to ensure explanations are not only accurate but also actionable and informative for users.

By actively addressing the limitations of XAI and promoting responsible development practices, we can foster a future where AI is not just powerful but also transparent, trustworthy, and beneficial to society. This will ensure that AI serves as a force for good, empowering humans to make informed decisions and solve complex challenges. By working towards more robust and comprehensive XAI techniques, we can unlock the full potential of AI and ensure its ethical and responsible implementation across various domains.

References

1. Adadi, A., Berrada, M.: Peeking inside the black box: a survey on explainable artificial intelligence (XAI). IEEE Access **6**, 52138–52160 (2018)
2. Char, D.S., et al.: Explainable AI (XAI): a review of the literature and opportunities for healthcare. Artific. Intell. Med. **102**, 103426 (2020)
3. Lime: explaining the predictions of any machine learning classifier. GitHub. https://github.com/marcotcr/lime
4. Hui, A., Ahn, S., Lye, C., Deng, J.: Ethical challenges of artificial intelligence in health care: a narrative review. Ethics Biol. Eng. Med.: An Int. J. (2022). https://doi.org/10.1615/EthicsBiologyEngMed.2022041580
5. Brożek, B., Furman, M., Jakubiec, M., Kucharzyk, B.: The black box problem revisited. Real and imaginary challenges for automated legal decision making. Artificial Intell. Law **32**, 1–14 (2023). https://doi.org/10.1007/s10506-023-09356-9

6. Langer, M., et al.: What do we want from explainable artificial intelligence (XAI)? A stakeholder perspective on xai and a conceptual model guiding interdisciplinary XAI research. Artif. Intell. (2021). https://doi.org/10.1016/j.artint.2021.103473
7. Lundberg, S.M., Lee, S.-I.: Search for "Lundberg SM, Lee SI. Towards a science of explainable artificial intelligence. In: Proceedings of the International Conference on Artificial Intelligence and Statistics. PMLR vol. **51**, pp. 200–208 (2016)
8. Miller, T.: Search for Miller T. A framework for understanding artificial intelligence explainability. arXiv preprint arXiv:1606.08849 (2016)
9. Amodei, D., Been Kim, J., Celeste, C.K., et al.: Concrete problems in AI safety. arXiv preprint arXiv:1606.06565 (2016)
10. Ribeiro, M.T., Sameer, S., Carlos, G.: Why should I trust you?: explaining the predictions of any classifier. In: Proceedings of the 22nd ACM SIGKDD International Conference on Knowledge Discovery and Data Mining, pp. 1135–1144 (2016)
11. Lundberg, S.M., Lee, S.-I.: Search for "Lundberg SM, Lee SI. A unified approach to interpretable machine learning models. arXiv preprint arXiv:1704.07974 (2017)
12. Pearl, J.: Causality: Models, Reasoning, and Inference. Cambridge University Press (2000). https://doi.org/10.1017/S0266466603004109
13. Vaswani, A., et al.: Attention is all you need, NIPS'17: Proceedings of the 31st International Conference on Neural Information Processing Systems December, pp. 6000–6010 (2017)
14. Loyola-González, O.: Black-box vs. white-box: understanding their advantages and weaknesses from a practical point of view. IEEE Access. **7**, 154096–154113 (2019). https://doi.org/10.1109/ACCESS.2019.2949286
15. Kumar, A.: Feature contribution to an in-depth understanding of the machine learning model interpretation. Przegląd Elektrotechniczny **1**(2), 147–150 (2024). https://doi.org/10.15199/48.2024.02.29
16. Rajjliwal, N.S., Chetty, G.: Cardiovascular disease detection based on interpretable and explainable AI. In: 2022 IEEE Asia-Pacific Conference on Computer Science and Data Engineering (CSDE), pp. 1–7. Gold Coast, Australia (2022). https://doi.org/10.1109/CSDE56538.2022.10089349
17. Guleria, P., Naga Srinivasu, P., Ahmed, S., Almusallam, N., Alarfaj, F.K.: XAI framework for cardiovascular disease prediction using classification techniques. Electronics **11**(24), 4086 (2022). https://doi.org/10.3390/electronics11244086
18. Hagan, R., Gillan, C.J., Mallett, F.: Comparison of machine learning methods for the classification of cardiovascular disease. Informat. Med. Unlocked **24**, 100606 (2021). https://doi.org/10.1016/j.imu.2021.100606
19. Mridha, K., Kuri, A.C., Saha, T., Jadeja, N., Shukla, M., Acharya, B.: Toward explainable cardiovascular disease diagnosis: a machine learning approach. In: Chaki, N., Roy, N.D., Debnath, P., Saeed, K. (eds.) Proceedings of International Conference on Data Analytics and Insights, ICDAI 2023, pp. 409–419. Springer Nature Singapore, Singapore (2023). https://doi.org/10.1007/978-981-99-3878-0_35
20. Moreno-Sanchez, P.A.: Development of an explainable prediction model of heart failure survival by using ensemble trees. In: 2020 IEEE International Conference on Big Data (Big Data), pp. 4902–4910. Atlanta, GA, USA (2020). https://doi.org/10.1109/BigData50022.2020.9378460
21. Bharati, S., Mondal, M.R.H., Podder, P.: A review on explainable artificial intelligence for healthcare: why, how, and when? IEEE Trans. Artific. Intell. **5**(4), 1429–1442 (2024). https://doi.org/10.1109/TAI.2023.3266418
22. Gautam, N., Mueller, J., Alqaisi, O., et al.: Machine learning in cardiovascular risk prediction and precision preventive approaches. Curr. Atheroscler. Rep. **25**, 1069–1081 (2023). https://doi.org/10.1007/s11883-023-01174-3

23. Gupta, S., Bharti, V., Kumar, A.: A survey on various machine learning algorithms for disease prediction. Int. J. Recent Technol. Eng. **7**, 84–87 (2019)
24. Samek, Wojciech, Artur Krasuski, and Thomas Roth. Explainable machine learning for data scientists. Springer Nature Switzerland AG, 2022
25. Pearl, J.: Causality: Models, Reasoning, and Inference. Cambridge University Press (2009). https://doi.org/10.1017/CBO9780511803161
26. Molnar, C.: Interpretable Machine Learning. Lulu.com (2019)
27. Friedman, J.: Greedy function approximation: a gradient boosting machine. Ann. Stat. (2000). https://doi.org/10.1214/aos/1013203451
28. Guidotti, R., Monreale, A., Ruggieri, S., Turini, F., Giannotti, F., Pedreschi, D.: A survey of methods for explaining black box models. ACM Comput. Surv. **51**(5), 1–42 (2018). https://doi.org/10.1145/3236009
29. Saeed, W., Omlin, C.: Explainable AI (XAI): A systematic meta-survey of current challenges and future opportunities
30. Chen, S., Zhao, Q.: REX: Reasoning-aware and grounded explanation. In: Proceedings of the IEEE/CVF Conference on Computer Vision and Pattern Recognition, pp. 15586–15595 (2022)
31. Guidotti et al.: A Survey of Methods for Explainable Artificial Intelligence (2023). https://arxiv.org/pdf/2101.09429
32. Liu, et al.: A Survey of Privacy-Preserving Explainable Artificial Intelligence (2023). https://arxiv.org/abs/2204.08237
33. Miller, et al.: Explainable AI: A User-Centric Perspective (2023). https://arxiv.org/abs/2204.03209
34. Arrieta, et al.: Explainable Artificial Intelligence (XAI): A Review. https://arxiv.org/abs/2006.00093 (2023)
35. Xu, et al.: Adversarial Attacks and Defenses in Explainable AI: A Survey (2023). https://arxiv.org/abs/2204.03232
36. Rudin, C.: Stop explaining black box machine learning models for high-stakes decisions and use interpretable models instead. Nat. Mach. Intell. **1**(5), 206–215 (2019)
37. Fisher, A., Rudin, C., Dominici, F.; Model class agnostic feature importance with monotonic causal effects. In: Advances in Neural Information Processing Systems, pp. 7835–7846 (2019)
38. Wachter, S., Mittelstadt, B., Russell, C.: Counterfactual explanations without opening the black box: Building trust in machine learning models. In: Proceedings of the 2017 ACM Conference on Computer and Communications Security, pp. 1179–1190 (2017)
39. Kahng, M., et al.: ActiVis: Visual exploration of deep learning models for improved understanding and debugging. IEEE Trans. Visualization Comput. Graphics **25**(8), 2602–2610 (2018)
40. Hui, A.T., Ahn, S.S., Lye, C.T., Deng, J.: Ethical challenges of artificial intelligence in health care: a narrative review. Ethics Biol. Eng. Med. **12**(1), 55–71 (2021)
41. Dave, D., Naik, H., Singhal, S., Patel, P.: Explainable AI meets healthcare: a study on heart disease dataset. Comput. Sci. **03195**(2011), 1–23 (2020)
42. Barocas, S., Hardt, M., Narayanan, A.: Fairness and machine learning (2019). arXiv preprint arXiv:1901.04039.

Design of PID Controller for Speed Control of Hybrid Electric Vehicle

Samir Bag[1], Rosy Pradhan[1](✉), and Santosh Kumar Majhi[2]

[1] Department of Electrical Engineering, Veer Surendra Sai University of Technology, Burla, Odisha 768018, India
rosy.pradhan88@gmail.com

[2] Department of Computer Science and Information Technology, Guru Ghasidas Vishwavidyalaya (Central University), Bilaspur, (C. G) 495009, India

Abstract. The enormous growth and demand for hybrid electric vehicle (HEV) technology leverage modern, efficient control methods in the on-demand HEV system. This work introduces a novel approach to controlling the speed of the HEV by implementing a PID controller with a distinct metaheuristic technique. The PID controller parameters are tuned using metaheuristic algorithms. Moreover, to optimize the parameters of the controller, the Honey Badger Algorithm (HBA) technique using PID controller is developed for regulating the speed of a HEV. Various objective functions such as IAE, ISE, ITAE, and ITSE are used to minimize the controller error. The efficacy of the proposed methods is measured in terms of the transient and steady-state analysis of the system, which is evaluated based on its performance index. The proposed controller gives optimum speed control of the HEV system using tuned values for the better result.

Keywords: PID Controller · IAE · ISE · ITAE · ITSE · HBA Algorithm

1 Introduction

Nowadays, the whole world is talking about the sustainability of vehicle fuel. The vehicles that have been in use for many decades are fossil fuels. The fossil computing of fossil fuels raises environmental concerns, and its sustainability will also be a challenge in the future. Therefore, the demand for alternatives to fossil fuels has necessitated the desire for electric vehicle (EV) technology in the market.

In recent days, EVs have gained momentum because of their high performance, relatively low maintenance cost, pollution-free operations, and ease of operation. EVs are preferred in most smart urban cities and semi-urban cities due to the very low pollution in terms of air and noise. In EV technology, the propulsion system plays a vital role in deciding the efficacy of the EV system. Currently, researchers both in industry and academia are focusing on the development of the control strategies of the EV propulsion system. Moreover, more dynamic performance and efficient energy management are the primary issues in the EV system, which is the focus of the researchers.

B. K. Mishra et al. (Eds.): ICT4GS 2024, ISEM 56, pp. 306–322, 2025.
https://doi.org/10.1007/978-3-031-94997-5_27

The EV system is highly non-linear, time-dependent, and unreliable because of fluctuating road conditions, motor settings, and outside disturbances. Scientists and researchers developed the idea of HEVs to address all these problems [1–6]. HEVs can be plug-in, series, parallel, series-parallel, or complicated, depending on how they are built. It is commonly an intricate system because it has a combination of mechanical and electrical components, making it a form of interdisciplinary technological advancement.Advanced control system technology is used based on diverse, linked elements that determine performance [7–9].

It is propelled by an electric motor and a traditional internal combustion engine (ICE). Hazardous exhaust pollutants and fuel consumption are guaranteed to decrease with the hybrid model. When compared to the car and EV, it offers several great advantages. It is employed in motor systems and electrical energy storage. ICE supplies the mechanical power needed to move the car. HEVs, on the other hand, need power produced by an engine in addition to another energy source [10, 11]. In the case of the HEV, the ICE has a throttle valve. An ICE manifold's throttle valve allows air to enter. The throttle position, bar, or plate manages the throttle valve's airflow. The Electronic Throttle Control System (ETCS) regulates this throttle position.

2 Objective of the Work

The main objective is to focus on the PID controller's efficient design and circuit realization for HEV speed control. In this context, many optimization approaches, including HBA, ALO, and GWO, are employed to fine-tune the settings of the PID controller. The evaluation of the suggested controller's performance for the speed control of HEV is showcased and compared.

This project offers a comprehensive understanding of the design and adjustment of the PID controller, employing various optimization strategies to minimize error and provide optimal speed control for HEVs.

The remaining part of this paper is organized as follows: Section 3 describes the system modeling of HEV. Section 4 explains the design and tuning of the proposed algorithm for the PID controller; Section 5 shows the results and summary, and Section 6 contains the conclusion.

3 System Overview

3.1 Model Description of HEV

Figure 1 illustrates the HEV architecture within a control system with a closed loop. Information flows through the following: reference input (Rd), output (y), error signal (e), controller, plant, disturbance (Kg), and the controlling line. The performance of HEVs is influenced by various factors such as the vehicle's weight, the load it carries, and the road conditions including the gradient and surface.

Newton's second law states that the change in vehicle speed may be predicted based on the force applied. The vehicle's dynamic equation is defined as:

$$\omega \frac{ds}{dt} = Ke(\theta) - \sigma s^2 - K_g \tag{1}$$

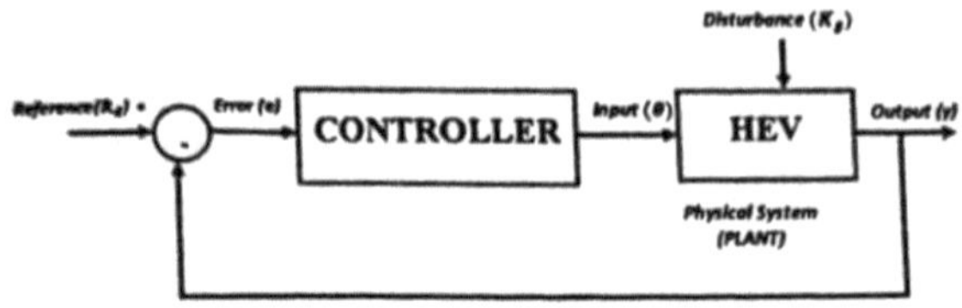

Fig. 1. Control diagram of HEV

$$\delta_e \frac{dKe(\theta)}{dt} = -Ke(\theta) + K_i(\theta)g \tag{2}$$

$$K_i(\theta) = -K_i + \gamma\sqrt{\theta} \tag{3}$$

where,

Ke = Position of the throttle value of ICE
K_g = Gravity induced force, road grade function
S = vehicle speed
θ = Throttle position
ω = Mass of the vehicle
δ_e = Time constant of engine

Table 1. Numerical Values of HEV.

Idle force of the engine (Ki)	6400N
Time Constant of the engine (δ_e)	0.2 s
Coefficient of the Engine force (γ)	12500N
Mass of the vehicle (ω)	1000kg
Drag coefficient (σ)	$\frac{4N}{(\frac{m}{s})^2}$

All the numerical values considered to develop the HEV are used in Table 1.

The mathematical equation is also nonlinear. The state-space analysis is one of the best methods to solve this nonlinear system. By using this method, we simplify and get the required parameters. Classical methods of a system analysis so far discussed are valid for LTIS, whereas State Space Analysis (SSA) is valid for nonlinear and time-variant systems and LTIS. So, the linearized model of system obtained is explained [18].

The transfer function of the system obtained as:

$$G(s) = \frac{\{0.01292s^3 + 0.005944s^2 + 0.0004034s + 1.836e - 05\}}{\{s^5 + 0.2985s^4 + 0.1139s^3 + 0.01532s^2 + 0.001381s + 4.641e - 05\}} \tag{4}$$

3.2 PID Controller

PID Controller is most widely used for industrial purposes because PID can be adjusted on-site. PID controller has several advantages, which give a combination of both PD and PI controller. It is employed to both increase stability and decrease steady-state error. There are two zeros and a pole at the origin. The stability is increased by one zero, while the pole is compensated for by the other. The proportional integral derivative controller's transfer function is

$$G_c(s) = \left(K_p + \frac{K_i}{S} + K_d S\right) = K_p\left(1 + \frac{1}{T_i S} + T_d S\right) \tag{5}$$

where,

$T_i = \frac{K_p}{K_i}$ is a rest or integral time
$T_d = \frac{K_d}{K_p}$ is a derivative or rate time
K_p is proportional gain, K_d is derivative gain

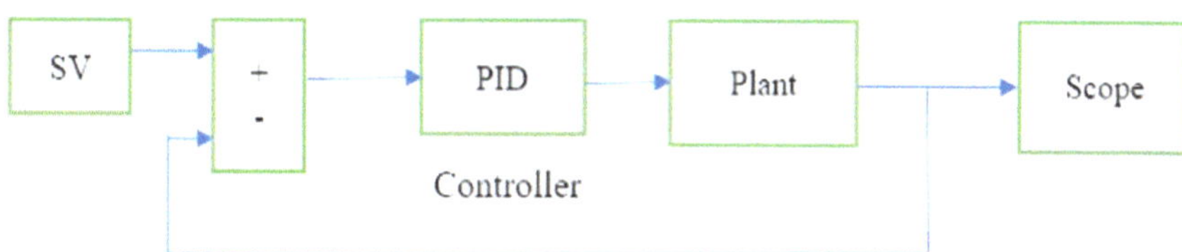

Fig. 2. Simulink model of PID Controller for HEV

In this work, we employ the HBA, ALO, and GWO Algorithms to determine the various PID controller parameters, suchas K_p, K_i *and* K_d while establishing an ideal outcome, we address it in a later section. Figure 2 illustrates the Simulink model of a HEV equipped with a PID controller.

4 Implementation and Optimization of the Proposed Technique for PID Controller

4.1 Honey Badger Algorithm (HBA)

HBA is a newly created metaheuristic approach that imitates the hunting habits of honey badgers. The honey badger adopts two distinct methods to locate its food: it either trails the honeyguide bird or utilizes its sense of smell to locate and excavate its prey, referred to as honey mode and digging mode, respectively. During the digging mode, honey badger utilizes its olfactory senses to identify its prey. Once it has found the prey, it proceeds to encircle it to determine the optimal location for digging and grabbing it. The honey badger utilizes the honeyguide bird's guidance in honey mode to locate the beehive with precision. All the key steps in the HBA method have a mathematical model available [12].

Mathematical Model of HBA. The technical details, including the mathematical formulation of the HBA algorithm, are discussed in this section. The HBA balances the exploration and exploitation phase to achieve global optimization value.

In HBA, the population of possible solutions is depicted as follows:

$$Population\ of\ candidate\ solutions = \begin{bmatrix} x_{11} & x_{12} & x_{13} & \cdots & x_{1D} \\ x_{21} & x_{22} & x_{23} & \cdots & x_{2D} \\ \vdots & \vdots & \vdots & \vdots & \vdots \\ x_{n1} & x_{n2} & x_{n3} & \cdots & x_{nD} \end{bmatrix}$$

$$i^{th}\ position\ of\ honey\ badger\ x_i = [x_i^1, x_i^2, \ldots\ldots\ldots x_i^D$$

The mathematical equations of the hunting nature of honey badgers are

$$x_i = lb_i + r_1 \times (ub_i - lb_i), r_1 \tag{6}$$

$$l_i = r_2 \times \frac{S}{4\pi d_i^2} \tag{7}$$

$$S = (x_i - x_{i+1})^2 \tag{8}$$

$$d_i = x_{prey} - x_i \tag{9}$$

$$x_{new} = x_{prey} + F \times \beta \times I \times x_{prey} + F \times r_3 \times a \times d_i \times |\cos(2\pi r_4) \times [1 - \cos(2\pi r_5)]| \tag{10}$$

$$x_{new} = x_{prey} + F \times r_7 \times a \times d_i \tag{11}$$

Here, x_i denotes the ith honey badger position, N is the population size, and r_2 is a random number between 0 and 1

S denotes the source and d_i denotes the distance between the prey and the ith badger
x_{prey} denotes the position of the prey
x_{new} denotes the flag that alters the search direction
F denotes the flag that alters the search direction

Dimension-wise diversity in the search process can be quantified as

$$Div_j = \frac{1}{N}\sum_{i=1}^{N} median\left(x^j\right) - x_i^j \tag{12}$$

$$Div^t = \frac{1}{D}\sum_{j=1}^{D} Div_j \tag{13}$$

$$Exploration\% = \frac{Div^t}{\max(Div)} \times 100 \tag{14}$$

$$Exploitation\% = \frac{\left|Div^t - \max(Div)\right|}{\max(Div)} \times 100 \tag{15}$$

where,

x_i^j denotes the jth dimension of ith honey badger position
$median(x^j)$ denotes the median value of jth dimension in the population of N candidate solutions
Div_j is mean diversity for dimension j max(Div) is the maximum diversity in t_{max} iterations

4.2 Ant Lion Optimization (ALO)

Ant Lion Optimization (ALO) is recently developed nature-inspired optimization method based on the chasing behavior of the grey ant lions. The prey of the ant lions is the ants. The hunting process of an ant lion is based on digging a conical pit and disguising the larva under it to catch the ant. The trap behavior of the ant lion is described and mathematically modeled in five steps: randomly walking of ants, construction capture, setup of ants in entrap, grab preys, and reconstruction of entraps. The mathematical model of all the important five steps of the ALO algorithm is given [13].

Random Walks of Ants: The prey-catching behavior of the ant lion is mathematically modeled by considering the interaction between the ant lion and the ant. The ants move towards the conical pit trapped by an ant lion to find food and shelter; in turn, ant lions hunt the ant. The movement of ants is captured using a random walk and mathematically represented in Eq. 16.

$$A(t) = \left[0, yadd(2ft_1), \ldots\ldots yadd(2ft_n - 1)\right] \tag{16}$$

where n and t represent max. Iteration and random walks of ants, yadd is average sum, f(t) is 1 when m greater than 0.5 and 0 if less than equals to 0.5.

Random walks of ant can be write in mathematically

$$A_i(t) = \frac{(A_i^t - P_i)x(S_i^t - f_i^t)}{(q_i - p_i)} + f_i^t \tag{17}$$

where, p_i and q_i are the min and max value of ant's walk respectively. $f_i\prime$ and $S_i\prime$ are the min. And max. Value of ith variable and t[th] iteration.

$$g_i^t = antlion_j^t + g^t \tag{18}$$

$$h_i^t = antlion_j^t + h^t \tag{19}$$

where, g' and h' are min. And max. Variables respectively at t^{th} iteration. j^{th} Position of an ant represents with $antlion_j^t$ with t^{th} iteration.

$$antlion_j^t \geq ant_j^t \tag{20}$$

Equation 3.15 satisfy when $ant_j^t \geq antlion_j^t$ where

$$ant_j^t = \frac{L_A' + L_E'}{2} \tag{21}$$

where L_A' *and* L_E' are random walk of antlion and elite chosen by roulette wheel at t^{th} iteration.

4.3 Grey Wolf Optimization (GWO)

Grey wolf's rank are decided w.r.t. their hunting and their responsibilities. Generally, there are four kinds of a wolf named $\propto$, β, δ *and* μ. Where α is the highest priority to the wolves or called group leaders. In a group, there are 10 to 15 members are there, α is generally a male and a female. For any group α, take the responsibilities such as the decision in the groups, where to halt, where to sleep, and how much time to sleep [14].

β is the second-level wolves of the group. These are subordinate and help the α for the different activities. These are maybe a male or female and the best candidate for the $\propto$. If $\propto$ becomes older, then the β is taking all the decisions with the help of lower groups of wolves. It plays an advisory role in the pack.

δ are exceptional case which does not belong to the $\propto$, β *and* μ. This category submits the $\propto$ *and* β but is dominant to the μ. This category takes the action of Scouts, sentinels, elders, hunters, and caretakers. Scouts play a major role in the border security of the Union Sentinels and re the safety pack of the group. Experienced wolves are elders help to the $\propto$ β. Hunters are responsible for hunting prey and helping the $\propto$ *and* β by providing food to the entire group.

Their group activities and their hunting nature are observed, and an optimization technique called GWO is observed. These optimization techniques were developed by S. Mirjalili in 2015. All mathematical equations are based on their hunting prey.

Pseudo Code for GWO Algorithm

Set population of grey wolf $X_i = (i = 1,2,3,4, \ldots \ldots n)$

Initialize the parameters a, A and C

Find the fitness status for every search agent

$X_{\propto} = best\ search\ agent$

$X_{\beta} = second\ best\ search\ agent$

$X_{\delta} = third\ best\ search\ agent$

$\boldsymbol{while}\ (t < Max\ number\ of\ repeatitions)$

$\boldsymbol{for}$ each search agent

Revise the location of the existing search agent.

end $\boldsymbol{for}$

Revise a, C and A

Evaluate the fitness status of all search agents.

Revise $\vec{X_{\alpha}}, \vec{X_{\beta}}\ and\ \vec{X_{\delta}}$

$i = t + 1$

end $\boldsymbol{while}$

return X_a

4.4 Problem Formulation and Performance Criteria

Objective functions are used to reduce the error and get optimum control in the controller. Equations (22)–(25) are objective functions used to minimize the error in the PID controller.

Integral Absolute Error(IAE)

$$F = \int_0^\infty |e(t)|dt \tag{22}$$

Integral Square Error (ISE)

$$F = \int_0^\infty \left|e^2(t)\right|dt \tag{23}$$

Integral of Time multiplied Absolute Error (ITAE)

$$F = \int_0^\infty t|e(t)|dt \tag{24}$$

Integral of Time multiplied Square Error (ITSE)

$$F = \int_0^\infty t\left|e^2(t)\right|dt \tag{25}$$

The optimization problem for the PID controller design is formulated as minimize the function F is subjected to the range

$$K_{pmin} < K_p < K_{pmax}$$
$$K_{imin} < K_i < K_{imax}$$
$$K_{dmin} < K_d < K_{dmax}$$

PID has three unknown parameters, which is used to find with three algorithms are HBA, ALO and GWO.

5 Result and Summary

This section provides an explanation and examination of the reaction and evaluation of speed control in HEVs. The model is designed and considered for their responses. Transient analysis of the system and their performance index such as rise time (*tr*), settling time (*ts*) and maximum overshoot (*Mp*) is described.

HBA-PID Controller: This section uses all objective functions, *i.e.*, IAE, ISE, ITAE and ITSE to minimize the error. The proposed model is developed with unit step. Parameters of the HBA-PID controlled with all response with different objective function and their performance index is mentioned (Figs. 3, 4, 5 and 6).

Figures 3, 4, 5 and 6 shows the response of HEV, where IAE, ISE, ITAE and ITSE are the objective functions. Obtained responses are very smooth so they can be achieved at the very smooth speed for HEV (Tables 2 and 3).

ALO-PID Controller: This section of PID is designed with an optimization technique called Ant Lion Optimization (ALO). In this case, IAE, ISE, ITAE and ITSE are the objective function used to reduce the error. Obtained parameters of the controller and response of the system are mentioned (Figs. 7, 8, 9 and 10 Tables 4 and 5).

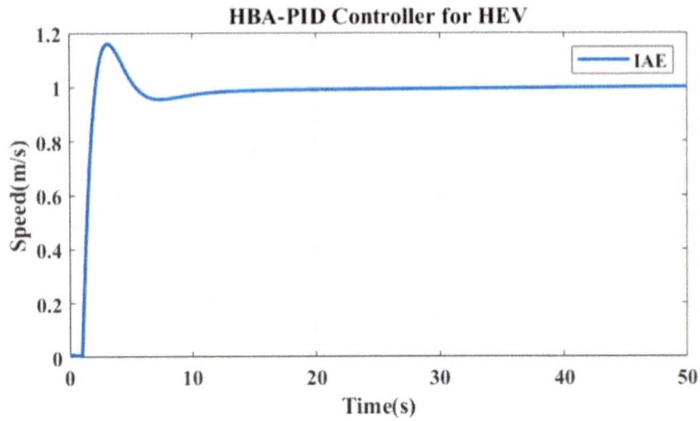

Fig. 3. Unit Step response with IAE

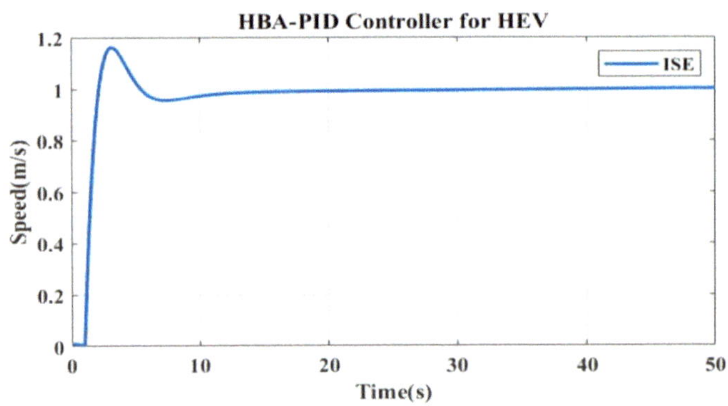

Fig. 4. Unit Step response with ISE

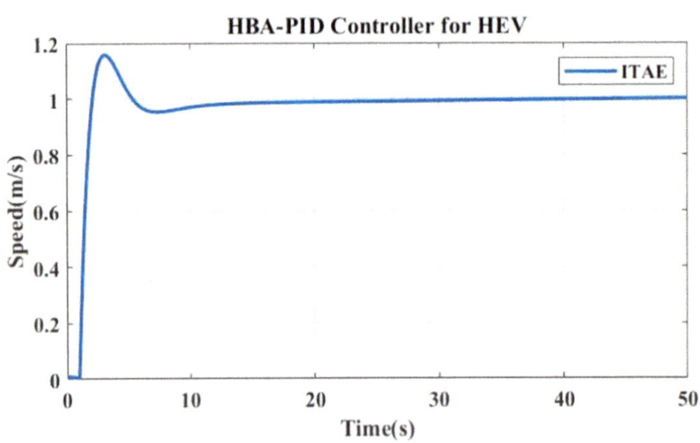

Fig. 5. Unit Step response with ITAE

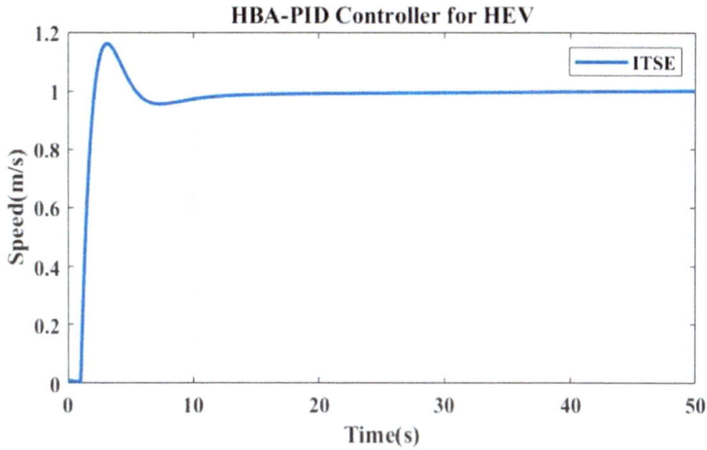

Fig. 6. Unit Step response with ITSE

Table 2. PID Controller parameters with all Objective function tuned by HBA

Parameters of PID Controller	IAE	ISE	ITAE	ITSE
K_p	51.882	51.417	51.814	51.764
K_i	4.298	4.780	3.157	4.967
K_d	117.597	117.234	115.655	117.862

Table 3. Performance index of HBA-PID Controller

HBA-PID with objective functions	Maximum Peak overshoot (%)	Rise Time(s)	Settling Time(s)
IAE	15.925	0.830	12.078
ISE	16.013	0.833	11.754
ITAE	15.327	0.806	11.209
ITSE	16.034	0.830	11.615

GWO-PID Controller: GWO is based on the wolves hunting nature towards the prey. This hunting nature is designed with algorithms. Generally, four-category wolves called $\propto, \beta, \delta$ *and* μ.

This section used population-based met heuristics optimization techniques; Grey Wolves Optimization tuned the PID controller. PID has three unknown parameters. Obtained parameters and responses are explained (Figs. 11, 12, 13 and 14, Tables 6 and 7).

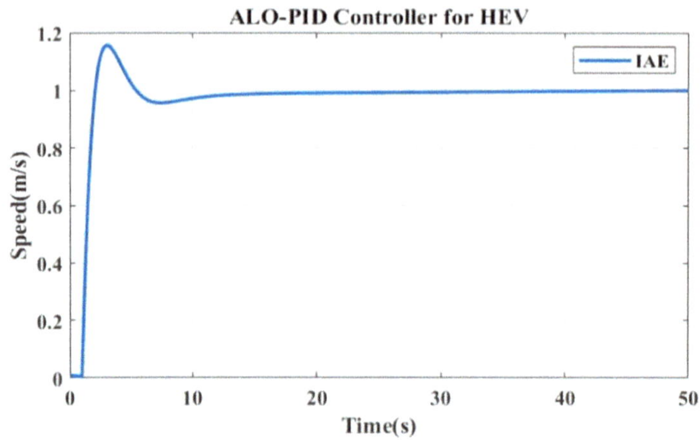

Fig. 7. Unit Step response with IAE

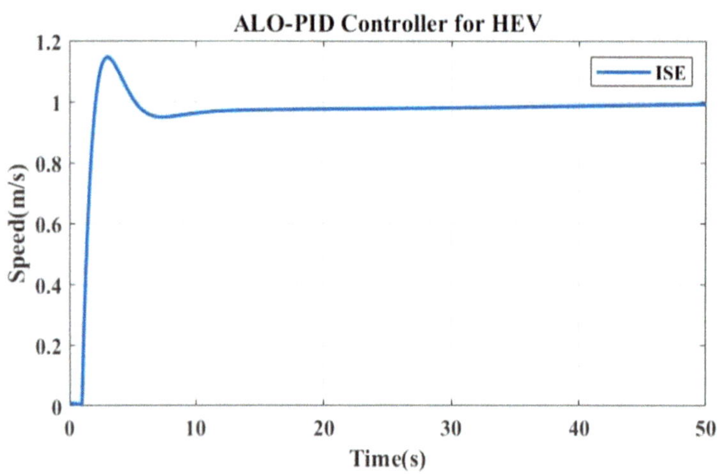

Fig. 8. Unit Step response with ISE

The proposed model and its responses are characterized. Finally, it can be concluded from the result Table 8 that for PID controller HBA with ITAE gives the better performance and optimum responses as compared to the other designed techniques.

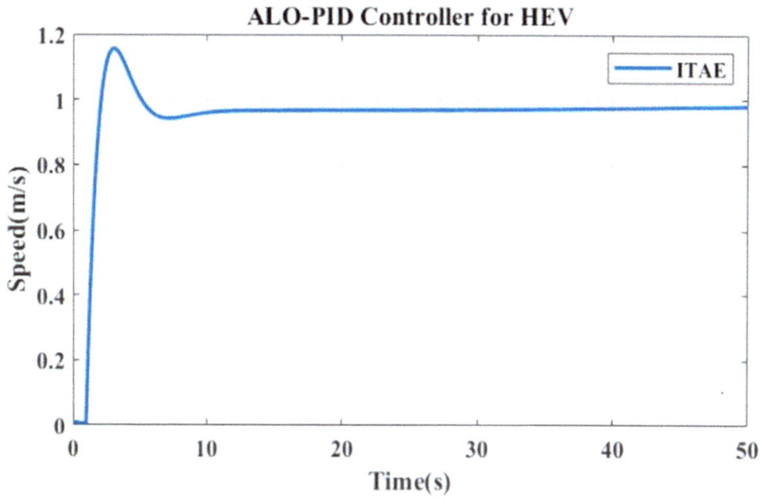

Fig. 9. Unit Step response with ITAE

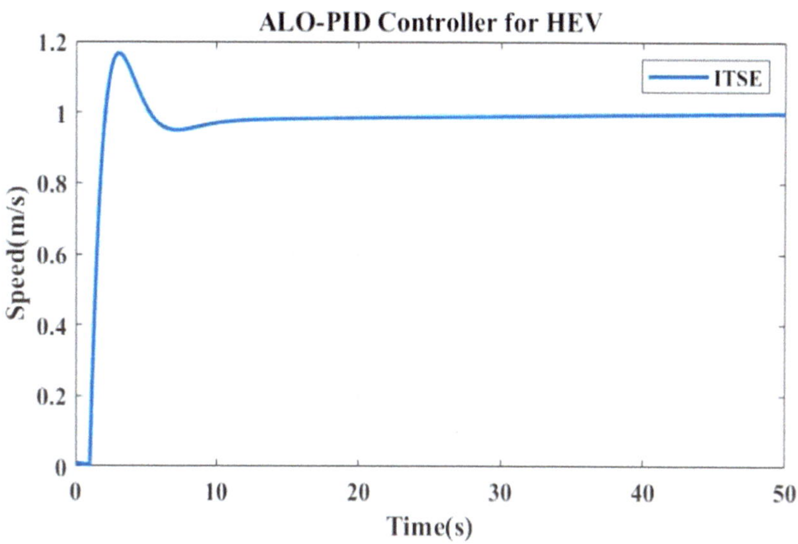

Fig. 10. Unit Step response with ITSE

Table 4. PID controller parameters with all Objective function by ALO

Parameters of PID Controller	IAE	ISE	ITAE	ITSE
K_p	51.091	50.877	54.349	54.945
K_i	4.999	2.037	1.215	3.536
K_d	120	119.770	115.052	115.244

Table 5. Performance index of ALO-PID Controller

ALO-PID with objective functions	Maximum Peak overshoot (%)	Rise Time(s)	Settling Time(s)
IAE	15.549	0.825	11.718
ISE	15.506	0.816	13.958
ITAE	17.859	0.806	10.319
ITSE	16.868	0.826	12.088

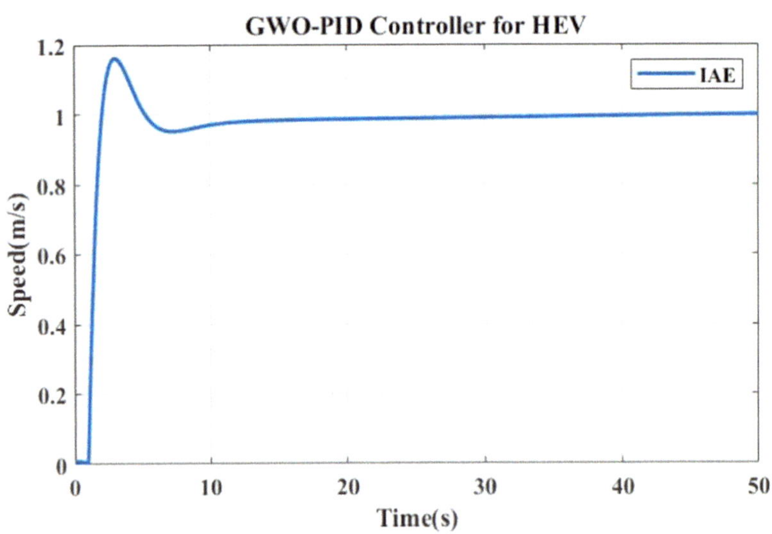

Fig. 11. Unit Step response with IAE

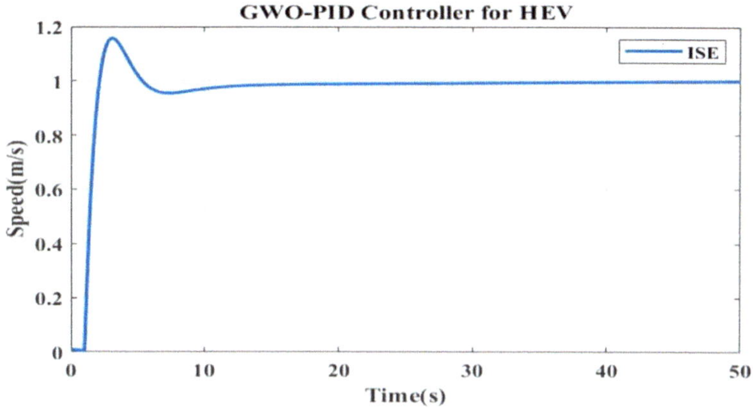

Fig. 12. Unit Step response with ISE

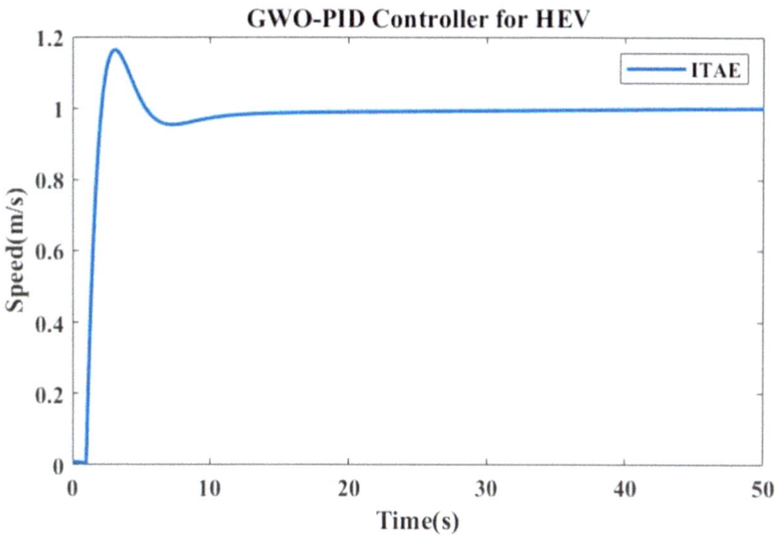

Fig. 13. Unit Step response with ITAE

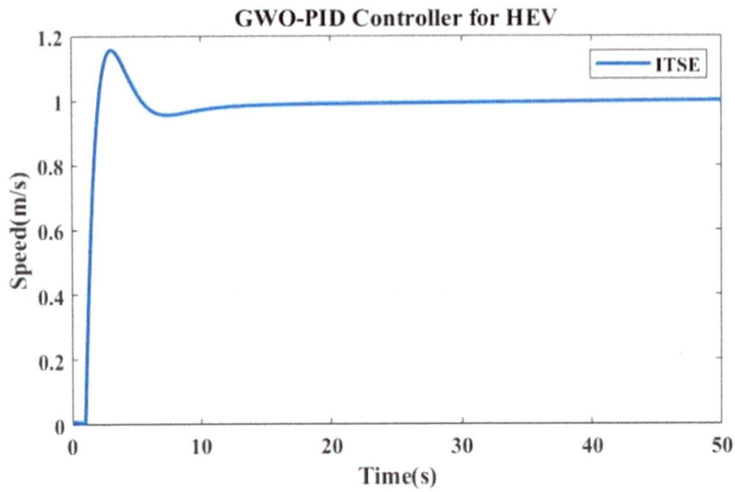

Fig. 14. Unit Step response with ITSE

Table 6. PID controller parameters with all Objective function tuned by GWO

Parameters of PID Controller	IAE	ISE	ITAE	ITSE
K_p	54.142	51.804	53.922	51.070
K_i	3.656	4.577	4.824	4.762
K_d	117.706	119.499	118.558	119.306

Table 7. Performance index of GWO-PID Controller

GWO-PID with objective functions	Maximum Peak overshoot(%)	Rise Time(s)	Settling Time(s)
IAE	16.283	0.820	12.278
ISE	15.672	0.824	11.930
ITAE	16.326	0.819	11.498
ITSE	15.589	0.827	11.874

Table 8. Parameter and Performance index of PID Controller

PID Paramerter					Performance index of controllers		
Algorithms	Objective Function	K_p	K_i	K_d	Max. Peak Overshoot (%)	Rise Time (s)	Setting Time (s)
HBA	IAE	51.882	4.298	117.597	15.925	0.83	**12.178**
	ISE	51.417	4.780	117.234	16.013	0.833	11.754
	ITAE	**51.814**	**3.157**	**115.655**	**15.370**	**0.806**	**11.209**
	ITSE	51.764	4.967	117.862	16.034	0.830	11.615
ALO	IAE	51.091	4.999	120	15.549	0.825	11.718
	ISE	50.877	2.037	119.770	15.506	0.816	13.958
	ITAE	54.349	1.215	115.052	17.859	0.806	10.319
	ITSE	54.945	3.536	115.244	16.868	0.826	12.088
GWO	IAE	54.142	3.656	117.706	16.283	0.820	12.278
	ISE	51.804	4.577	119.449	15.672	0.824	11.930
	ITAE	53.922	4.824	118.558	16.326	0.819	11.498
	ITSE	51.070	4.762	119.306	15.589	0.827	11.874

6 Conclusion

Through this research, a PID controller is developed to regulate the velocity of a HEV. The controller with a PID is constructed with metaheuristics-based techniques, which are very popular to solve real-life and engineering problems.

IAE, ISE, ITAE and ITSE, are used to reduce the error. PID controller designed with ITAE gives a better performance index (0.806 s. rise time, 11.209 s. settling time, and 15.370% max. Peak overshoot). By observing Table 8, the proposed model for the HEV PID controller has optimum performance in PID tuned with HBA, which designed with the objective function ITAE. According to the concept of theory it is practically acceptable.

Ultimately, after examining the complete suggested model, i.e., the PID controller tuned with HBA and constructed with the goal function ITAE demonstrated the most optimal response for HEV equipment when compared to alternative methods of control. As a result, the suggested controller regulates the throttle angle and is capable of achieving a more seamless speed, thereby controlling the velocity of a HEV.

References

1. Veysi, M., Aghaei, J., Shasadeghi, M., Razzaghi, R., Bahrani, B., Ryan, D.J.: Energy-efficient speed control of electric vehicles: linear matrix inequality approach. IEEE Trans. Veh. Technol. **69**(10), 10469–10483 (2020). https://doi.org/10.1109/tvt.2020.3008500

2. Kumar, V., Rana, K.P.S., Mishra, P.: Robust speed control of hybrid electric vehicle using fractional order fuzzy PD and PI controllers in cascade control loop. J. Franklin Inst. **353**(8), 1713–1741 (2016). https://doi.org/10.1016/j.jfranklin.2016.02.018
3. Khooban, M.H., Gheisarnejad, M., Vafamand, N., Boudjadar, J.: Electric vehicle power propulsion system control based on time-varying fractional calculus: implementation and experimental results. IEEE Trans. Intell. Veh. **4**(2), 255–264 (2019). https://doi.org/10.1109/TIV.2019.2904415
4. Sanjaka, G.W., Ali, E.: Classification and review of control strategies for plug-in hybrid electric vehicles. In: IEEE Transactions on Vehicular Technology, vol. 60, no. 1 (2011)
5. Kumar Yadav, A., Gaur, P., Kant Jha, S., Gupta, J.R.P., Mittal, A.P.: Optimal speed control of hybrid electric vehicles. J. Power Electron. **11**(4), 393–400 (2011)
6. Bishit, P., Yadav, J.: Optimal speed control of hybrid electric vehicle using GWO based fuzzy-PID controller. In: Proceedings-2020 International Conference on Advances in Computing, Communication and Materials, ICACCM 2020. pp. 115120 (2020). https://doi.org/10.1109/JICACCM50413.2020.9212985
7. Yang, C., Jiao, X., Li, L., Zhang, Y., Chen, Z.: A robust H. control-based hierarchical mode transition control system for plug-in hybrid electric vehicle. Mech. Syst. Signal Process. **99**, 326–344 (2018). https://doi.org/10.1016/j.ymssp.2017.06.023
8. Luo, C., Shen, Z., Evangelou, S., Xiong, G., Wang, F.Y.: The combination of two control strategies for series hybrid electric vehicles. IEEE/CAA J. AutomaticaSinica **6**(2), 596–608 (2019). https://doi.org/10.1109/JAS.2019.1911420
9. Conatsera, R., Wangerb, J., Gantab, S., Walker, I.: Diagnosis of automotive electronic throttle control system. Eng. Pract. **12**, 23–30 (2004)
10. Weng, C., Wang, Y., Tsourapas, V., Patil, C., Sun, J.: Optimal control of hybrid electric vehicles with power split and torque split strategies: a comparative case study. Eng. J. (2011). 10. 0/Linux-x86_64
11. Morteza, M.-G., Poursamad, A., Ghalichi, B.: Application of genetic algorithm for optimization of control strategy in parallel hybrid electric vehicles. J. Franklin Instit. **343**, 420–435 (2006)
12. Yuchen, L., Hess, H.L., Edwards, D.B.: Adaptive control of an ultracapacitor energy storage system for hybrid electric vehicles. IEEE Int. Electr. Mach. Drives Conf. **1**, 129–133 (2007)
13. Fatma, A.H., Essam, H.H., Hussain, K., Mabrouk, M.S., Al-Atabany, W.: Honey badger algorithm: new metaheuristic algorithm for solving optimization problems. Mathemat. Comput. Simulation **192**, 84–110 (2022)
14. Seyedali, M.: Ant Lion Optimization, In: Advances in Engineering Software, vol. 83, pp. 80–98 (2015). https://doi.org/10.1016/j.advengsoft.2015.01.010
15. Mirjalili, S., Mohammad Mirjalili, S., Lewis, A.: Grey wolf optimizer. In: Advances in Engineering Software, vol. 69, pp. 46-61 (2014).
16. Naik, K.P., Pradhan, R., Majhi, S.K.: Speed control of hybrid electric vehicle by using moth flame optimization. In: Kumar, A., Mozar, S., Haase, J. (eds.) Advances in Cognitive Science and Communications. ICCCE 2023. Cognitive Science and Technology. Springer, Singapore (2023). https://doi.org/10.1007/978-981-19-8086-2_81

Comparative Study of Thyroid Classification Using Deep Learning Tools

Swarnima Shrivastava[1](✉), Varsha Thakur[2], and Surendra Kumar Patel[1]

[1] Department of Information Technology, Government N.P.G. College of Science, Raipur, India
swarnima.shrivastava@gmail.com

[2] Department of Computer Science, Government N.P.G. College of Science, Raipur, India

Abstract. Nowadays thyroid disease has become very common but it is important to check it at the right time otherwise it can turn into some serious disease. It is easy to do a blood test to check thyroid. But it can also contain some errors like noisy data. Therefore, to find out the exact disease related to thyroid, deep learning algorithm should be used. In today's modern era, special contribution of deep learning algorithms is made to process vast data. There are some deep learning tools that can process large datasets accurately and efficiently. Some such attempts are made to optimize classification models by using deep learning tools with small datasets. In this paper thyroid dataset from Kaggle is executed on data mining tool WEKA and deep learning tool TensorFlow, Keras. WEKA gives an accuracy of 94.77% and TensorFlow 92.70%. Our objective is to experiment with both the tools and find a better tool for future research.

Keywords: Keras · PyTorch · MXNet · WEKA · TensorFlow

1 Introduction

The thyroid is an endocrine gland that regulates the body's iodine balance in addition to secreting thyroid hormones. The anterior, inferior neck is where it is located. The thyroid produces 90% of the passive thyroid hormone thyroxine (T4) and 10% of the working thyroid hormone triiodothyronine (T3). Peripherally, passive thyroid hormone can be converted into either activated thyroid hormone or alternative inactive thyroid hormone [1].

Since it controls the body's metabolism, among the most significant glands is the thyroid gland in the person body. It does this by releasing hormones into the bloodstream that regulate the body's functions. Hypothyroidism and hyperthyroidism are the two distinct hormone diseases. A specific hormone that controls the body's metabolism is released into the bloodstream by the thyroid gland in response to various illnesses. Thyroid problems can be exacerbated by autoimmune diseases, inflammation, and iodine shortage. A blood test is used to diagnose the illness, however noise and disturbance are often present. Data cleansing techniques can be applied to make analytics indicating the patient's risk of thyroid malady very easy to do [2]. Deep learning and machine learning are essential for identifying thyroid conditions.

B. K. Mishra et al. (Eds.): ICT4GS 2024, ISEM 56, pp. 323–334, 2025.
https://doi.org/10.1007/978-3-031-94997-5_28

Machine learning is the process of creating programs that can access and use data on their own. As a result, computers can improve at what they do and learn from their experiences without requiring explicit programming [3].

The foundation of deep learning is artificial neural networks. Within machine learning, deep learning is a subset that gradually extracts higher level characteristics from raw input by using multiple layers. Artificial Neural Network get their name from their artificial representation of the functioning of a human nervous system [4–6].

Deep learning is one of the newest and most interesting areas belongs to machine learning. The most efficient machine learning method in terms of value and time is deep learning, which operates under supervision. Instead of being a limited learning strategy, deep learning adheres to a diversity of topographies and procedures through which wide range of challenging issues can be used to solve. The method acquires the differentiating and illustrating properties in a highly layered manner. Deep learning tools achieved notable growth in a range of applications, demonstrating noteworthy performance and offering valuable security features. It is believed to be the best alternative used for finding intricate design in elevated dimensions data when imposing the back propagation. Broadly applied area deep learning, in the fields of science, trade and government [7].

Deep learning tools are helping in prediction and detection of thyroid disease more accurately. This paper gives a brief review about novel tools of deep learning technology. From these tools WEKA and TensorFlow is chosen to perform experimental work. Paper is divided in three sections I. Brief reviews of deep learning tools, Literature review and Experimental work using WEKA and TensorFlow, Keras.

Some popular deep learning tools are TensorFlow, Keras, PyTorch, MXNet, Theano, CNTK and WEKA.

2 Brief Reviews of Deep Learning Tools

TensorFlow The "Google Brain Team" developed the freely available deep learning package TensorFlow. It is among the most significant library. It functions with deep learning models such as CNNs, RNNs, and GANs [8]. An implementation for carrying out machine learning algorithms also an interface in support of expressing them are TensorFlow [9]. Small, portable gadget like mobile phones and tabs to enormous, dispersed systems with many of machines and many more peripherals say Graphic Processing Unit cards—all of these can conduct TensorFlow-expressed computations with little to no change. Deep neural network model training and inference algorithms are among the many algorithms that the system is able to represent. The TensorFlow API and reference implementation are accessible at www.tensorflow.org and were released as an Apache 2.0 license open-source package in November 2015 [9].

Keras Keras is python-written and employed with TensorFlow and other lower-level frameworks. Its user-friendliness, modularity, and extensibility were key design considerations in order to facilitate rapid prototyping and experimentation [10]. Keras aids numerous kinds of neural network modules, like convolution layer, thick layer, recurring layers, dropout layers, and their variants. To make the best employ of possessions like Graphics Processing Unit (GPU) and Central Processing Unit (CPU) and the programming manages them dynamically. Python users can access TensorFlow functions through a Deep Learning plug-in that combines features from Keras libraries [11].

Keras is a neural network API with a Python foundation that powers TensorFlow, CNTK, and Theano. Python uses the notation tf.keras when referring to Keras. It operates on four tenets [12]:-

1. Easy-to-use
2. Flexibility
3. Simple extension
4. Utilize Python

PyTorch Facebook's AI Research Lab (FAIR) has first created the freely available machine learning platform PyTorch and it is mostly used for deep learning and natural language processing applications. It's currently included in the Linux Foundation. It is renowned for having a continually evolving computational graph that allows for real-time code testing and adaptability. Models made using PyTorch are specified by taking a class derivative of torch. PyTorch's tensor computing adds automatic discrimination for GPU acceleration and neural network training [8]. PyTorch is a Python programming library designed to support deep learning applications. It is used for many different purposes and possesses several applications [13].

WEKA The University of Waikato in New Zealand created the Waikato Environment for Knowledge Analysis (WEKA) computer program with the goal of learning from unprocessed data collected from agricultural areas. WEKA is a cutting-edge laboratory for creating machine learning methods and using them to solve practical data mining issues. These machine learning techniques are designed to be used with data mining applications. The methods are immediately applied to the dataset. Numerous common data mining activities, including selection of features, initial processing, classification, clustering, regression, and visualization, are supported by WEKA. You be able to download WEKA for free, as long as you abide by the GNU common persons license. It is quick and simple to set up and run because to its graphical user interface [14].

MXNet A multilingual machine learning (ML) package called MXNet makes it easier to create ML algorithms, particularly for deep neural networks. It combines imperative tensor computation with declarative symbolic expression, all embedded within the host language. To derive gradients, it provides auto differentiation. MXNet operates on a variety of heterogeneous systems, from distributed GPU clusters to mobile devices, and is memory and computation efficient [15].

In contrast to other open-source machine learning systems, MXNet supports additional systems, including GPU clusters, and offers a superset programming interface to Torch7 [16]. While MXNet may also include imperative tensor operations, it is more similar to TensorFlow [9, 15].

3 Literature Review

Prathibha et al. [17] proposed a model which detect thyroid diseases. It uses Keras library as part of TensorFlow. In this paper CNN based modified ResNet is used with Adam and SGD optimizers are used to improve the performance. Multiple datasets are used to detect thyroid disease.The modified ResNet with dual optimizer gives better performance than basic ResNet.

J. Cao et al. [18] paper presented Tnc-Net for Automatic categorization of thyroid nodule. It uses 606 thyroid Doppler ultrasound images from 537 patients from Nanjiang TCM Hospital, Bazhong City, Sichuan Province. Deep learning library PyTorch is used. Model can perform well in processing small and unevenly classified ultrasonic image datasets. Tnc-Net model gaining an accuracy of 90.2% on the test set.

Gothane et al. [19] Used the WEKA tool 3.8, and classified the dataset Hypothroid having 30 attributes. For classification Zero R algorithm is used. As a consequence 3481 properly classified instances and 291 incorrectly classified instances found. The categorization done with labels named primary hypothyroid, secondary hypothyroid, compensated hypothyroid and negative. This paper analyses, predicted and classified thyroid data. Also the result shows that women are more affected than men with hypothyroidism.

Kumar et al. [2] Purpose of the paper is utilizing hypothyroid disease data and classifies the data in WEKA environment. For this dimensionality reduction method is used to opt for the subpart of instances from the unique data. For classification dataset from UCI machine learning library Thyroid dataset is opted. Decision stump and J48 data mining classification algorithms are applied. To find accuracy of algorithm and error rate confusion matrix is used. Output of J48 algorithm is 99.58%, and 95.38% for decision stump tree. Consequently, J48 performs better than decision stump. Error rate of J48 is much lower than Decision stump. That is 4.61 for decision stump and 0.424 for J48.

A model that builds deep learning (DL) models was proposed by Qiao et al. [23] to advance the accurateness and effectiveness of thyroid scintigraphy as a diagnostic tool for thyroid illness. A Discovery NM/CT 670 scanner (GE Healthcare, Chicago, IL, USA) was used to obtain the images. PyTorch 1.4.0 is used to diagnose thyroid illness accurately and quickly. For the three models, the range of values for precision, recall, NPV, specificity, accurateness, and F1-score were 73.33% to 97.00%.

Chaganti et al. [20] by using machine learning approach this paper suggested a model which predict thyroid disease. UCI thyroid disease datasets is used with different machine learning algorithms like SVM, RF, ADA, LR, and GBM. In which Random Forest gives best result 98% over other algorithms.

R et al. [21] Proposed a model for prediction of Thyroid disease which approaches machine learning. KNN, SVM, ANN, Decision Tree,LR algorithms are used with new_thyroid dataset from UCI machine learning repository. Logistic Regression (LR) model gives highest accuracy with 96.92%.

Arepalli et al. [22] Proposed a CNN model which has CT thyroid scans to assess, identify, and categorize thyroid cancer. This paper's primary goal is to enhance the finding accuracy in Computed Tomography images. The dataset on thyroid disease was obtained from the Garavan Institute in Australia. Dataset acquired 3000 scans out of which 2800 scans used as training dataset and 200 scans as testing dataset. Two other datasets are also used for cross validation. In detecting thyroid cancer proposed CNN model exhibits better accuracy than other existing models (Table 1).

Table 1. Literature Review

Author Name	Year & Reference	Problem	Dataset	Tool	Result	Accuracy
Prathibha et al	2023 [17]	Detection of thyroid diseases	Thyroid dataset taken from kaggle and UCI repository	Keras,modified ResNet based on CNN	Modified ResNet gives better accuracy than basic ResNet	Modified ResNet accuracy is 97%
J. CAO et al	2024 [18]	Thyroid nodule classification	606 thyroid Doppler ultrasound images	PyTorch	Model can perform well in processing small and unevenly classified ultrasonic image datasets	Tnc-Net model gaining an accuracy of 90.2% on the test set
Gothane et al	2020 [19]	Diagnosis hypothyroid	Hypothyroid dataset taken from github library	WEKA	Women are more affected with thyroid than men	WEKA 92% Precision
Kumar et al	2023 [2]	Thyroid detection and diagnosis	Thyroid dataset from UCI machine learning library	WEKA	J48 algorithm gives better performance than decision stump	J48- 99.57% Decision Stump--95.38%
Qiao et al	2021 [23]	Accurately and efficiently diagnosing thyroid disease	Discovery NM/CT 670 scanner (GE Healthcare, Chicago, IL, USA) was used to obtain the images	PyTorch 1.4.0	ResNet exhibits best performance	ResNet-92.78% Alexnet-88.56% VGGNet-89.89%
Chaganti et al	2022 [20]	Thyroid disease prediction	UCI thyroid disease datasets	RF, LR, SVM, ADA, and GBM	Random Forest gives best result over other algorithms	RF - 98% GBM - 97% ADA – 97% LR - 85% SVM – 85%

(*continued*)

Table 1. (*continued*)

Author Name	Year & Reference	Problem	Dataset	Tool	Result	Accuracy
R et al	2021 [21]	Thyroid disease prediction	new_thyroid dataset from UCI machine learning repository	KNN,SVM,ANN,Decision Tree,LR	Logistic Regression(LR) model gives highest accuracy	KNN-93.84% ANN- 75.38% Decision Tree-92.3% SVM- 95.38% LR-96.92%
Arepalli et al	2023 [22]	To assess, identify, and classify thyroid cancer in CT scans	Thyroid Disease dataset (D1)collected from Garavan Institute, Australia	CNN	CNN exhibits better accuracy than existing methods	Proposed CNN-99.94% Existing methods-98.26%

4 Experimental Work Using WEKA

Various deep learning tools are used by different authors namely Tensorflow, Keras, PyTorch, MXNet, and Caffe. TensorFlow, Keras, PyTorch, MXNet, Caffe are using Python as its primary language. WEKA uses java library. Based on the learning of tools from literature review this paper taken basically three deep learning tools **TensorFlow, Keras and WEKA**. To explore the features and functionality of these three tools Thyroid_diff dataset from Kaggle is opt. In this work same set of data in three different tools were used, comparative analysis of tool is done to find out which tool is best suited and gives most accurate result.

This paper used WEKA 3.8.6 interface. The classification model using WEKA has following steps:-

4.1 Loading Dataset in WEKA

Dataset used is Thyroid_Diff from Kaggle [24] loaded into WEKA data mining tool. Dataset has 17 attributes and 383 instances.

Dataset has 2 categories say No, Yes.

NAME: CLASS

LABEL	COUNT	WEIGHT
NO	275	275
YES	108	108

DATASET NAME:-Thyroid_Diff
INSTANCES:-383
ATTRIBUTE:-17

Serial Number	ATTRIBUTE NAME
1	AGE
2	GENDER
3	SMOKING
4	HX SMOKING
5	HX RADIOTHREAPY
6	THYROID FUNCTION
7	PHYSICAL EXAMINATION
8	ADENOPATHY
9	PATHOLOGY
10	FOCALITY
11	RISK
12	T
13	N
14	M
15	STAGE
16	RESPONSE
17	CLASS

4.2 Dataset Preprocessing

Every attribute in the dataset has been preprocessed. It is the fundamental phase of data mining.

4.3 Classification

Data mining techniques like classification are used to aggregate instances that fall into the same category. By grouping data according to training data, supervised learning is being done. Neural nets and decision trees are two examples of data mining classification methods [2]. In this experiment classification done through decision tree and J48 algorithm.

4.4 Performance Measurement

Performance of algorithmis measured through confusion matrix.

Confusion Matrix: - An algorithm for classifying data is evaluated using a confusion matrix, which is a 2D table. A confusion matrix describes the performance of the classification method [2].

Accuracy:-- Accuracy is defined as the number of correct predictions division by the sum of datasets. The values of 1.0 and 0.0 represent the best and worst cases of accuracy, respectively [2].

4.5 Result Analysis

Confusion matrix for J48 algorithm

```
=== Confusion Matrix ===

   a   b   <-- classified as
 269   6 |   a = No
  14  94 |   b = Yes
```

363 occurrences were successfully classified and 20 instances were wrongly classified by the J48 algorithm. The algorithm's accuracy is 94.7781.

5 Experimental Work using TensorFlow, Keras

This paper used TensorFlow, Keras framework in Google colab. The model has following steps:-

5.1 Importing Libraries

Importing important libraries such as NumPy, Pandas, TensorFlow, and Keras.

5.2 Data Preprocessing

One crucial step in the data mining process is data preparation. This procedure prepares raw data for additional analysis by converting it into a comprehensible format. The goal is to raise the caliber of the data and make it suitable for the intended uses. Preprocessing data aids in converting it into a format that may be used. Data reduction, cleansing, integration, transformation, and transformation are tasks that aid in data preprocessing. By managing missing values and reducing noise through the use of binning, regression, and clustering, data cleaning eliminates incomplete data. To create a single dataset, data integration brings together information from several sources [25].

Loading dataset:-- Dataset used is Thyroid_Diff from Kaggle [24]. Dataset has 17 attributes and 383 instances.

The dataset has two classes No and Yes. No indicates that after primary treatment thyroid cancer is not recurred. Yes indicates that after primary treatment thyroid cancer is recurred. Here 275 cases are not recurred ie they belong to no class and 108 cases are recurred ie they belong to Yes class.

Renaming the Columns:- Columns are renamed for clarity of fields. Renaming of columns makes it more understandable.

5.3 Data Visualization

Figure 1 exhibits percentage of each stage of cancer. Here, First Stage cancer is detected in more than 85% cases. Remaining 15% distributed in Second, Third and Fourth stage cancer.

Figure 2 indicates the age distribution of thyroid cancer patients. Most of the patients belong to the age range 28–35 years.

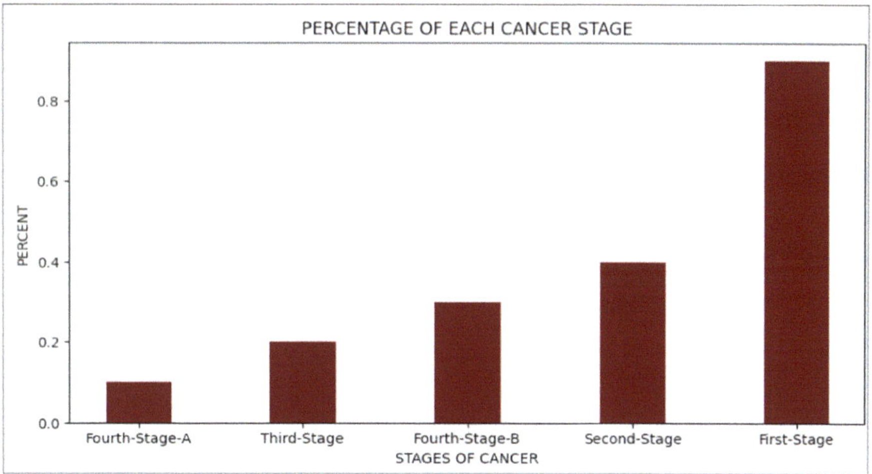

Fig. 1. Percentages of Each thyroid Cancer Stages

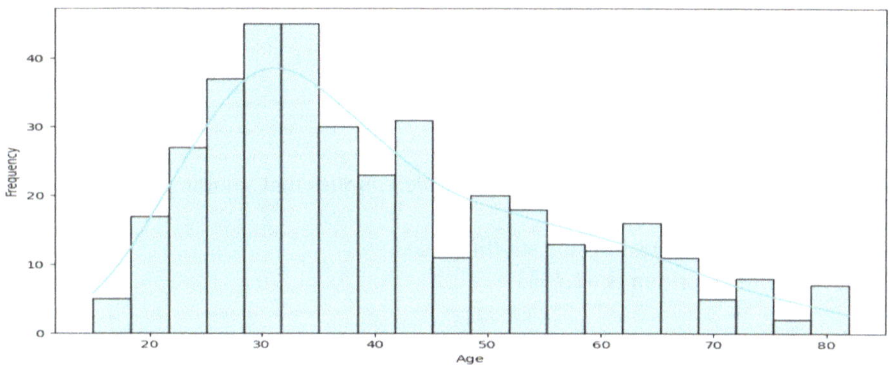

Fig. 2. Distribution of Age

5.4 Additional Data Preprocessing

This includes following task-

Dividing the Set of Data in Train and Test Sets: Train data and test data division ratio is 75:25

Feature Scaling: It is a way for standardizing the range of autonomous variables or data attributes.

5.5 Building the ANN

An important part is creating artificial neural network model using keras sequential model. Model building contains following steps:-

i. Initializing the ANN
ii. Adding the input layer and hidden layers

5.6 Training the ANN

Training the ANN on the Training set for epochs = 30

5.7 Building the Predictions and Assessing the Model

Making the Confusion Matrix

```
[[65   1]
 [6   24]]
```

0.927083333333
Correctly classified instances:-89
Incorrectly classified instances:-07

5.8 Result Analysis

.

1. The majority of patients do not smoke, which implies that smoking has a minor risk of causing thyroid cancer.
2. Mass of patients is healing at a startling rate.
3. Accuracy of algorithm is 92.7083%.

6 Conclusion

This paper provides a brief review about some deep learning tools most of the tools are python based. Two most promising tools WEKA and TensorFlow, experiment one dataset named Thyroid_diff on them using classification technique. Under classification, in WEKA we have chosen decision tree J48. In TensorFlow framework, data undergoes a number a processing steps including preprocessing, training and testing of data. Both the environment has classified data in with a good accuracy rate. WEKA exhibits 94.7781 accurate where as TensorFlow shows 92.7083%. The dataset has very less number of instances numbered 383. In this small experiment of tool we found WEKA more accurate than TensorFlow but there are some limitations using WEKA that it has very limited set of algorithm. It works with a small number of dataset. It takes more time to build and execute model than TensorFlow. Whereas instead of having lesser accuracy than WEKA with this dataset TensorFlow has a big range of algorithm more sophisticated processing of data. In future TensorFlow with Keras library may be explored to process big dataset. By applying certain constraint the accuracy of TensorFlow and Keras may be increases.

References

1. Armstrong, M., Asuka, E., Fingeret, A.: Physiology, Thyroid Function. in *StatPearls*, Treasure Island (FL): StatPearls Publishing (2024). Accessed: 27 Jul. 2024. http://www.ncbi.nlm.nih.gov/books/NBK537039/
2. Kumar, R.P.R., Lakshmi, M.S., Ashwak, B.S., Rajeshwari, K., Zaid, S.M.: Thyroid Disease Classification using Machine Learning Algorithms. E3S Web Conf. **391**, 01141 (2023). https://doi.org/10.1051/e3sconf/202339101141
3. Agbaraji, E.: Deep Learning Technology: A Vital Tool for National Development. Int. J. Eng. Res. **V8** (2019). https://doi.org/10.17577/IJERTV8IS070161
4. Nag, S., et al.: Deep learning tools for advancing drug discovery and development. 3 Biotech **12**(5), 110 (2022). https://doi.org/10.1007/s13205-022-03165-8
5. LeCun, Y., Bengio, Y., Hinton, G.: Deep learning. Nature **521**(7553), 436–444 (2015). https://doi.org/10.1038/nature14539
6. Min, S., Lee, B., Yoon, S.: Deep Learning in Bioinformatics. Brief. Bioinform. **18** (2016). https://doi.org/10.1093/bib/bbw068
7. Dargan, S., Kumar, M., Ayyagari, M.R., Kumar, G.: A survey of deep learning and its applications: a new paradigm to machine learning. Arch. Comput. Methods Eng. **27**(4), 1071–1092 (2020). https://doi.org/10.1007/s11831-019-09344-w
8. Mohialden, Y.M., Kadhim, R.W., Hussien, N.M., Hussain, S.A.K.: Top python-based deep learning packages: a comprehensive review. Int. J. Pap. Adv. Sci. Rev. **5**(1) (2024). https://doi.org/10.47667/ijpasr.v5i1.283
9. Abadi, M., et al.: TensorFlow: Large-Scale Machine Learning on Heterogeneous Distributed Systems (2016). arXiv: arXiv:1603.04467. https://doi.org/10.48550/arXiv.1603.04467
10. Moolayil, J. An Introduction to Deep Learning and Keras: A Fast-Track Approach to Modern Deep Learning with Python, pp. 1–16 (2019). https://doi.org/10.1007/978-1-4842-4240-7_1
11. Chicho, B.T., Sallow, A.B.: A Comprehensive Survey of Deep Learning Models Based on Keras Framework. J. Soft Comput. Data Min. **2**(2) (2021)
12. Joseph, F.J.J., Nonsiri, S., Monsakul, A.: Keras and TensorFlow: A Hands-On Experience. In: Prakash, K.B., Kannan, R., Alexander, S.A., Kanagachidambaresan, G.R. (eds.) Advanced Deep Learning for Engineers and Scientists. EICC, pp. 85–111. Springer, Cham (2021). https://doi.org/10.1007/978-3-030-66519-7_4
13. Imambi, S., Prakash, K.B., Kanagachidambaresan, G.R.: PyTorch. In: Prakash, K.B., Kanagachidambaresan, G.R. (eds.) Programming with TensorFlow: Solution for Edge Computing Applications. Springer International Publishing, Cham, pp. 87–104 (2021). https://doi.org/10.1007/978-3-030-57077-4_10
14. Asha Kiranmai, S., Jaya Laxmi, A.: Data mining for classification of power quality problems using WEKA and the effect of attributes on classification accuracy. Prot. Control Mod. Power Syst. **3**(1), 29 (2018). https://doi.org/10.1186/s41601-018-0103-3
15. Chen, T., et al.: MXNet: A Flexible and Efficient Machine Learning Library for Heterogeneous Distributed Systems (2015). arXiv. https://doi.org/10.48550/ARXIV.1512.01274
16. Collobert, R., Kavukcuoglu, K., Farabet, C.: Torch7: A Matlab-like Environment for Machine Learning (2011)
17. Prathibha, S., Dahiya, D., Robin, R., Nishkala, C., Swedha, S.: A novel technique for detecting various thyroid diseases using deep learning. Intell. Autom. Soft Comput. **35**, 199–214 (2023). https://doi.org/10.32604/iasc.2023.025819
18. Cao, J., Zhu, Y., Tian, X., Wang, J.: Tnc-Net: automatic classification for thyroid nodules lesions using convolutional neural network. IEEE Access **12**, 84567–84578 (2024). https://doi.org/10.1109/ACCESS.2024.3414675

19. Gothane, S.: Data mining classification on hypo thyroids detection: association women outnumber men. Int. J. Recent Technol. Eng. **8**, 2277–3878 (2020). https://doi.org/10.35940/ijrte.F7261.038620
20. Chaganti, R., et al.: Thyroid disease prediction using selective features and machine learning techniques. Cancers **14**, 3914 (2022). https://doi.org/10.3390/cancers14163914
21. Chandan, R., Vasan, C., Chethan, M.S., Devikarani, S.: Thyroid detection using machine learning. Int. J. Eng. Appl. Sci. Technol. **5** (2021). https://doi.org/10.33564/IJEAST.2021.v05i09.028
22. Arepalli, L., Rao, K., Dabbiru, M.: Deep learning approach for analyzing and predicting thyroid cancer in computed tomography images: review. SN Comput. Sci. **4** (2023). https://doi.org/10.1007/s42979-023-02010-w
23. Qiao, T., et al.: Deep learning for intelligent diagnosis in thyroid scintigraphy. J. Int. Med. Res. **49**(1), 0300060520982842 (2021). https://doi.org/10.1177/0300060520982842
24. Thyroid Disease Data. Accessed: 15 Jun. 2024. https://www.kaggle.com/datasets/jainaru/thyroid-disease-data
25. Data Preprocessing in Data Mining: Accessed: 17 Jun. 2024. https://www.tutorialspoint.com/data-preprocessing-in-data-mining

Role of Emerging Technologies for Sustainable Smart Cities

Anjum Nazir Qureshi(✉)

Rajiv Gandhi College of Engineering Research and Technology, Chandrapur, India
anjnaznus@gmail.com

Abstract. Smart city is one of the topics most discussed by the researchers. As the world is becoming a global village and more people are moving towards cities it is essential to know the requirements of the cities and upgrade the facilities to make life comfortable. Various technologies like IoT, AI, Blockchain, and Big Data have helped to improve smart city environments. The municipal corporations should provide facilities for residents to make living comfortable. Besides this, the cities should be protected from the impact of increasing environmental pollution, climate change, and global warming. Recent technologies can ensure better living conditions in smart cities through real-time monitoring. The current cities can be converted into smart cities by incorporating the latest infrastructural development and adopting sustainable practices. This paper will study the importance and role of a few emerging technologies in converting existing cities into sustainable smart cities. It will be helpful to learn about how the technologies are being used currently and how they can be applied to make smart cities more sustainable.

Keywords: Smart City · IoT · AI · Blockchain · Sustainability

1 Introduction

Urbanization is the shifting of humans from rural to urban areas. Over the past few centuries, people liked living in small communities. The last few decades have witnessed a mass migration of people from small villages to big cities.

According to a report by the World Bank, more than half of the global population resides in cities. It is predicted that by 2050, approximately 70% of people worldwide will live in urban areas. Urbanization therefore seems to be one of the most dominant factors for global development currently. Urbanization is likely to generate favorable circumstances for a better life, providing opportunities for reducing poverty and playing a significant role in economic growth globally [1]. Furthermore, 43 megacities are expected to emerge by 2030 that will have more than 10 million residents. Most of these cities will be in the developing region. One in eight people lives in 33 megacities globally; half of the urban inhabitants live in smaller establishments with less than 500,000 inhabitants. As the world continues to urbanize, understanding the key trends in urbanization will be one of the valuable steps toward achieving a sustainable development agenda by 2030. The attainment of sustainable development will depend on fulfilling the basic needs of

B. K. Mishra et al. (Eds.): ICT4GS 2024, ISEM 56, pp. 335–351, 2025.
https://doi.org/10.1007/978-3-031-94997-5_29

the growing urban population and meeting their needs related to housing, infrastructure, transportation, employment, education, and healthcare [2].

Many cities are on the path of becoming smart and sustainable. However, a few challenges have arisen with the rapid urbanization. Traffic congestion, water contamination, carbon emissions, energy consumption, and related health issues have increased drastically in the cities. Though many cities are becoming smart and sustainable, the versatility of Information and communication (ICT) technologies should be harnessed to build smarter and more sustainable cities. ICTs in smart and sustainable cities can help improve the quality of life of the people by making urban services more efficient, improving waste management, and enhancing housing and healthcare facilities. Recent advancements in technologies like the Internet of Things (IoT), Artificial Intelligence (AI), robotics, and digital twins drive the development of smart sustainable cities worldwide [3]. There is no single agreed definition for a smart city worldwide, many experts use the term coined by IBM and Cisco in 2007 [4]. As the word smart describes a human ability to deal effectively with situations, cities need to become smart enough to deal with the problems of the cities with increased urbanization. A study of the background of smart cities, as shown in Fig. 1 indicates that every global financial or energy crisis has triggered a technical revolution. The Industrial Revolution, the Electrical Revolution, and the Information Revolution are the few factors that have left an impact to a great extent on the cities of the present day. These changes came into existence to cope with resource shortages and the low potential of urban functions [5].

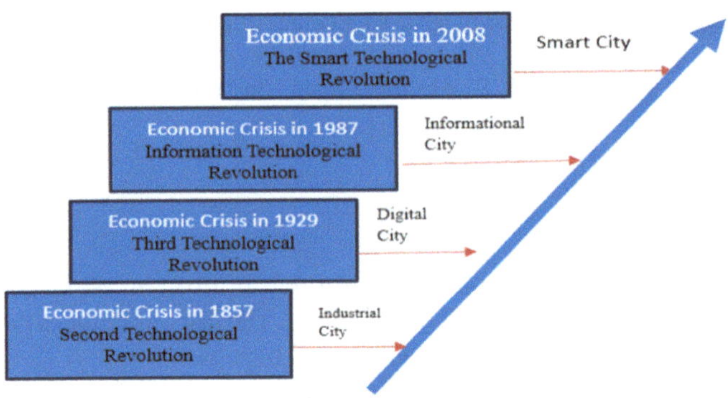

Fig. 1. The Background of Smart City [5].

It has been observed that urbanization and income are closely related. As the countries grow richer, they become more urbanized [6]. The standard of living of the urban residents is always better than the rural people. They have better access to electricity, safe drinking water, sanitation, and clean fuels and have lower malnourishment. However, a rapid increase in the urban slum population will cause increased inequalities in the urban localities [8]. The migration of people from villages to cities has shifted employment trends from agriculture to industry or manufacturing services [7]. Though technology has a greater role in the evolution of smart cities, these cities should not be limited to market-building strategies for emerging industries and impassioned billionaires. Amidst the various characteristics and expectations of smart cities, it is believed that a smart city should be a safe, livable, and sustainable place to live in [9].

The study on sustainable smart cities has been done to attain the following objectives: (i) study the concept of smart cities and the basic requirements to convert a city into a smart city, (ii) have an overview of how the technologies are contributing to developing new applications of a smart city and (iii) make like more comfortable and to know the challenges faced while using these technologies and possible solutions to overcome it. Scientific literature analysis is used in this study to reveal the theoretical concepts of how the technologies are being used to implement smart city applications.

2 An Overview of Smart Cities

Smart City is commonly discussed among researchers and common people. A smart city of the 21st century is expected to have many features to make life comfortable and easier. A few characteristics and features of smart cities are mentioned in Fig. 2 below.

A Smart city unites stakeholders from various sectors to build interdisciplinary partnerships. It facilitates entrepreneurial relationships, creates opportunities for new challenges, and serves its residents in multiple aspects to make life smart, attractive, happier, and smart. The world's ecosystem is changing and is being impacted by technologies like virtual reality (VR), AI, IoT, and the behavior of human relationships in innovative surroundings.

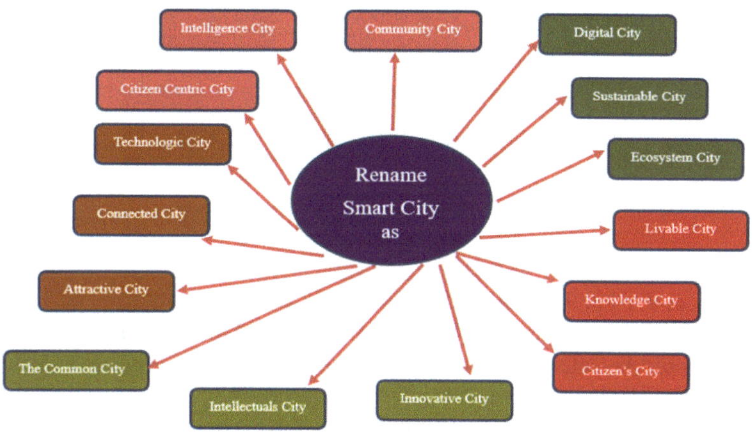

Fig. 2. Features and Characteristics of Smart City [4]

The European Commission proposed that smart cities should tackle common challenges, develop innovative solutions, recognize the demands of the cities, and attract business opportunities. The smart cities are systems with people, resources, finance, and services that foster sustainable economic development. These cities ensure a high quality of life through the smart use of recent technologies and distinguished urban planning to promote better economic and financial activities of the communities. The Concept of smart cities has been developed along six main dimensions: Economy, mobility, environment, people, living, and governance [10]. In technical aspects, the word "smart" refers to an automated mechanism capable of performing the desired task with less or no human intervention or a task that can be controlled with some handheld devices. The main entities of the smart city's ecosystem are given in Fig. 3 [11]. Technology, organization, policy, economy, community, governance, infrastructure, and natural environment are eight critical factors for an integrative framework of a smart city. These factors depend on each other and provide a basis for comparing how the cities envision their smart initiatives, execute their shared services, and address the challenges. As seen in Fig. 4., these factors are represented in two levels of influence. Technology has been placed as one of the inner influential factors and is considered to have an important share in implementing smart city initiatives. As most initiatives of smart cities rely on technology, it can heavily influence the success of other factors mentioned in the framework [12].

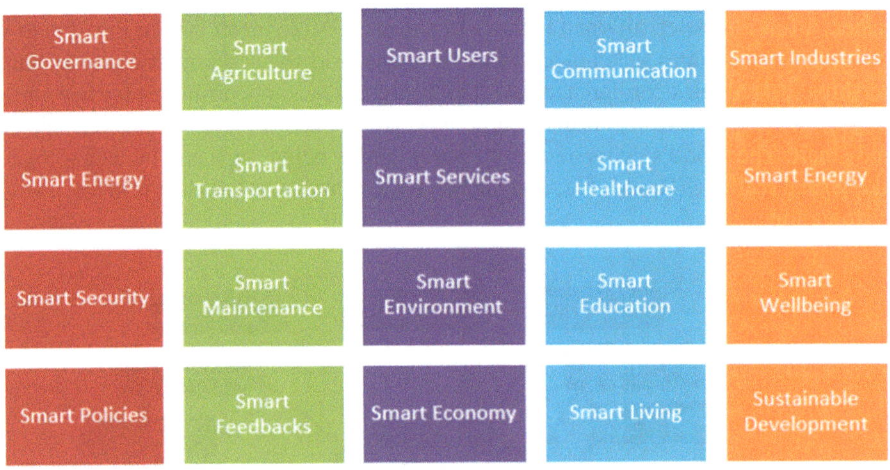

Fig. 3. Core Entities of Smart Cities [11]

The devices used in the smart cities are embedded with sensors and actuators. They are automated and self-sustainable. These devices need seamless connectivity and at the same time should be safe from security and privacy breaches. All these requirements of the smart city devices can be fulfilled by using enabling technologies like the IoT, Blockchain, AI, Cloud Computing, Wireless Sensor Networks, cyber-physical systems, security protocols, and Big Data. These technologies can be integrated to perform various tasks using the interconnected devices of smart cities [13]. The smart cities that will emerge in the coming years will combine excellent features of the existing technologies and harness the potential of collaborative technologies. The city Governments can encourage people's participation to utilize the full potential of technology. Four basic steps in this regard are: (i) Using the internet and digital technologies to connect distributed groups of people to ensure better utilization of resources especially in cities with limited resources. (ii) Crowdsourcing data from social media sites and through sensors in mobile phones that will act as a supplement for the IoT networks in the city. (iii)

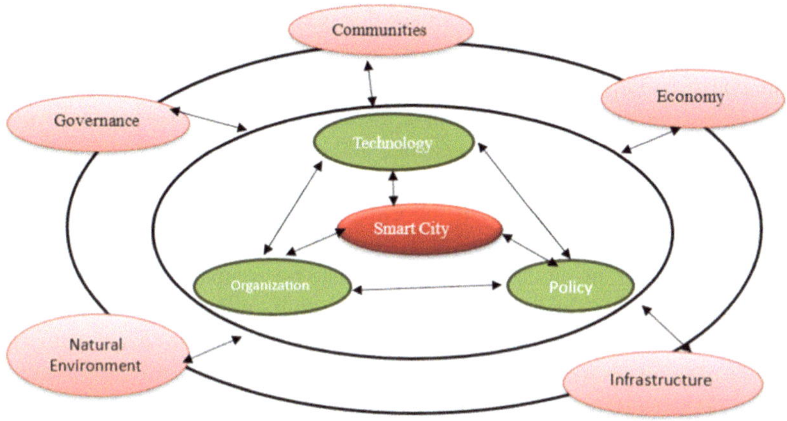

Fig. 4. Smart City Integrity Framework [12]

Using new digital tools to involve people in policy-making and planning to ensure better cities through democratic decisions. (iv) Encourage crowdfunding by connecting people online to collect funds for community projects. The city Governments can channel these funds for projects that match the people's wishes and needs [14].

3 Application of Emerging Technologies in Smart Cities

Various researchers have highlighted using technology for the sustainable development of smart cities. Technology connects devices and people to make processes easier, faster, and smarter. The local Governments should utilize existing technologies to ensure better facilities for people and better living standards. Recognizing the right technologies for a given application is a primary step toward achieving smart and sustainable cities. Several emerging technologies can be combined to yield better results. From agriculture to waste management, transportation to health care, all the entities can be made better and affordable in a smart city by combining technologies.

3.1 Smart Waste Management

The waste generation in cities is increasing proportionally with the expanding population. Most of the waste generated in the cities in either dumped in landfills or incinerated. These methods are unsafe as they cause the emission of harmful gases into the atmosphere. Moreover, the traditional waste-collecting methods in the cities are outdated. The waste collection team may have unnecessary or a late visit in some cases. Integrating technologies like IoT, AI, Cloud Computing, Geographic Information Systems, Data Access networks, and Blockchain helps better waste management and realize the goals of 3R reduce, reuse, and recycle.

IoT-based smart bins use sensors to monitor garbage levels and locations of the bins. The level of waste collected in these bins can be known using an application on your smartphone. Once the smart bin is almost full the waste management staff receives an alert along with its location. This saves time that is sometimes wasted in visiting places with no waste and saves fuel and time as the waste management team reaches the exact location. It also avoids the spilling of trash outside the bins and saves from foul smell and emission of harmful chemicals in the waste. These sensor-enabled bins can detect smoke to avoid fire eruptions caused by the presence of flammable substances in the waste [46]. These smart bins contain cameras that will capture images of the waste that comes into the dustbin. ML and Image processing algorithms help identify different types of wastes and classify them. Convolution Neural Networks (CNN) identifies and classifies the items as glass, plastic, metal, and undetected waste by comparing the captured images with the stored database. The dustbin has different compartments and the waste enters the respective section after classification [47]. This approach helps to save the environment from the disposal of plastic wastes in the surroundings and especially into water. As the metals and plastics are recyclable, the waste that goes to landfills will be reduced and it will save animals from facing severe health consequences owing to the swallowing of plastics. Robots and computer vision techniques are additional options for recognizing and classifying waste. AI-based algorithms are used as prediction models for predicting waste generation. The bins are equipped with weight sensors. These sensors calculate and estimate the waste collected during a fixed period. This will help in deploying sufficient vehicles and staff for waste collection. The data collection, processing, and analysis are done through cloud computing platforms. Blockchain enhances the waste management process by enabling transparency and traceability. Blockchain reduces the risks of hacking. The Ethereum Blockchain enables updating the disposal of e-waste and this can be used to reward the people who help in recycling. These technologies can be used for household, medical wastes, and industrial waste.

3.2 Smart Agriculture

AI and IoT are considered favorable technologies leading towards the transitioning from traditional farming techniques to smart agriculture. This can help in crop monitoring, weed handling, disease management, irrigation, and harvesting. These technologies can protect the crops against environmental changes, monitor the water level in fields, and avoid excessive use of pesticides and fertilizers [26]. The automation of agriculture can expand soil addition and enhance soil wealth. The AI-based algorithms can be used to strengthen the security of IoT [27]. As digitization has become more attainable the agriculture sector has started realizing the benefits of edge computing. Edge computing in agriculture protects the stakeholders of the agrifood value chain. It provides them unlimited access to a large dataset in different categories for tracking their farms. Cloud, fog, and edge computing techniques are used in agriculture for livestock management, crop farming, and greenhouses. These applications are generally combined with IoT and wireless sensor networks to achieve the monitoring and analyzing objectives [28].

3.3 Smart Healthcare

AI, IoT, and robotics are becoming the future of smart healthcare systems. IoT has enabled remote monitoring of patients. The doctors can monitor the health of their patients through wearables that will continuously monitor health parameters like sugar level, blood pressure, oxygen level, and body temperature. The doctors and relatives or the caretakers receive alerts and notifications on mobile and email when the health parameters change drastically. These notifications help the doctors immediately prescribe first aid and medicines to the patient. This technique is helpful for patients living in remote and acute geographical locations and for those having locomotive disabilities. IoT-enabled homes, equipped with sensors can be used for monitoring the physical and mental health of the elderly. These homes will have different sensors to monitor their motion in their house, the time when their lights are on, fall detection, and other health parameters. IoT-enabled wheelchairs are one more promising feature by which it is easy to detect the location of a patient and monitor fall detection. Smart cities will have the facilities of IoT-enabled ambulances that will help to take care of the health of a person with chronic illness on the way to the hospital in an ambulance. The doctors will continuously monitor the patient before reaching the hospital and provide suggestions of treatments to the accompanies. This initiative will save many lives as many do not reach hospitals on time due to the localized multispecialty and critical health care centers in urban areas.

AI and robotics are also considered emerging technologies for the healthcare sector. Robots can assist in surgery, streamline hospital logistics, and conduct routine checkups efficiently in hospitals. AI ensures safe working conditions for healthcare workers. Robots for surgeries and sanitization in hospitals will reduce the risks of infections for the doctors and hospital staff. Some future applications of AI and robotics include drug discovery, diagnosis of disease, remote monitoring of patients, and prediction of epidemic outbreaks. However, high costs, lack of skilled labor, and fixed mindset of people are some challenges to the wide adoption of these technologies in healthcare [29]. The Machine Learning (ML) and Deep Learning (DL) models use neural networks and feature classification techniques to help healthcare experts diagnose and detect medical conditions from medical images [30]. The various DL algorithms can be used for X-ray images, Computerized tomography, mammography, and digital histopathology images. These algorithms are faster than manual detection and provide accuracy in detection [31]. The AI and IoT-integrated applications are vulnerable to attacks. The change in health parameters by hackers can be fatal to a patient. Blockchain technology is used for securing these healthcare applications. It helps to deal with the privacy and security issues associated with computer vision and acoustic AI in healthcare [32].

3.4 Smart Transportation

Smart transportation integrates the latest technologies into transportation systems to enhance traveling and manage traffic congestion. The adoption of AI and IoT improves the logistical performances of transport systems and facilitates connecting cities. Smart transport systems improve the services provided to visitors and residents across the transport facilities, connecting roads, railways, and airports. Automation enhances transportation inside and outside the cities. AI, IoT, data, and Blockchain are fundamental pillars of reducing traffic congestion, allowing optimization of paths in the city and traffic flow optimization [33]. Efficient parking systems and enhanced safety, street lighting, accident prevention, predicting jams due to traffic or some climatic changes, and maintenance of roads are additional benefits of smart transportation systems. Wireless Sensor networks. Cloud computing, 4G/5G protocols, and ML algorithms help to implement and boost the performance of smart transportation systems [34]. The Traffic Incident Management systems in a smart city will be automated using CCTV cameras, computer vision capabilities for image detection, edge chips for video processing and cloud computing, and GPS for receiving appropriate information and location. This will avoid car rashes, road blockages, illegal parking, careless drivers, and commuters. These systems would be beneficial by initiating fast emergency services, adjusting traffic controls, and sending notifications to the drivers. The intelligent transportation system will accelerate the designing of greener environments by enabling real-time collection of data related to air pollution, carbon dioxide emissions, damage caused to the roads due to floods, cyclones, etc., and traffic throughput during various weather conditions. This real-time data collected by sensors, pre-processed on edged devices is sent to the cloud for further analysis. The obtained intelligence helps to frame policies for improving transportation in the cities. Another interesting feature is the Electronic Road Pricing system that uses the entities of smart transportation to issue automatic bills to the drivers and encourage people to use public transport. Smart transportation combines connectivity, hardware, and software technologies to make the traffic system more organized [35].

3.5 Smart Education

Future education should be compatible with the technological advancements of the smart city. Modern learning methods should cooperate with collaborative and interactive learning methods and provide more access to learning methods. Smart cities would improve the quality of life and smart education will be required to produce smart citizens who appreciate the values of these cities [36]. Educators will have an important role in smart education. They should upskill themselves and provide technological support to the students when needed. Some of the constituents of a smart education are depicted in Fig. 5 below.

Web 2.0 provides enhanced connectivity between the teacher and the learner. Internet and mobile connectivity will be needed to provide universal access by which educators and learners can access educational tools irrespective of time and location. Smart education uses innovative technologies and adaptive features to enhance perception and performance. It will provide better learning opportunities to students living in remote locations and those with some disabilities [38].

IoT and AI in education make things interesting through personalized and immersive learning experiences for learners. IoT-enabled cameras and microphones are used for real-time monitoring of the students. IoT-enabled laptops and tablets assist in generating automatic grading of assessments and assignments. This will help with remote monitoring for the students who cannot attend the classes due to tough geographical locations. AI uses the data collected by IoT-enabled devices to develop learning experiences in real-time. It understands students' perception levels and interests, generates pedagogical material, generates quizzes, and provides instant feedback [39]. The combination of AI and IoT in the education system benefits smart cities by advancing the application and infrastructure levels to provide efficient and effective learning scenarios. Table 2 below demonstrates the role of IoT and AI in the smart education system.

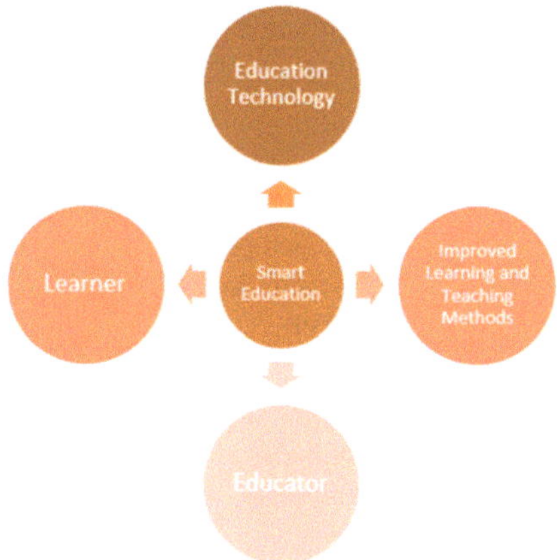

Fig. 5. Elements of Smart Education [37]

3.6 Smart Energy Usage

Energy consumption and its management are challenging for smart cities. Technological advancements optimize energy consumption and reduce costs to support a sustainable environment. The five major areas related to energy are its generation, transmission, distribution, utilities, and storage. The hardware layer consisting of the components and devices, the communication layer, and the intelligence layer are three layers that support energy systems. Smart grids, renewable energy sources, energy storage, smart buildings, electric vehicles, smart homes, and data analytics are a few strategies for the energy management of smart cities [40]. Integrating technology and energy management ensures meeting the increasing energy demands of urban areas. It optimizes energy production, distribution, and consumption to make it livable, cost-effective, and sustainable [41].

A Blockchain-based peer-to-peer energy transaction model can be used to meet the growing energy demands of a smart city. This Smart-Energy-Trading model allows independent transactions while communicating with home automation applications. This application controls the generation, consumption, and distribution of clean energy using renewable energy sources. Buildings and homes using either solar or wind energy will trade off the excess energy to other homes. Each house or building is treated as a Smart Energy-Chain digital transaction block. The validated blocks collect information on excess power generation by the buildings and power requirements of those who do not have power generation arrangements. All these blocks identify the energy consumers, their producers, and excess power available for trade [42]. IoT, cloud computing, and 5G technologies can be used to optimize energy usage in smart cities. All the electrical appliances are equipped with sensors. The 5G technologies are deployed to transfer data within the communication channels and cloud computing is used to store and retrieve data. It has been observed that lighting and billboards consume a lot of energy in the cities. IoT assists in the detecting motion or arrival of human or vehicle in the streets or roads to switch on or off the lights and billboards in the street. This model saves energy by minimizing energy waste [43].

Table 1. Contribution of AI and IoT towards bettering of the education system [39].

Application Infrastructure	Usability of AI	Usability of IoT
Classroom management	Monitor the behavior of the students in real-time, and provide support to the teachers to improve classroom engagement	Monitor temperature, and air quality in the classrooms to create more comfortable learning environments in class
Campus Management	Optimize campus operations, like scheduling and resource allocation for efficient utilization of resources and reduce costs	Monitor and control access to classrooms and labs, analyze and interpret the data to improve safety and security on the school premises
Remote Learning	Personalize and optimize remote learning experiences, to improve the engagement and performance of learners	Provide access to remote learning facilities like online libraries, and educational apps for monitoring the participation and progress of learners
Parent-Teacher Communication	Facilitate the involvement of parents in the education of wards by using chatbots and virtual assistants	Providing real-time updates on homework and assignment completion by the students and facilitating communication between parents and students
Health and Safety of Students	Identifying symptoms of illness and providing early warning for probable health risks	Track the movement of students on the campus and their interaction with others, and analyze the data to revise safety protocols
Learning Analytics	Analyze students' performance to contribute insights on upgrading the teaching-learning process	Identify the pattern of the participation of the students, and track how the students utilize the learning platforms, to collect data on their engagement and behavior

3.7 Smart Buildings

Smart buildings are essential building blocks of a smart city. A smart building has minimum human intervention, uses automation, connects the sensor-enabled smart devices in the building, and manages the data for better functioning of the entire system. The supporting technologies widely used for smart buildings are big data, IoT, sensors, 5G connectivity, geospatial technology, and robotics [44]. Smart buildings have facilities like smart water, energy, mobility, smart security, and smart waste management. They use automated control for providing heating, ventilation, air conditioning, lighting, and shading. Real-time data from the sensors is used to adjust settings based on human occupancy in the building, weather conditions, and energy usage patterns. This makes the building energy efficient and ensures enhanced living for its residents [45]. Integration of technologies into smart buildings has spread its applicability to industries apart from residential spaces. IoT in smart buildings collects data from the devices and buildings by deploying sensors. AI and ML are integrated with the IoT sensors to understand objects and environments and make adjustments based on analysis of user preferences and past trends. For example, the AI-based service automation system will study the normal functioning of machines and predict upcoming maintenance with minimal human intervention. Centralized building control using the building automation system, centrally controls the temperature, ventilation, electricity, humidity, and water pressure. The building therefore becomes energy efficient and has its operating costs decreased. These smart building technologies create a pleasant ambiance for people at the workplace and improve their productivity. Building Information Modeling (BIM) and Augmented Reality (AR) are useful for facilities management. For example, an AR-enabled glass can examine a plant with unfamiliar equipment and machines. Integrating BIM allows viewing digital representations of the equipment: their identification, installation dates, and warnings. It will help during emergencies. Virtual Reality (VR) is one more popular tool used in the construction and design industries. VR can be used during the mockup phase to allow a client to understand the proposed building as if entering the space. The client can walk around, and check for the windows, doors, etc., and the entire structure before any construction has begun. Aerial drones can be used inside buildings and retail spaces to take over time-consuming tasks like scanning shelves for expired goods, detecting intruders in the building, and reaching out to the rooftop machinery. These drones are like mobile eyes that detect minute details and activities of the building. Smart building technologies will provide safe and healthy environments for its occupants, reduce costs, improve security, and make the buildings more energy efficient.

4 Challenges in the Realization of Smart Cities

There are various issues and challenges faced during the realization of smart cities. These challenges can be broadly classified into three categories: technical, socio-economic, and environmental. Technical challenges that arise due to technology are security and privacy threats, cyber security threats, interoperability, scalability, internet connectivity, etc. Socio-economic issues arise due to the mindset of people and the lack of trained personnel to implement smart city initiatives. Sometimes people are not ready to accept the changes being done by the city government. Successful implementation of an initiative depends on how people anticipate and use it for their benefit. Along with this as technology gets involved, there is a lack of trained and qualified people to handle the smart city projects. Sometimes city governments and municipalities face difficulties in executing their initiatives due to the deficiency of a competent and skillful workforce. The environmental concerns of a smart city include tackling natural disasters and managing Greenhouse Gas emissions and e-waste. The significance of technologies in smart city implementation and the challenges faced during the process are given in Table 1 below [13]. The occurrence of natural disasters is affected by factors like global warming and climate change. Natural disasters like floods, earthquakes, cyclones, and droughts wipe away the efforts to develop a smart city. They cause death, injury, economic losses, and degradation of urban systems. Proper utilization of ICT for smart cities when combined with urban planning and engagement of the community in decision making can improve urban systems. It will help people to be prepared to face disasters, assist in disaster resilience, and help communities to recover from disasters [15]. The rapid technological advancements have caused an increase in e-waste. Many devices become obsolete after a certain period and cannot be upgraded. Disposing of outdated hardware becomes difficult due to the lack of specialized recycling infrastructure in many cities [16].

Table 2. Emerging Technologies, Their Significances, and Challenges [13]

Technology	Challenge	Significance
Cloud Computing	Managing enormous data in a centralized manner	Efficient data storage for the smart city application
IoT	Security and Privacy threats, interoperability, low storage	Enables connecting devices over a large network
Blockchain	Lack of infrastructural system and skilled workforce	Trusted transactions to deal with the security breaches arising in IoT and AI
AI	Implementation may be difficult sometimes due to a lack of technical knowledge and needed facilities	Complicated procedures in some applications
ICT	Security breaches	Enables the IoT applications
Geospatial Systems	Lack of standardization, More processing times, and high costs	Smart navigation, and remote sensing, enhance urban mobility
Wireless Networks	Safety and security issues	Real-time monitoring and collection of data
Big Data	Harness large amounts of data, enable data-driven decision-making	Privacy, security, and ethical concerns

A few Challenges related to using technologies for smart cities are:

4.1 Security Threats

Security and privacy are key issues in handling the working of smart cities. These issues should be addressed during the initial stages of implementation. A smart city performs real-time monitoring of its residents to provide intelligent services for health care, transportation, education, etc. Ensuring security and privacy are important in retaining trust and confidence among the people towards the smart city initiatives and increasing its adaptability [17]. The highly interconnected nature of the services, components, and technologies increases the security risks in smart cities. Compromising any of the elements in the interdependent networks can disrupt the whole system and have devastating effects on the entire system [18]. Smart cities come up with new applications and new devices, they may have some different security issues that are difficult to handle by the existing security policies and authentication or encryption methods [19]. IoT applications and devices become targets for cyber attackers due to a lack of safety mechanisms [23]. The data collection and preprocessing stages in AI-based systems are vulnerable to spoofing and scaling attacks, the training phases are subjected to poisoning attacks and inference phases at adversarial attacks [24].

4.2 Interoperability

Interconnected devices and sensors use self-defined formats to store the data. However, the schemes and data formats are sometimes incompatible for some devices. Data generated from various sources and multiple formats create interoperability issues, causing problems for integrating and analyzing information. Interoperability problems can be minimized by adopting semantic models, languages, technologies, and standards for modeling the information or data [20]. Interoperability issues are found while using technologies like IoT [21]. Interoperability across different problems is needed to increase the flexibility of AI and ML models for use by various vendors. The development teams can use the best framework and several features if the AI and ML models are interoperable and flexible [22].

4.3 Shortage of Supporting Infrastructure

The smart infrastructure in a smart city consists of energy-efficient buildings, smart transportation systems, smart grids, and smart waste management systems. Moreover, the advanced technologies should be integrated into the existing infrastructure. The infrastructure of cities should be upgraded to support IoT devices, data centers, and high-speed internet to connect smart city applications. Integrating new intelligent systems in old buildings without affecting their original structure can be challenging sometimes. Implementing the smart city infrastructure is expensive. Several infrastructure development projects face problems in completion on time due to a lack of funds and uncertainties in return on investment. Transformation of urban infrastructure to smart infrastructure will need skilled labor and this will increase investment and time requirements. A smart city will be irrelevant until the people are digitally empowered and learn to utilize technology. The Information Technology (IT) infrastructure supports the demands of solutions from the technology and application providers. Unless the infrastructure is well connected it is not possible to fully utilize the physical infrastructure. Traffic congestion and carbon emissions will make lives miserable. People may feel that the present infrastructure does not match their needs.

4.4 Unstructured Data Management

Digital technologies like the IoT involve storing, processing, analyzing, and transferring a large volume of real-time data. Sensors of different types are used to gather various types of data. This collected data can be structured, unstructured, and semi-structured in real time. The data can be text, video, image, geospatial, etc. The success of a smart city application depends on the proper analysis of the data and delivering the right data or information on time to its users. The storage and processing of large data is beyond the limits of a normal computer. The traditional database management systems are insufficient to handle this data. Cloud computing and big data help to store the voluminous data generated in real-time and meet the necessary arrangements for processing it.

4.5 Lack of Universal Standards

Smart city standards should be defined to encourage the large-scale adoption of smart cities and successfully implement their products and services. Official recognition by national and international organizations is needed for the societal advances in smart cities. Lack of uniformity across metrics like smart technologies, physical infrastructure, connectivity standards, etc. increases the complexities in maintaining the standards. The IoT companies are reluctant to agree upon interoperability standards for sharing data among the different types of sensors. There is no globally acclaimed standard for smart cities. The various kinds of standards make it difficult for any standard to gain global acceptance. The International Organization for Standardization (ISO), the International Telecommunication Union (ITU), and the British Standard Institute (BSI) are the most widely accepted standardization bodies for smart cities. Many other standardization agencies are working in specific nations. National standard-setting organizations for various countries are trying to develop smart city standards for their countries. But there is no such standard that is universally accepted. Joint efforts by the various standardization agencies will help to build a unified understanding of smart cities and ensure cooperation and collaboration among the nations to benefit the citizens.

5 Conclusion

This paper has tried to study how technologies like IoT, AI, Blockchain, Big Data, and Blockchain can enhance the living experiences of smart cities. Some smart city applications discussed in this paper are waste management, health care, agriculture, transportation, buildings, and education. Security and privacy threats remain great concerns while using technology and interconnecting devices and applications in a smart city. The technologies help to make the cities smarter and more sustainable. Technical, socio-economic, and environmental challenges may slow the development and adaptability of sustainable smart cities. At the same time, more research should be done to overcome the challenges observed while integrating technologies into smart city applications.

References

1. World Bank (2023). https://www.worldbank.org/en/topic/urbandevelopment/overview
2. Population Division of United Nations Department of Economic and Social Affairs (UN DESA), (2018). World Urbanization Prospects, https://www.un.org/sw/desa/68-world-population-projected-live-urban-areas-2050-says
3. The International Telecommunication Union (ITU), (2021). Smart Sustainable Cities, https://www.itu.int/en/mediacentre/backgrounders/Pages/smart-sustainable-cities.aspx#gsc.tab=0

4. Weinstein, Z.: OPED: Why it is Called A Smart City?, Bee Smart City (2023). https://www.beesmart.city/en/smart-city-blog/why-is-it-called-a-smart-city#:~:text=Despite%20the%20lack%20of%20a,quality%20of%20life%20for%20citizens
5. Zhou, Y.: The Path Towards Smart Cities in China: From the Case of Shanghai Expo 2010. Proceedings REAL CORP 2014 Tagungsband, pp. 1032–1027 (2014). ISBN: 978-3-9503110-7-5 (Print), https://repository.corp.at/314/1/CORP2014_132.pdf
6. Bloom, D.E., Canning, D., Fink, G.: Urbanization and the wealth of nations. Science **319**(5864), 772–775 (2008). http://science.sciencemag.org/content/319/5864/772.short
7. Michaels, G., Rauch, F., Redding, S.J.: Urbanization and structural transformation. The Quarterly Journal of Economics **127**(2), 535–586 (2012). https://academic.oup.com/qje/article/127/2/535/1824278
8. Ritchie, H., Samborska, V., Rose, M.: Urbanization, Our World in Data (2024). https://ourworldindata.org/urbanization
9. Anderson, M.: Smart Cities Examples: The Smart, The Not-So-Smart and The Ugly, One Key (2022). https://onekeyresources.milwaukeetool.com/en/smart-cities
10. European Parliament: Mapping Smart Cities in the EU. Directorate General for Internal Policies Policy. Department A: Economic and Scientific Policy. Brussels: European Parliament (2014)
11. Guelzim, T., Obaidat, M.S., Sadoun, B.: Chapter 1 - Introduction and overview of key enabling technologies for smart cities and homes. In: Obaidat, M.S., Nicopolitidis, P. (eds.) Smart Cities and Homes, Morgan Kaufmann, pp. 1–16 (2016). ISBN 9780128034545, https://doi.org/10.1016/B978-0-12-803454-5.00001-8
12. Chourabi, H., et al.: Understanding Smart Cities: An Integrative Framework. 2012 45th Hawaii International Conference on System Sciences (2012). https://doi.org/10.1109/hicss.2012.615
13. Ahad, M.A., Paiva, S., Tripathi, G., Feroz, N.: Enabling technologies and sustainable smart cities. Sustainable Cities and Society **61**(2020) 102301 (2020), https://doi.org/10.1016/j.scs.2020.102301
14. Saunders, T., Baeck, P.: Rethinking smart cities from the ground up. Nesta, public reports (2015)
15. Neha, B., Mukherjee, M., Gairola, A.: Smart Cities and Disaster Resilience. In: Book "From Poverty, Inequality to Smart City" (2017). https://doi.org/10.1007/978-981-10-2141-1_8
16. Babii, E., Goals, Z.W., Challenges, T.: (2024) https://aesg.com/perspective/zero-waste-goals-and-true-challenges/
17. Ahmed, S., Abbas, S.M., Zia, H. (eds.): Smart Cities—Opportunities and Challenges. Lecture Notes in Civil Engineering (2020). https://doi.org/10.1007/978-981-15-2545-2
18. Alibasic, A., Al Junaibi, R., Aung, Z., Woon, W.L., Omar, M.A.: Cybersecurity for smart cities: a brief review. In: Data analytics for renewable energy integration. Springer International Publishing, pp. 22–30 (2017)
19. Martínez-Ballesté, A., Pérez-Martínez, P.A., Solanas, A.: The pursuit of citizens' privacy: a privacy-aware smart city is possible. IEEE Commun. Mag. **51**(6), 136–141 (2013)
20. Kumar, K.P., Kumar, V.R., Raghuveer, K.: A survey on semantic Web technologies for the Internet of Things. International Conference on Current Trends in Computer, Electrical, Electronics and Communication (CTCEEC), IEEE (2017), pp. 316–322 (2017). https://doi.org/10.1109/CTCEEC.2017.8454974
21. Rahman, H., Hussain, M.I.: A comprehensive survey on semantic interoperability for Internet of Things: state-of-the-art and research challenges. Trans. Emerg. Telecommun. Technol. **31**(12), e3902 (2020). https://doi.org/10.1002/ett.3902
22. Vinton, G.C.: Thoughts of AI Interoperability, Communications of the ACM (2024). https://cacm.acm.org/opinion/thoughts-on-ai-interoperability/

23. Dorobantu, O.G., Halunga, S.: Security threats in IoT, 2020 International Symposium on Electronics and Telecommunications (ISETC). Timisoara, Romania **2020**, 1–4 (2020). https://doi.org/10.1109/ISETC50328.2020.9301127
24. Hu, Y., et al.: Artificial Intelligence Security: Threats and Countermeasures. ACM Computing Surveys **55**(1), Article No.20, 1–36 (2021)
25. Sarkar, S.: Smart Cities Rely on Smart Infrastructure To Make People's Lives Better. https://climate-change.ieee.org/news/yp-smart-cities/
26. Katiyar, S., Farhana, A.: Smart Agriculture: The Future of Agriculture using AI and IoT. J. Comput. Sci. **17**(10), 984–999 (2021). https://doi.org/10.3844/jcssp.2021.984.999
27. Parasuraman, K., Anandan, U., Anbarasan, A.: IoT Based Smart Agriculture Automation in Artificial Intelligence. 2021 Third International Conference on Intelligent Communication Technologies and Virtual Mobile Networks (ICICV) (2021). https://doi.org/10.1109/icicv50876.2021.9388578
28. David, P.E., Chelliah, P.R., Anandhakumar, P.: Chapter Nine - Reshaping agriculture using intelligent edge computing. In: David, P.E., Anandhakumar, P. (eds.) Advances in Computers, Volume 132, pp. 167–204. Elsevier (2024). ISSN 0065-2458, nISBN 9780323885447, https://doi.org/10.1016/bs.adcom.2023.08.007
29. Deo, N., Anjankar, A.: Artificial Intelligence With Robotics in Healthcare: A Narrative Review of Its Viability in India. Cureus. **15**(5), e39416 (2023). https://doi.org/10.7759/cureus.39416. PMID:37362504;PMCID:PMC10287569
30. Rana, M., Bhushan, M.: Machine learning and deep learning approach for medical image analysis: diagnosis to detection. Multimed Tools Appl. 1–39 (2022). https://doi.org/10.1007/s11042-022-14305-w. Epub ahead of print. PMID: 36588765; PMCID: PMC9788870
31. Puttagunta, M., Ravi, S.: Medical image analysis based on deep learning approach. Multimed. Tools Appl. **80**, 24365–24398 (2021). https://doi.org/10.1007/s11042-021-10707-4
32. Shinde, R., Patil, S., Kotecha, K., Potdar, V., Selvachandran, G., Abraham, A.: Securing AI-based healthcare systems using blockchain technology: A state-of-the-art systematic literature review and future research directions. Trans. Emerg. Telecomm. Technol. (2023). https://doi.org/10.1002/ett.4884
33. Anwar, A.H.M.M., Oakil, A.T.: Smart Transportation Systems in Smart Cities: Practices, Challenges, and Opportunities for Saudi Cities. In: Belaïd, F., Arora, A. (eds.) Smart Cities. Studies in Energy, Resource and Environmental Economics. Springer, Cham (2024). https://doi.org/10.1007/978-3-031-35664-3_17
34. Oladimeji, D., Gupta, K., Kose, N.A., Gundogan, K., Ge, L., Liang, F.: Smart transportation: an overview of technologies and applications. Sensors (Basel). **23**(8), 3880 (2023). https://doi.org/10.3390/s23083880.PMID:37112221;PMCID:PMC10143476
35. Zavadko, V.: Intelligent Traffic Management Systems: A Lowdown of Software & Hardware Components, Intellias (2024). https://intellias.com/intelligent-traffic-management/
36. Totoitich, P.Y.: Smart Education for Smart Cities. IEEE Smart Cities (2022). https://smartcities.ieee.org/newsletter/april-2022/smart-education-for-smart-cities
37. Demir, K.A.: Smart education framework. Smart Learn. Environ. **8**, 29 (2021). https://doi.org/10.1186/s40561-021-00170-x
38. El Janati, S., Maach, A., El Ghanami, D.: SMART education framework for adaptation content presentation. Procedia Computer Science **127**, 436–443 (2018). https://doi.org/10.1016/j.procs.2018.01.141
39. Kamruzzaman, M.M., et al.: AI- and IoT-Assisted Sustainable Education Systems during Pandemics, such as COVID-19, for Smart Cities. Sustainability **15**, 8354 (2023). https://doi.org/10.3390/su15108354
40. Pandiyan, P., et al.: Technological advancements toward smart energy management in smart cities. Energy Reports **10**, 648–677(2023). ISSN 2352-4847, https://doi.org/10.1016/j.egyr.2023.07.021

41. Appio, F.P., Lima, M., Paroutis, S.: Understanding smart cities: innovation ecosystems, technological advancements, and societal challenges. Technol. Forecast. Soc. Change **142**(2019), 1–14 (2019)
42. Pangracious, V.: Smart Energy for Smart Cities, IEEE Smart Cities (2022). https://smartcities.ieee.org/newsletter/february-2022/smart-energy-for-smart-cities
43. Humayun, M., Alsaqer, M.S., Jhanjh, N.: Energy Optimization for Smart Cities using IoT. Applied Artificial Intelligence **36**(1), e2037255 (2022). Taylor and Francis
44. Singh, T., Solanki, A., Sharma, S.K.: Role of Smart Buildings in Smart City—Components, Technology, Indicators, Challenges, Future Research Opportunities. Digital Cities Roadmap, 449–476 (2021). https://doi.org/10.1002/9781119792079.ch14
45. Wang, W., Hong, T., Li, N., Wang, R.Q., Chen, J.: Linking energy-cyber-physical systems with occupancy prediction and interpretation through WiFi probe-based ensemble classification. Appl. Energy **236**, 55–69 (2019)
46. Awawdeh, M., Bashir, A., Faisal, T., Ahmad, I., Shahid, M.K.: IoT-based Intelligent Waste Bin, Advances in Science and Engineering Technology International Conferences (ASET), pp. 1–6. Dubai, United Arab Emirates (2019). https://doi.org/10.1109/ICASET.2019.8714406
47. Abeygunawardhana, A.G.D.T., et al.: AI-Driven Smart Bin for Waste Management. 2nd International Conference on Advancements in Computing (ICAC). IEEE (2020). https://doi.org/10.1109/icac51239.2020.9357151

Designing and Developing Bio-Lattice Structures Using 3D Printing with ABS Carbon Fiber Inspired by Foxtail Palm Seed Fiber

Sivasankara Raju Rallabandi[1](✉), Harikiran Vuddagiri[2], Ashok Darsigunta[3], and Anandhini Chikkam[4]

[1] Department of Mechanical Engineering, Aditya Institute of Technology and Management, Tekkali, Andhra Pradesh 532201, India
sivaraju80@gmail.com

[2] Department of Mechanical Engineering, Avanthi Institute of Engineering and Technology, Makavarapalem, Andhra Pradesh 531113, India

[3] Department of Mechanical Engineering, RISE Krishna Sai Prakasam Group of Institutions, Ongole, Andhra Pradesh, India

[4] Department of Mechanical Engineering, Andhra University, Visakhapatnam, Andhra Pradesh, India

Abstract. The goal of this article is to design and develop bio-lattice structures utilizing 3D printing technology with ABS carbon fiber and nTopology software, which are inspired by the fibrous pattern of foxtail palm seed fiber. The aim is to replicate the complex structure of foxtail palm seeds while maximizing biocompatibility and mechanical robustness. ABS carbon fiber will be incorporated into the lattice structures using cutting-edge manufacturing processes to increase their durability. This work combines bio-inspired design with cutting-edge materials and testing methods to advance bio-mimetic structures in biomechanics and sustainable materials development. The analytical examination of lattice structures, namely Iso-Truss and Fluorite lattice structures entailed assessing crucial characteristics such as displacement, tension, and strain. The results showed that the Iso-Truss lattice structure had a maximum displacement of 1.45e-04 and a maximum stress of 7.49e + 06, whereas the Fluorite lattice structure had a maximum displacement of 1.54e-04 and a maximum stress of 4.83e + 06. These findings offer useful knowledge about the mechanical behaviour and performance of the lattice structures, emphasizing their strengths and peculiarities when subjected to various loading circumstances.

Keywords: Bio-lattice structures · Foxtail palm seeds · Structural Analysis · nTopology · 3D printing

1 Introduction

Biomedical applications as well as engineering have shown a great deal of interest in the design and development of biomimetic structures. The objective of this study is to investigate the complex fibrous pattern found in the fibers of foxtail palm seeds and uses

B. K. Mishra et al. (Eds.): ICT4GS 2024, ISEM 56, pp. 352–362, 2025.
https://doi.org/10.1007/978-3-031-94997-5_30

it as a model for bio-lattice constructions [1, 2]. With additive manufacturing methods like 3D printing and sophisticated design tools like nTopology software, the project aims to create these bio-lattice structures out of ABS carbon fiber. An interesting model for the creation of bio-inspired materials with improved mechanical qualities and biocompatibility is provided by the distinctive fibrous arrangement of foxtail palm seed fibers. By utilizing nTopology software, parametric design techniques will be used to replicate the structural features of foxtail palm seed fibers while maximizing for specific mechanical requirements. A potential approach for easily and stably constructing these complex lattice structures is the use of 3D printing technology with ABS carbon fiber reinforcement [3–5]. Evaluating the performance of the produced bio-lattice structures by extensive mechanical testing, such as tensile and compression tests are one of the main goals of this study. These tests will offer essential data about the strength, stability, and mechanical behavior of the structures under various loading scenarios [6, 7]. To verify the effectiveness of the bio-inspired design methodology and evaluate the possible uses of these structures in the engineering and biomedical domains by carefully analyzing the test results. In general, this project is an example of an integrated effort that connects advanced design tools, additive manufacturing processes, mechanical testing, and biomimicry [8–10]. The development of bio-inspired materials with customizable properties for a range of engineering applications by creating and analyzing bio-lattice structures patterned after the seed fibers of the foxtail palm, vide in Fig. 1. In the realm of botany, the intricate lattice structure of seeds plays a crucial role in seed dispersal, protection, and germination. One compelling example of such a botanical marvel is found in the seeds of the foxtail palm (shown in Fig. 1(c)).The only known toxic palm fruit is the seed of the foxtail palm, which makes it special. When fully grown, the seeds are orange-red, and they must be removed from the seed pod while still inside the crown shaft. Because the seeds are poisonous if consumed, it's critical to handle them carefully. Using foxtail palm seed fibers as inspiration for creating lattice structures is one of the main points of the literature review. The researchers want to mimic the complex pattern of the seed fibers using parametric design tools and additive manufacturing techniques. In order to further the development of biomimetic materials, this strategy aims to improve mechanical qualities and biocompatibility. The usefulness of this bio-inspired engineering process is investigated through extensive testing [11, 12].

(a) (b) (c)

Fig. 1. (a) foxtail palm tree(Wodyetiabifuracate) (b) Palm tree fruit (c) foxtail palm dry seed

Ramakrishna and bala [13] investigates 3D-printed bio-inspired lattice structures for energy absorption applications.. The review summarizes recent advances in the development of bio-inspired structures, focusing on the mechanical properties of natural structures like bones, teeth, and dermal layers, which can be biomimicked to design economical energy absorbers. Büşra and Ali Rıza [14] optimizes lattice architectures utilizing advanced additive manufacturing. The authors improved structure mechanical performance and material efficiency using computational calculations and experimental testing. The optimized lattice architectures increased tensile and compressive strength while using less material, suggesting lightweight and strong engineering components. Li et al. [15] utilized in super capacitors, batteries, oil/water separation, self-cleaning, water collection, and droplet manipulation. The review examines the present condition of this research and the potential for utilizing 3D-printed biomimetic structures to tackle difficulties in the realms of energy and the environment.

This work uses 3D printing using ABS carbon fibre and nTopology software to design and build bio-lattice structures. Foxtail palm seed fibre inspired these bio-lattice constructions. The goal is to imitate foxtail palm seed complexity while maximizing biocompatibility and mechanical strength. Biomimetic structures in biomechanics and sustainable materials development will be studied by merging bio-inspired design with sophisticated materials and testing methods.

2 Material

The combination of ABS (Acrylonitrile Butadiene Styrene) carbon fiber composites makes such a potential material for 3D printing. This is toughness, heat resistance and impact resistance is just a few of the characteristics that make ABS a popular thermoplastic [3, 16]. The addition of carbon fibers to ABS carbon fiber composites improves these qualities even more and making them appropriate for a variety of uses. Significant gains in strength, stiffness, and dimensional stability are achieved when carbon fibers are added to ABS. When compared to conventional ABS, the resulting composite can bear larger loads and stresses since carbon fibers are known for having high tensile strength and modulus [17–19]. For applications demanding structural integrity and longevity, ABS carbon fiber composites are therefore perfect.

ABS carbon fiber composites also have outstanding thermal characteristics, such as a low coefficient of thermal expansion and a high heat deflection temperature. This qualifies them for use in high-temperature applications such industrial gear, aerospace parts, and automobile components.ABS carbon fiber composites provide advantages in terms of printability and surface finish in addition to mechanical and thermal qualities. During 3D printing, the carbon fibers enhance the material's flow characteristics, producing prints that are smoother, more accurate, and experience less shrinkage and warping. This allows high dimensional accuracy production of complicated designs. Moreover, ABS carbon fiber composites can be further refined and customized for the printed parts through post-processing utilizing conventional machining methods including milling, drilling, and tapping. For 3D printing applications where strength, stiffness, thermal stability, and dimensional precision are essential criteria, ABS carbon fiber composites offer a high-performance and adaptable material alternative. The author presents an overview of bio-lattice architectures, which are inspired by natural materials, particularly the fibers of foxtail palm seeds. The author provides an overview of CSA composites, highlighting the benefits of integrating carbon fibre reinforcement into ABS matrices [20–22]. They are appropriate for a variety of industrial, automotive, aerospace, and consumer product applications due to their compatibility with additive manufacturing methods.

2.1 Why Choose ABS Carbon Fiber Material

ABS carbon fiber materials have great strength, stiffness, durability and making them highly appealing for 3D printing. These composites provide improved mechanical qualities by combining carbon fibers with ABS thermoplastic; as a result, they are perfect for producing parts that need to have a high degree of structural integrity and strength. ABS carbon fiber composites additionally offer exceptional dimensional stability and thermal resistance, guaranteeing top performance even in difficult conditions. Their lightweight design strikes a compromise between strength and weight reduction, increasing their appeal even further. ABS carbon fiber composites are a flexible and affordable option for creating functional prototypes, finished parts, and tooling components using additive manufacturing techniques. They can be compatible with a broad range of industries and applications [23, 24].

2.2 nTopology for Designing Bio-Lattice Structures

nTopology provides a wide range of tools and features that greatly improve the design and development of bio-lattice structures. The software's significant features in geometric control, optimization, simulation, and connection with 3D printing make it extremely beneficial for designing strong and biocompatible structures that are inspired by natural patterns. The new algorithms enable precise manipulation of lattice parameters, ensuring faithful reproduction of the delicate patterns found in foxtail palm seed fibres. The software demonstrates exceptional proficiency in optimizing designs to achieve precise mechanical attributes, including strength, stiffness, and weight. Designers can utilize nTopology to customize the bio-lattice structures in order to get the appropriate equilibrium between these characteristics, hence improving their overall performance and utility [25, 26].

However, it is important to acknowledge that there are also certain limits associated with this software. Tackling these obstacles and harnessing the software's functionalities can result in significant progress in the fields of material science, engineering, and sustainable development.

3 Results and Discussion

3.1 FE-SEM Analysis on Foxtail Palm Seed Fiber

Field Emission Scanning Electron Microscopy (FE-SEM) is used to analyse the structure and properties of the fibre in foxtail palm seed fibre, shown in Fig. 2. This research offers comprehensive insights into the fiber's shape, content, and surface characteristics. The efficacy of utilizing foxtail palm fruits to produce activated carbon for water purification applications is demonstrated by the study on foxtail palm fruits as prospective activated carbon for eliminating pollutants like metamifop and methylene blue. The study highlights the value of using natural resources, such as foxtail palm fruits, to create environmentally friendly solutions for problems facing the environment and highlights the potential uses of these materials in long-term projects.

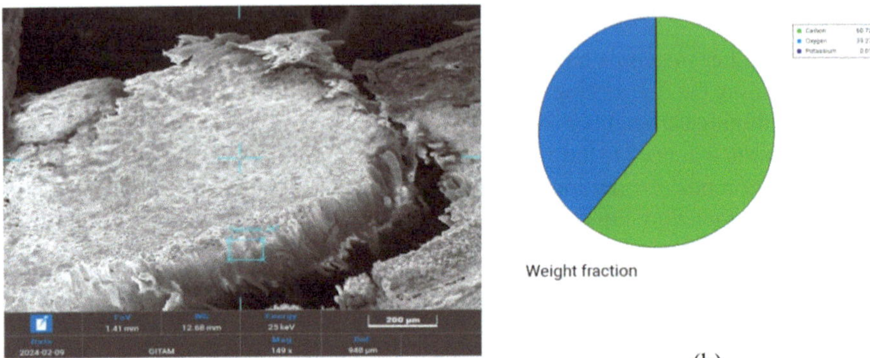

(a) (b)

Fig. 2. (a) SEM of foxtail palm seed fiber (b) Atomic Fraction of foxtail palm seed fiber

Examining the foxtail palm seed's composition and physical attributes is part of the structural analysis process, shown in Fig. 3. The surface of the foxtail palm seed fiber was closing similar to iso-truss and fluorite lattice structures. These lattice structures are designed and developed with help of nTopology software.

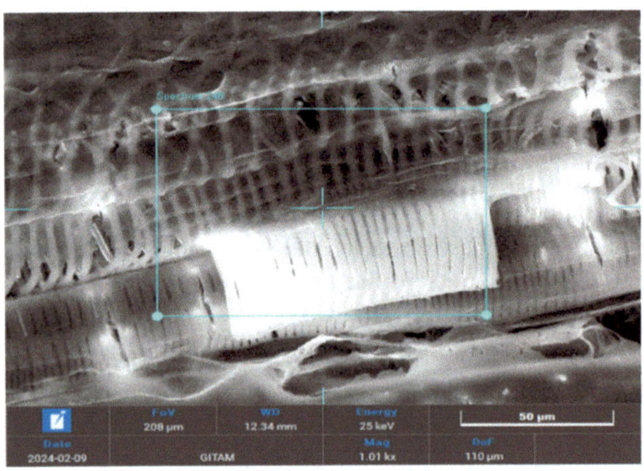

Fig. 3. Structure analysis of foxtail palm seed

Furthermore, Fig. 4 showed that they might be used to make activated carbon, which would be utilized to clean up pollutants like metamifop from water sources. The method included utilizing nitric acid for chemical activation to produce activated carbon from the fruits of the foxtail palm. This procedure showed encouraging outcomes in lowering the amounts of metamifop in tainted water. This study highlights the value of foxtail palm seeds in sustainable solutions by showcasing their adaptability and efficacy in environmental applications. There are several ways to analyze the structure of foxtail palm seeds, such as by looking at their germination patterns, composition, and physical traits. Studying the morphology of the seed, which entails assessing its dimensions, form, and surface characteristics, is a popular strategy. Furthermore, the seed's interior components the embryo, endosperm, and seed coat offers clues into the seed's growth and possible germination process. At high magnification, the seed's surface can be seen using methods like scanning electron microscopy (SEM), enabling a thorough analysis of its morphology and texture.

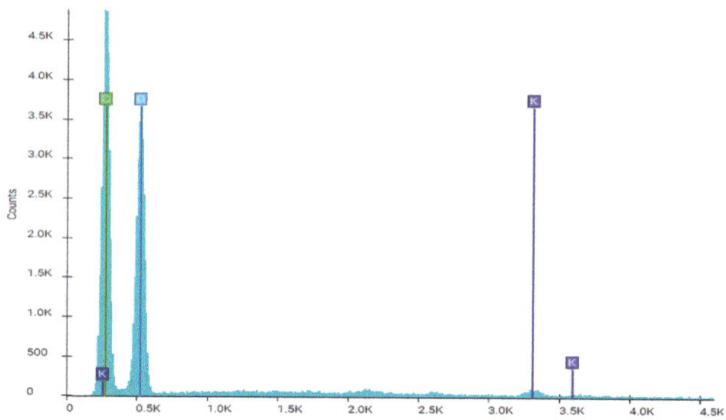

Fig. 4. EDX Analysis of foxtail palm seed fiber

Additionally, examining the seed's histochemistry can provide important insights on the cellular components and chemical makeup of the seed, which can help comprehend the processes involved in its germination and growth. Researchers can examine the structure of foxtail palm seeds in detail to learn more about their traits and germination behavior by combining various analytical techniques.

3.2 Design and Analysis of Lattice Structures

Making intricate models to examine the structure and properties of foxtail palm seeds can be a part of the design and analysis process utilizing nTopology software. Geometric morphometric analysis is a component of topology software that can be used to examine the characteristics and form of seeds, such as foxtail palm seeds. The authors accurately investigate the grain shape and other morphological properties of the seeds by using the software's geometric morphometric analysis capabilities, which can provide important insights into the seeds' functionality and design. Furthermore, the nTopology software facilitates the development of automated simulation workflows for design analysis. These workflows can be used to model numerous situations pertaining to the behavior and structure of the seed, assisting in the comprehension of its performance in various circumstances. With nTopology software's sophisticated functionalities, scientists may do comprehensive design and analysis of foxtail palm seeds, thereby augmenting their comprehension of the seeds' morphology and its uses.

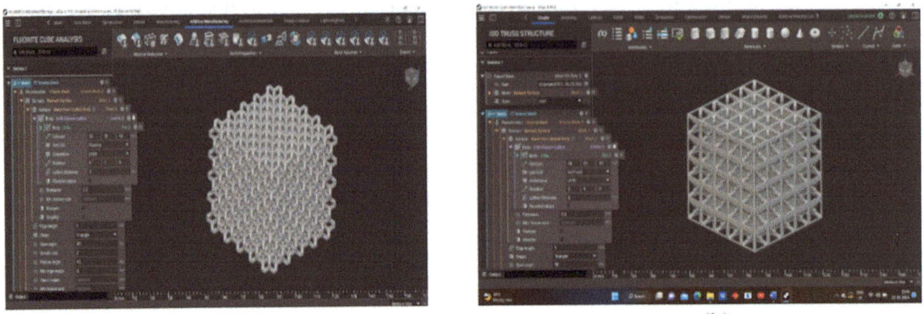

(a) (b)

Fig. 5. Design of bio lattice structures (a) Fluorite (b) Iso-Truss

To fully grasp the features and actions of materials that have fluorite structures, it is necessary to carefully study their structures (Vide in Fig. 5a), investigation is essential. It entails examining how crystal defects, lattice factors, and ion coordination affect the structural stability and functional characteristics of fluorite compounds.

The utilization of nTopology software in the design and analysis of Iso-Truss structures entails employing sophisticated tools for geometric morphometric analysis and simulation operations. Iso-Truss structures are distinguished by its lattice-like configuration shown in Fig. 5b, which provides exceptional strength-to-weight ratios and structural efficiency. nTopology offers functionalities for enhancing the design of truss structures by taking into account elements such as topology, member sizing, and load-bearing requirements. Engineers can use nTopology's simulation processes to analyze the performance of Iso-Truss structures under different loading circumstances, guaranteeing structural integrity and efficiency.

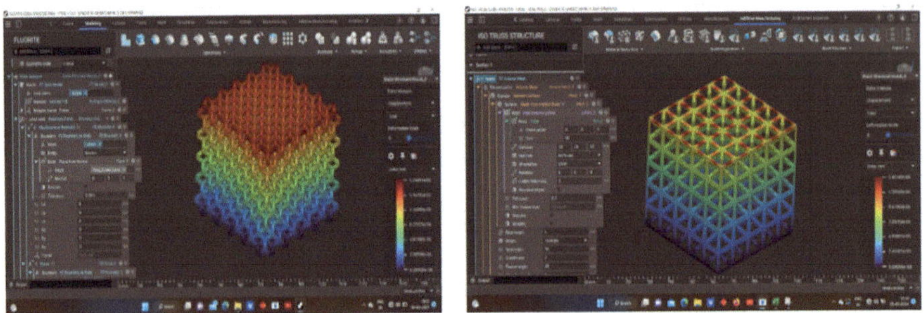

(a) (b)

Fig. 6. Displacement in Static analysis (a) Fluorite (b) Iso-Truss

3.3 Discussion of Static Analysis on Lattice Structure

The Displacement result has shown the model deformed under the given boundary conditions as displacement resistant and load 100N in z- axis. The obtained maximum and minimum of total deformation, as shown in Fig. 6(a) and 6(b).

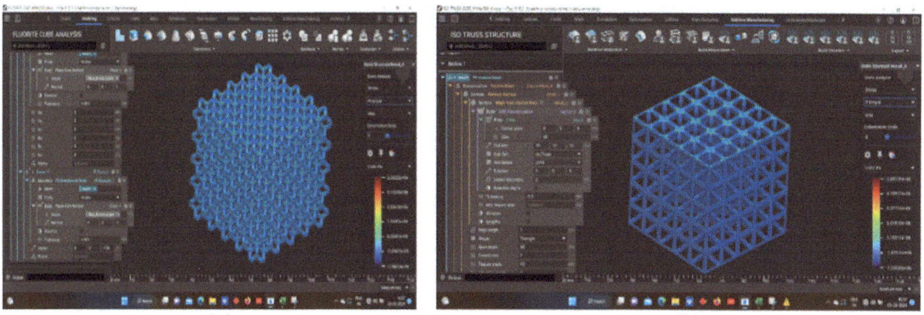

(a) (b)

Fig. 7. Principal stress in Static analysis(a) Fluorite (b) Iso-Truss

The Stress results has shown, the model stressed under the given boundary conditions as displacement resistant (0, 0, 0) and load (0, 0, −100). Here the principal stresses are developed as shown in Fig. 7(a) and 7(b).

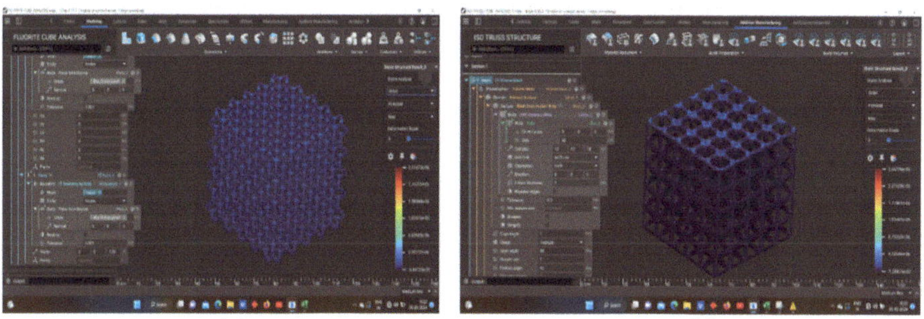

(a) (b)

Fig. 8. Strain in Static analysis(a) Fluorite (b) Iso-Truss

The strain results has shown, the model deformed under the given boundary conditions as displacement resistant (0,0,0) and load (0, 0, −100). The maximum and minimum values of strain developed as shown in Fig. 8(a) and Fig. 8(b). Figure 8(a) depicts the original shape of the modelled item, illustrating the beginning conditions prior to the application of any weight or displacement. On the other hand, Fig. 8(b) shows the distorted state of the model after implementing the stated boundary conditions, thus demonstrating the structure's reaction to the imposed load. These visualizations are frequently used in engineering simulations to comprehend the mechanical properties of materials and structures when subjected to different loading conditions. Analyzing data in this manner aids in forecasting potential places of failure, displacement, and the distribution of stress. These factors are crucial for guaranteeing the safety and dependability of designs in various domains, including civil, mechanical, and aeronautical engineering. The results are summarized in Table 1 of lattice structures. It clearly representing the maximum displacement occurred in Iso-Truss. Similarly found the maximum Stress and maximum Stain in Fluorite lattice structure.

Table 1. Analytical Analysis of lattice structure

S.No	Structure	Displacement		Stress		Strain	
		Max	Min	Max	Min	Max	Min
1	Iso Truss	1.45e-04	2.42e-05	7.49e + 06	1.21e + 01	2.67e-05	−1.216e-07
2	Fluorite	1.54e-04	2.56e-05	4.83e + 06	1.67e + 04	2.12e-05	−6.44e-07

3.4 Validation of Properties

Figure 9 compares the performance of two constructions, Fluorite and Iso-Truss. Iso-Truss outperforms in all criteria, including higher peak load (20 KN vs. 15 KN), elongation at peak (6.6 mm vs. 5.0 mm), and compression strength (7.2 KN/m2 vs. 6.9 KN/m2). This shows that Iso-Truss is stronger and more flexible than Fluorite. In the three metrics of load at peak, elongation at peak, and compression strength, iso-truss performs better than fluorite. This shows that Iso-Truss could be a stronger and more flexible structure than Fluorite.

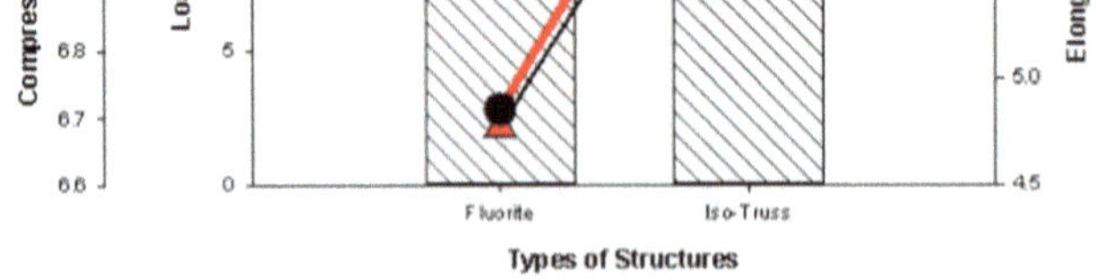

Fig. 9. Mechanical Properties of Fluorite vs. Iso-Truss

4 Conclusion

In this work, nTopology software used to create lattice structures, drawing inspiration from the complex microstructure of foxtail palm seeds. The foxtail palm seed fibers' microstructure using FE-SEM analysis, revealing their amazing intricacy. Next, two lattice structures were created that mirrored the fundamental design principles of the seed: fluorite and iso-truss. Based on the data designed lattice shapes that were optimized for strength and resilience by utilizing the adaptability of nTopology. By bridging biomimicry and computational design, this multidisciplinary method opens up new avenues for solving engineering problems. The research highlights how biomimetic-inspired design can be used to develop cutting-edge materials and structures. By working on this work, support innovation at the nexus of nature and technology by advancing computational tools and biomimetic design processes.

References

1. Novikov, A.S.: Self-Healing Polymers, https://typeset.io/papers/self-healing-polymers-1njvbswx (2022). https://doi.org/10.3390/polym14112261
2. Malekmohammadi, S., et al.: Smart and biomimetic 3d and 4d printed composite hydrogels: Opportunities for different biomedical applications (2021). https://doi.org/10.3390/biomedicines9111537
3. Aimar, A., Palermo, A., Innocenti, B.: The Role of 3D Printing in Medical Applications: A State of the Art (2019). https://doi.org/10.1155/2019/5340616
4. Prakash, A., Bag, R.K., Raju, S.S.: Parametric Optimization of Tungsten Inert Gas (TIG) Welding by using Taguchi. Int. J. Innov. Res. Sci. Eng. Technol. **5**, 3630–3638 (2016). https://doi.org/10.15680/IJIRSET.2016.0503164
5. Prakash, A., Bag, R.K., Ohdar, P., R, S.S.R.: Parametric Optimization of Metal Inert Gas Welding By Using Taguchi Approach. Int. J. Res. Eng. Technol. **05**, 176–182 (2016). https://doi.org/10.15623/ijret.2016.0502030
6. Seharing, A., Azman, A.H., Abdullah, S.: A review on integration of lightweight gradient lattice structures in additive manufacturing parts. Adv. Mech. Eng. **12**, 1–21 (2020). https://doi.org/10.1177/1687814020916951
7. Basoalto, H.C., et al.: A computational study on the three-dimensional printability of precipitate-strengthened nickel-based superalloys. Proc. R. Soc. A Math. Phys. Eng. Sci. **474** (2018). https://doi.org/10.1098/rspa.2018.0295
8. Stevens, C.V., Callum, A.S.H.: Industrial Applications of Natural Fibres. Wiley (2010). https://doi.org/10.1002/9780470660324
9. Atmakuri, A., Palevicius, A., Janusas, G., Eimontas, J.: Investigation of Hemp and Flax Fiber-Reinforced EcoPoxy Matrix Biocomposites: Morphological, Mechanical, and Hydrophilic Properties. Polym. **14**, 4530 (2022). https://doi.org/10.3390/POLYM14214530
10. Mzili, T., Mzili, I., Riffi, M.E.: Artificial rat optimization with decision- making: a bio-inspired metaheuristic algorithm for solving the traveling salesman problem. Decis. Mak. Appl. Manag. Eng. **6**, 150–176 (2023). https://doi.org/10.31181/dmame622023644
11. Wan, H., Ohtaki, H., Kotosaka, S., Hu, G.: A study of negative Poisson's ratios in auxetic honeycombs based on a large deflection model. Eur. J. Mech. A/Solids. **23**, 95–106 (2004). https://doi.org/10.1016/j.euromechsol.2003.10.006
12. Moroni, L., De Wijn, J.R., Van Blitterswijk, C.A.: 3D fiber-deposited scaffolds for tissue engineering: Influence of pores geometry and architecture on dynamic mechanical properties. Biomaterials **27**, 974–985 (2006). https://doi.org/10.1016/j.biomaterials.2005.07.023

13. Ramakrishna, D., Bala Murali, G.: Bio-inspired 3D-printed lattice structures for energy absorption applications: a review. Proc. Inst. Mech. Eng. Part L J. Mater. Des. Appl. **237**, 503–542 (2023). https://doi.org/10.1177/14644207221121948
14. Aslan, B., Yildiz, A.R.: Optimum design of automobile components using lattice structures for additive manufacturing. Mater. Test. **62**, 633–639 (2020). https://doi.org/10.3139/120.111527
15. Li, J., Li, M., Koh, J.J., Wang, J., Lyu, Z.: 3D-printed biomimetic structures for energy and environmental applications. DeCarbon. **3**, 100026 (2024). https://doi.org/10.1016/j.decarb.2023.100026
16. Chiang, K., Chang, F.: Application of grey-fuzzy logic on the optimal process design of an injection-molded part with a thin shell feature. Int. Commun. Heat Mass Transf. **33**, 94–101 (2006). https://doi.org/10.1016/j.icheatmasstransfer.2005.08.006
17. Vuddagiri, H.K., Rallabandi, S.R., Vadapalli, S., Pandi, T.: Assessment of mechanical and tribological performance of hybrid Al/MoS2/Al2O3 composite by GFRA. Metall. Mater. Eng. **28**, 79–102 (2022). https://doi.org/10.30544/764
18. Rao, D.A., Raju, M.V.J., Raju, R.S.: Evolution of mechanical, microstructural and tribological performance of DMLS deposited maraging steel. Int. J. Mech. Eng. **7**, 2998–3010 (2022)
19. Raju, R.S.S., Venkata Siva, B.: Fabrication and tribological studies of Al-CSA composite using RSM. Int. J. Mater. Eng. Innov. **12**, 83–102 (2021). https://doi.org/10.1504/IJMATEI.2021.115596
20. Rallabandi, S.R., Adapa, S.K., Jagadish, Rao, C.J., Yanda, S.: Evaluation of thermo-mechanical behavior of hemp fiber polymer. Compos. Mech. Comput. Appl. An Int. J. **13**, 113–132 (2022). https://doi.org/10.1615/CompMechComputApplIntJ.2022043103
21. Vikash Kumar, K., R, S.S.: raju: Statistical modeling and optimization of Al-MMCs reinforced with coconut shell ash particulates. In: Deepak, B., DRK, P., Jena, P.C. (eds.) Innovative Product Design and Intelligent Manufacturing Systems: Lecture Notes in Mechanical Engineering, pp. 703–712. Springer Nature Singapore Pte Ltd., Rourkela, India (2020). https://doi.org/10.1007/978-981-15-2696-1_67
22. Rallabandi, S.R., et al.: A contrastive characterization of pure Mg and AZ91D alloy based on the testing of mechanical, corrosion, wear, and erosion properties. Eng. Res. Express. (2023). https://doi.org/10.1088/2631-8695/ad16a2
23. Raju, S., Palli, S., Devi Prasad, P., Menda, V., Ramakrishna, B.: A hybrid AHP-TOPSIS, MOORA technique for multi-objective optimization of thermal, mechanical, and water absorption behavior of epoxy/hemp, pine apple, and palm fiber composites. J. Chinese Inst. Eng. **47**, 23–36 (2023). https://doi.org/10.1080/02533839.2023.2274092
24. Raju Rallabandi, S., et al.: Critical evaluation of epoxy-hemp-pineapple-palm fiber composites using hybrid AHM-TOPSIS technique for sustainable structural applications. J. Chinese Inst. Eng. Trans. Chinese Inst. Eng. A. **47**, 325–336 (2024). https://doi.org/10.1080/02533839.2024.2308250
25. Jafferson, J.M., Sharma, H.: Design of 3D printable airless tyres using NTopology. In: Materials Today: Proceedings, pp. 1147–1160. Elsevier (2021). https://doi.org/10.1016/j.matpr.2021.02.058
26. Sulaymon, T.A., Petri, H.T., Rayko, T.: Towards a complex geometry manufacturing: a case study on metal 3D printing of topology optimised bicycle parts with lattices. In: IFAC-PapersOnLine, pp. 1515–1520. Elsevier (2022). https://doi.org/10.1016/j.ifacol.2022.09.605

Performance Analysis of Simulation Techniques Used in Internet of Things

Rinki Kaur[1] and Surendra Kumar Patel[2(✉)]

[1] Information Technology, Government Nagarjuna P.G. College of Science, Raipur, Chhattisgarh, India

[2] Department of Information Technology, Government Nagarjuna P.G. College of Science, Raipur, Chhattisgarh, India

surendrapatelit2004@gmail.com

Abstract. Now a day, the Internet of things are become a most essential component of the human life. This important role to extensive growth of world. To become a smart, update and give the less human interaction for doing task which save our time. IoT is everywhere, Internet of Things changing the way of human living because is a portable, wearable, implantable, making a worldwide and linked universe, and converting physical things that surround us into an environment to the information. This research paper presents a comparative analysis of various IoT simulator, their working behavior. The paper goal is to study of the existing simulation technique to make enhancement in the IoT device system performance, enhance security, testing of theories, their modeling methods. This paper also present the IoT architecture and their domains, Device of IoT, Types of IoT communications and challenges with existing prospective and scenario.

Keywords: IoT · Architecture of IoT · Simulators · IoT Challenges and Communication · Domain

1 Introduction

Internet of things become the most used research area for the researchers in today era. IoT implies as Internet of Things which allow user to connect everyday thing surrounded with electronics, software, and sensors to internet allowing to collect and interchange data without human interaction. The word Things in Internet of Things termed as to anything and everything in today life which is processed or linked through the internet. IoT is automation and analytics structure which interrelate with artificial intelligence, sensor, networking, electronic, cloud messaging to give complete systems for the product or service. The system formed by IoT has greater transparency, control, and performance [1]. Purpose of IoT is to make a smart system by using smart things, objects and devices that use sensor and communicate autonomously to make data and convey it through the Internet for decision making [2] (Fig. 1).

The paper goal is to give comparative analysis about IoT simulator and describe the architecture of IoT, domain and challenges in IoT. This paper has been systematized as

B. K. Mishra et al. (Eds.): ICT4GS 2024, ISEM 56, pp. 363–374, 2025.
https://doi.org/10.1007/978-3-031-94997-5_31

Fig. 1. Internet of Things [3]

follows: - In section 2 we briefly discuss about the architecture and domain, IoT device, types of communications in IoT. In section 3 we present the comparative analysis of simulator of IoT and their working patterns. In section 4 the study of algorithms and their testing theories in detail. In section 5 we discussed about the challenges in IoT are discussed. Finally, in section 6 conclusion and future work are given.

2 IoT Architecture

According to the need of different application area of the IoT it works based on its designed and development. So there is no any specific defined architecture of working which is strictly followed worldwide. Here we will discuss the basic fundamental architecture of IoT which can be divided into 4 layers [1].

2.1 Layer/stage of IoT Architecture

Sensing Layer: - First layer of IoT, it is responsible for gathering data from the many sensor and actuators to gathering the information related to physical parameters. Theses device are connected to network layer via wired and wireless communication protocol.

Network Layer: - Second layer of IoT set the connectivity between the IoT devices. Uses protocol and technology, gateways and router which allow the device to link and communicate with respectively other and with the Internet. Cryptography algorithms protect against unauthorized access [4].

Data Processing Layer: - Third layer of IoT which collect raw data and facts from the device for analyzing and available for further investigation. It include number of technologies and tools, such as data administration systems, logical platforms, and machine learning algorithms those are used to get useful insights from the data and make conclusions based on that data [4].

Application Layer: - Forth layer of IoT which have S/W and application that are intended to communicate with the IoT infrastructure. Provide intermediate services that permit various IoT devices and systems to connect and share data seamlessly and includes analytics and processing abilities that allow data to be analyzed and transformed into meaningful understanding [4] (Fig. 2).

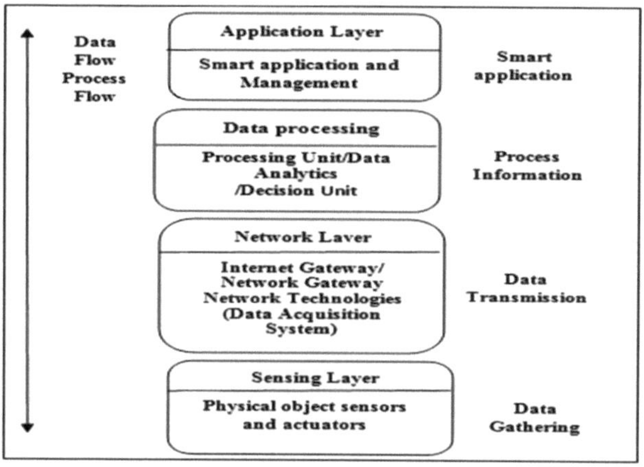

Fig. 2. IoT Architecture [3]

2.2 Domain

IoT system are different dependent upon the situation and work for themselves. Here we provide the some implementing area of the Internet of Things. IoT application can give some advantage like: -

- Smartly control the homes and cities via mobile phone to enhance the security and offer personal protection this automation saves the lots of time.
- Access of information being easy if we have distant from the main location and updated continuously in current time. It gives you a personal assistant for alerting with your regular plan.
- IoT application minimize the human effort because without human need IoT device communicate with each other to perform a variety of task. It is useful for safety because it senses any potential danger and warn user (Fig. 3).

2.3 IoT Device

For emerging any IoT infrastructure we need some physical objects which are sensors, actuators, electronic gadgets or machines. These Programmed device used for certain applications and can convey data over the internet or other networks and can be remotely monitored and controlled (Fig. 4).

2.4 Types of Communication in IoT

Communication play the important role in IoT because it creates a connection between devices to help to exchange the data, perform some task without any human interaction over the internet. Types are:-

Human to Machine(H2M): - Human interact with machine to perform some task. IoT device (Sensor, actuators, etc.) takes the input from the user like text, image, voice to

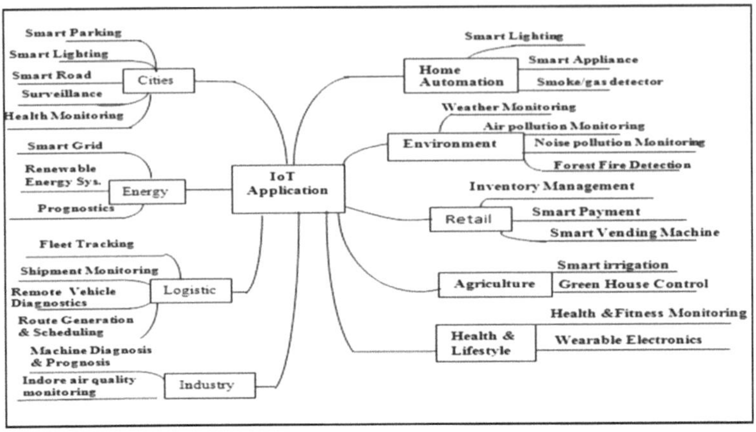

Fig. 3. IoT Domain [5]

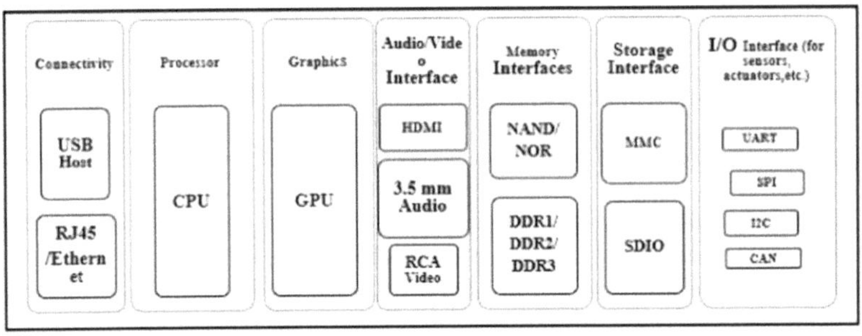

Fig. 4. Generic Block diagram of IoT Device [6].

understand the input to perform analysis and give some response to user in the form of text and visual display. **Example: -** Facial Recognition, Bio-metric attendance, Voice recognition (Fig. 5)

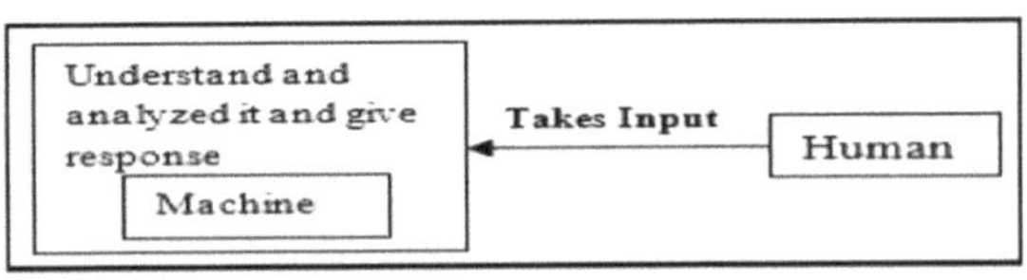

Fig. 5. Human to Machine(H2M)

Machine to Machine (M2M): - Two or more machine are interacting with each other to exchange the information. In this machine level instruction are needed for communication. Machines are connected with wire or wireless connection. IoT uses the M2M concept to create a large cloud of network devices that communicate with another network through the cloud networking platform (Fig. 6).

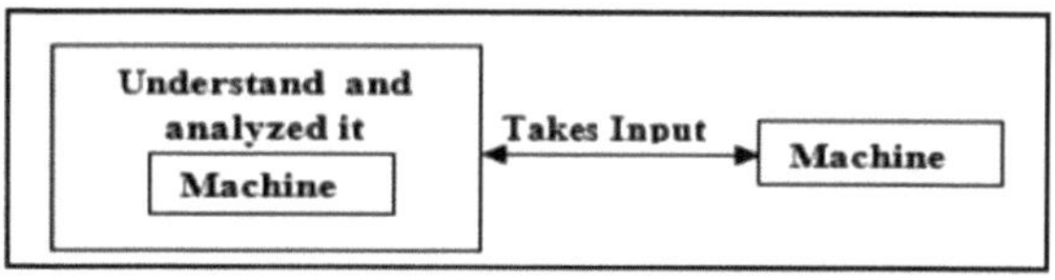

Fig. 6. Machine to Machine(M2M)

Example: - Smart Washing machine shows alerts to the owners' smart devices after completion of washing or drying of clothes [7]. Smart meters tracks amount of energy used in household or in companies and automatically alert the owner [7].

Machine to Human (M2H): - In this communication machine interact with human. Machine set off information like text, images, voice respectively or irrespectively in the presence of human. Human collaborate with machine by using tools or device to complete the task. It is most commonly used example of human daily life. **Example: -** Fire Alarms, Traffic Light, Smart gadgets, Health monitoring devices (Fig. 7)

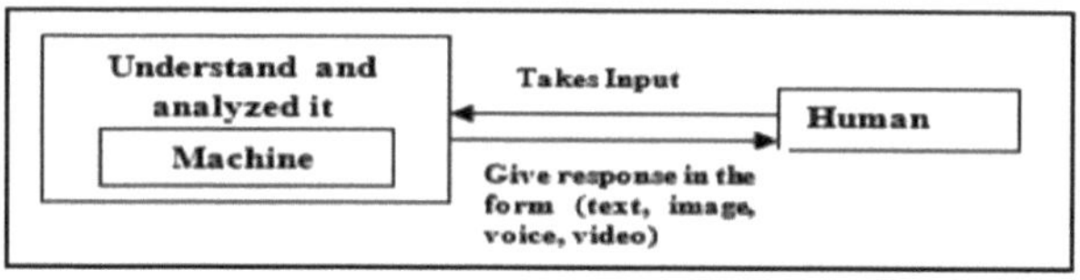

Fig. 7. Machine to Human (M2H)

Human to Human: - In this communication interchange of information are between human to human. Without H2H, M2M applications cannot produce the expected benefits unless humans can immediately fix issues, solve challenges, and manage scenarios [7]. **Example: -** Bluetooth, WiFi, Zigbee (Fig. 8)

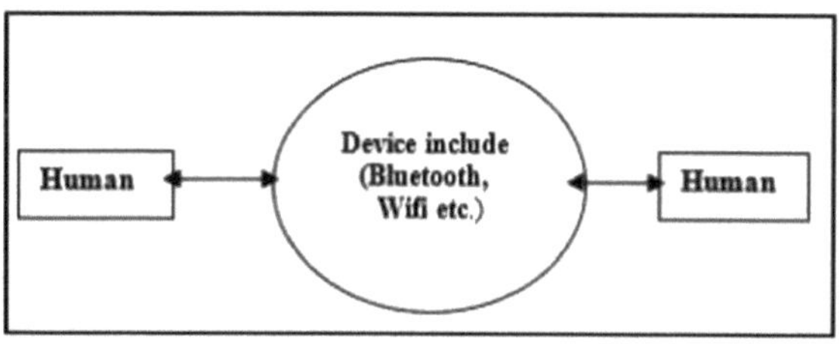

Fig. 8. Human to Human (H2H)

3 Comparison of Simulator

To implementing IoT platform many simulators are used to analysis the performance of network. **A Simulation is a techniques or specific program model to predict the behavior and performance of computer network and wireless network**. Many Commercial simulator are GUI driven and network simulator are CLI driven. IoT simulator provide a way to user to simulate hundreds of linked devices that are defined by the users without making any problems [8]. Here we present list of many IoT simulator and Emulator

MATLAB and Simulink: - This tools can help the user to design, prototype, and deploy IoT applications such as analytical preservation, operations optimization, controlling [9].

Iotify: - Provide IoT network simulation on the cloud platform , which is perfect large-scale Cloud-based IoT device simulation .Using this tools we can easily and quickly prototype, scale and manage IoT application [10].

OMNeT++:- A non-commercial simulation and discrete event simulator model based on a C++ simulation method and library for building network simulators and free in educational institutions [8].

NetSim:- It is advance network simulator used for protocol modeling or simulation, network devices, and security applications and used to investigate computer networks. A discrete event simulator covered an extensive range of wired, mobile networks, wireless, IoT devices, and sensor networks and covers 50 plus types of research globally for IoT, IIoT, WSN, LTE, MANET [8].

Cooja Emulator/Simulator: - Cooja is ajava based sensor network simulators and used for the limited source of devices such as the sensor nodes and it is created in the event- driven kernel [11].

IoTNetSim: - In IoT simulation is an innovative simulation and modeling tool that is flexible to end-to-end service. It keeps both the heterogeneous nodes with their organized details and its many models built on network connectivity and application logic . The simulation of the IoTNetSim can be divided into three groups as IoT layer, Edge layer, Cloud layer [12].

MIMIC IoT Simulator: - This simulator involves managing the several gateways, sensors, and further connected devices to build a real test lab to simulate the IoT system of Smart cities, factories, agriculture, Event driven architecture, Industry 4.0 [12] (Table 1).

4 Analysis of Some Algorithm and Testing Theories in Current Research Scenario

In this section we examine the previous work to find their implementation strategies and solution provided by the researchers To implement low-power IoT device Contiki operating system were used . Contiki cooja is a network simulator to simulate large and small networks of Contiki motes [13].

Table 1. Comparison of selected IoT Simulation.

Simulator	Type of Simula tor	Scope	Simulation Type	Programming language	Scale of operati on	Built in IoT Standards	Service Domain	Security Measure
MATL	IoT	Data	Discret	C, C++,	Large	Real Time	Engineering and Technica l service	High
AB & Simulin k	Simulat or	Analysis and Design	e-event	Java	Scale			
Iotify	IoT	Hardwar	Mobile	Python	Large	Real Time	Smart	High
	Simulator	e Connection	App	Java	Scale		City	
OMNe	IoT	Network		C++	Large	Manual	Generic	Medium
T++	Simulator		Discret e-event		Scale	Extension		
NetSim	IoT	IoT		C Code	Campu	802.15.4	Military	High
	Emulator	Network	Researc		s-Wide	LTE	Utilities	
		s	h Based		Use	MANETs		
Cooja	IoT	Network	Discret	C/Java	Small	Protocols	Enables	High
Simulator	Emulator		e-event		scale	supports by Contiki OS	Real World	
IoTNet	IoT	IoT	Discret	Java	Large	Real Time	Enables	High
Sim	Simulator	system and Network	e-event		Scale		real world	
MIMIC	IoT	IoT	CLI	C++, Java,	Large	Real Time	Real	High
IoT Simulator	Simulator	system and Network	Interface	PHP, Python	Scale		World	

4.1 Case Study of Some Related Work on Cooja Simulator for Design of IoT Scenario

Case Study 1: - In this article **Encryption Algorithms for WSN IOT: A Software based Analysis with Contiki Cooja** [14] they evaluating the number of clock ticks required to perform encryption and decryption on wireless sensor nodes using different modes (ECB, CBC, CTR) for AES, DES and 3DES with different key lengths and block size of data (16, 32, 48 and 64). For evaluation they use Contiki 3.0 and Contiki 2.7 as method 1 and method 2.Using this simulator they created nodes implementing on different encryption algorithms with different block sizes. They tested different sizes of blocks (16, 32,48 and 64 bytes) as well as various operational modes (ECB, CTR, CBC for AES in instance), and different key lengths (48, 128, 192, 256) depending on the parameters available for the algorithm. Each node implemented an encryption algorithm with specific parameters. For each type of node, 100 rounds were executed and the mean clock ticks of the round were used in order to compare the algorithms [14]. The Mode of operation, cipher block size, and key length for available algorithm based on simulation measurement [14] (Table 2).

Table 2. Simulation parameter scenario.

Algorithm	Mode	Key length	Block size
	CBC	256,192,128	16,32,48,64
AES	ECB	256,192,128	16
	CTR	256,192,128	16,32,48,64
DES	Not	48	16,32,48,64
	available		
3DES	Not available	48	16,32,48,64

In the method 1 measurement of each mode AES CTR,AES ECB,AES CBC, for each block size 16,32,48,64 bytes with each key length where they find that for the same block size 16and key length128, encryption with CBC, ECB and CTR use almost the same number of clock sticks and for decryption of blocks, CTR Mode tends to use almost half of the value needed by CBC and ECB which were relatively close and DES and 3DES consume a far greater number of clock ticks than AES in various modes (16 bytes blocks).They find the same conclusion in method 2 that give huge difference between DES,3DES and AES with the same size of block was observed so that these two algorithms (DES, 3DES)are not suitable for Wireless Sensor Networks and IoT devices as they will consume more energy and subsequently reduce the lifespan of nodes in the network [14]. This figure show the comparison of method 1 and method 2 for encryption and decryption (Figs. 9 and 10).

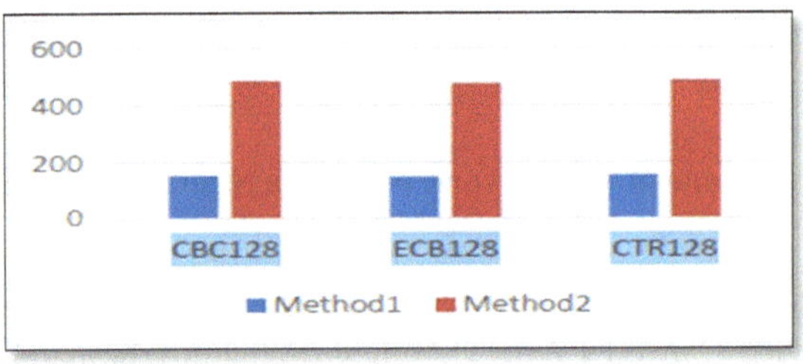

Fig. 9. Comparison of Method 1and Method 2 for encryption [14].

After analysis find with the studies DES and 3DES consume more energy compared with AES.CTR mode proved to be more efficient when it comes to decryption of data, compared with other AES modes.

Case Study 2: - In this article Anomaly Detection in IoT with Cooja Simulator the main purpose of this study is to design it is expected to be able to build simulation about the detecting the anomaly of the IoT network using Cooja Simulator and how the attack work and how to defend them [15].They design the system for the simulation (Fig. 11)

The simulation will take place in Contiki Operating System with Cooja Simulator using Rime stack to communicate each node [15]. For evaluation they create some IoT

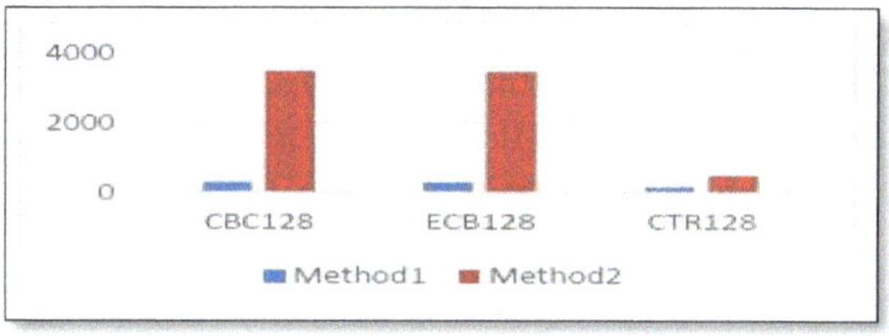

Fig. 10. Comparison of Method 1and Method 2 for decryption [14].

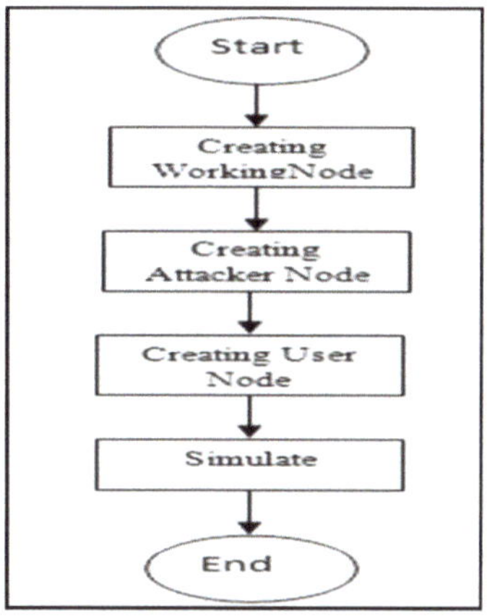

Fig. 11. Design of simulation process [15].

network scenario for four type of node, interactions between each pair of two different types of nodes are individually tested with and without the defending algorithm.

Scenario 1 & 2

In this scenario, one attacker node and one working node are placed in an IoT local network, attacker node requests for service every 1 to 2 seconds and will not stop until the end of simulation, after analysis of this scenario generate the result without defending algorithm and then applied with defending algorithm, the working node could distinguish the malicious peers and reject their requests after serving them for the first time. The records of malicious nodes are archived in the record list, which is indicated by the growing length in the record list [15].

Scenario 3

In this scenario, one user node and one working node are placed in an IoT local network, user node starts asking for service after the simulation begins for 10 seconds. The working node is expected to service the request and output the job status. If the request is served, the working node returns the served status with an enum in a unicast message to the user node. Then, the user node will print the message about its request has been served by the node of the responder to indicate the completion. Otherwise, it will send a rejected message back to the user node to notify it being unable to fulfill the request [15]. They used Bayesian network, which are suited for anomaly detection because they can handle high dimensional data which humans find difficult to interpret. Based on the simulation result they conclude that the proposed studies could effectively help the working node in an IoT network to distinguish malicious requests from attacker ones and process them differently.

4.2 Case Study of Some Related Work on OmNet++ simulator for Design of IoT Scenario: -

Case Study 1: -

In this article **A Beacon-Based BLE Mesh-Routing Algorithm for Smart Homes** [16] they present the beacon-based BLE routing algorithm for smart home applications. The reason for selecting beacons is to reduce the computational complexity and required resources for the implementation of the BLE routing algorithm while achieving compatibility with the BLE-enabled mobile devices. They proposed a BLE packet structure formatted with Eddystone protocol specification and BLE routing algorithm to slotted medium-access mechanism, where the time slots are divided into three regions called info, data and sleep (Figs. 12 and 13).

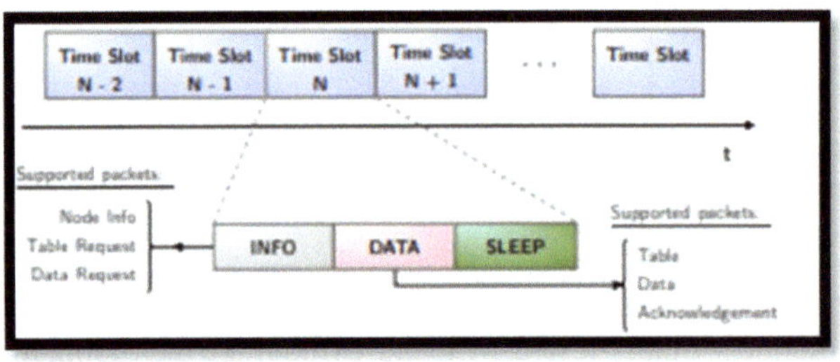

Fig. 12. BLEER algorithm time slotted medium access [16].

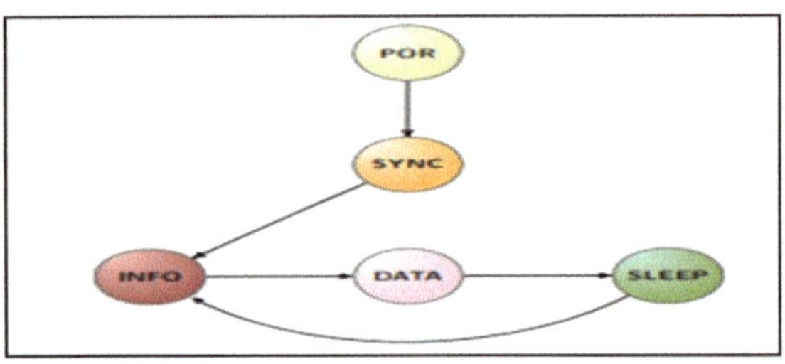

Fig. 13. Routing algorithm BLEER state machine [16].

The slotted medium-access approach adopted to classify the traffic by its function so that the nodes which do not have any data to send can utilize the sleep mode for longer periods of time during a slot. As a result, the power consumption is reduced and longer battery life is achieved. For the simulations of algorithm, they use OMNeT++

discrete simulation environment. The obtained results are compared with a real-life test network consisting of nRF52 BLE SoC's. The simulation results slightly exceed those from the real-life network due to the inherited limitations of the OMNeT++ simulator [16]. Result indicate that the proposed algorithm is good for implementation for IoT application where the energy efficiency is a key priority and propagation delay is not critical.

5 Challenges in Internet of Things

Security is a major challenge in IoT because it contains the protection of the data and system from the unauthorized access. It Device security , Network security ,Data security, Privacy The organization or designer should pay the attention of implementing security measures like encryption, firewall and software updates. To applying the security as priority basis, it protects the data and system in IoT environment and reduce the risk of cyber-attacks. **Privacy** is critical issue in IoT because connected IoT devices collect personal sensitive data, store it and exchange them in between. Taking privacy as prioritizing, organizations ensure that individuals' rights and freedoms are respected, and that sensitive information is protected from unauthorized access or misuse [17]. In IoT **scalability** is major issue in term of increasing number of connected devices, increased size of data and communication. To set scalability as priority organizations can confirm that their IoT systems can handle the growing number of connected devices and continue to deliver high performance and efficiency [17]. **Reliability** is system ability to perform the work consistently without any failure. IoT challenges related to reliability device failure, network connectivity, data accuracy. To addressing these challenges the organization and designer of the system should implement robust and reliable hardware and software and to perform regular testing and resolve any device issue to maintain it [17]. **Consumption of Power** the amount of energy used by the IoT devices and system. In IoT is an critical issue because many devices are intended to be small, low power and operated using batteries. It include battery life, energy efficiency, and power management to handling this adopt low power technologies and energy efficient design for IoT device The power consumption minimization increase the system battery life and reduce cost [17]. **Deployment challenges in IoT** related to connectivity ,Cross platform capability, Data collection and processing, Lack of skill set, Integration, Network infrastructure, Device management, Data management, Security, cost. IoT deployment in an organized and well-structured manner, organizations ensure that their IoT systems bring the intended benefits and results with a cost-effective result [17].

6 Conclusion

This paper briefly covers IoT concept, their architecture, their domain area and types of communication. This paper give shows the comparison of some present simulators and analysis of algorithms and their implementation techniques on Contiki Cooja and OmNet++ on the basis of previous work. After study of previous related work, we found that Cooja emulator is best for debugging and OmNet+ perform discrete event simulation with some inherited limitation. Future work will continue to with the security testing of the tools with the other scenario.

References

1. Internet of things: Wikipedia. (2024). Accessed: 29 May 2024. https://en.wikipedia.org/w/index.php?title=Internet_of_things&oldid=1232787992
2. Majid, M., et al.: Applications of Wireless Sensor Networks and Internet of Things Frameworks in the Industry Revolution 4.0: A Systematic Literature Review. Sensors **22**(6), (2022). https://doi.org/10.3390/s22062087
3. Malik, F.: So, What Exactly Is The Internet Of Things (IoT)? FinTechExplained. Accessed: 30 May 2024. https://medium.com/fintechexplained/so-what-exactly-is-the-internet-of-things-iot-89e165597b1d
4. "Architecture of Internet of Things (IoT)," GeeksforGeeks. Accessed: 30 May 2024. https://www.geeksforgeeks.org/architecture-of-internet-of-things-iot/
5. Pericherla, S.: Domain Specific IoT Applications - IoT Tutorial for Beginners. Startertutorials Blog. Accessed: 1 June 2024. https://www.startertutorials.com/blog/domain-specific-iot-applications.html
6. Mishra, H.: Physical Design of IoT. IoTbyHVM. Accessed: 1 June 2024. https://iotbyhvm.ooo/physical-design-of-iot/
7. "Types of Communications in IOT," GeeksforGeeks. Accessed: 2 June 2024. https://www.geeksforgeeks.org/types-of-communications-in-iot/
8. Patel, N., Mehtre, B., Wankar, R.: Simulators, Emulators, and Test-beds for Internet of Things: A Comparison, p. 145 (2019). https://doi.org/10.1109/I-SMAC47947.2019.9032519
9. "Internet of Things." Accessed: 2 June 2024. https://in.mathworks.com/solutions/internet-of-things.html
10. "Home | IoTIFY Network Simulator." Accessed: 3 June 2024. https://docs.iotify.io
11. "Contiki Cooja Simulator : Specially Designed Wireless Sensor Networks," Ns3 Projects. Accessed: 3 June 2024. https://ns3simulation.com/contiki-cooja-simulator/
12. "Top 5 List of IoT network simulator | Simulation Parameters for IoT network," Network Simulation Tools. Accessed: 21 Jul. 2024. https://networksimulationtools.com/iot-network-simulator/
13. Gaurav, D.K.: Programming Internet of Things using Contiki and Cooja." Accessed: 3 June 2024. https://www.opensourceforu.com/2017/06/programming-internet-things-using-contiki-cooja/
14. Ebobissé Djéné, Y., Berrazzouk, A., Brahim, E.B., Fakhri, Y.: Encryption Algorithms for WSN - IOT: A Software based Analysis with Contiki Cooja, p. 141 (2020). https://doi.org/10.5220/0010430601350141
15. Sanjaya, M.I.R., Putrada, A.G., Suryani, V.: Anomaly Detection In Iot Using Cooja Simulator. EProceedings Eng. **8**(2) (2021). Accessed: 3 June 2024. https://openlibrarypublications.telkomuniversity.ac.id/index.php/engineering/article/view/14696
16. Memic, S., Ibric, S.H.: A Beacon-Based BLE Mesh-Routing Algorithm for Smart Homes
17. "Challenges in Internet of things (IoT)," GeeksforGeeks. Accessed: 3 June 2024. https://www.geeksforgeeks.org/challenges-in-internet-of-things-iot/

Reliable Broadcast and Multi-channel Access in Hybrid Traffic Scenarios

Sunil Kumar, Lopamudra Hota, Biraja Prasad Nayak, and Arun Kumar(✉)

Department of Computer Science and Engineering, NIT Rourkela, Rourkela, India
{222cs3385,519cs1015,519cs1005,kumararun}@nitrkl.ac.in

Abstract. Advancements in self-driving technology, facilitated by artificial intelligence and smart transportation systems, have significantly enhanced safety and efficiency in autonomous vehicles, particularly through the concept of platooning. Recent studies highlight the potential of vehicular adhoc networks (VANETs) in optimizing the performance of self-driving cars when they travel closely together. Nevertheless, these studies often neglect the coexistence of self-driving and human-driven vehicles on the road. Addressing this, WAVE emerges as a crucial technology for communication in such mixed traffic scenarios. The WAVE standard has one CCH and six SCHs. To ensure efficient communication the proposed work considers platoon based multi-channel access for broadcast. The broadcast mechanism incorporates RTS/CTS for a multicast mode of communication. This paper provides a comprehensive throughput performance analysis of both Simple and Platoon-Centric broadcast techniques.

Keywords: MAC · WAVE · multi-channel · platoon · contention window

1 Introduction

The rapid advancement of technology and smart computers are making self-driving cars get better faster. Soon, lots of people will be able to use cars that drive themselves. But, these cars still have some problems. They might not see things far away, or bad weather can make it hard for them. Also, obstacles can be tricky for them to notice, especially in busy traffic. To help with this, there's a thing called vehicular platooning [1]. It means cars can drive really close, which makes roads less crowded, helps with wind resistance, and lets the cars share information to stay safe.

For platoons, efficiency and reliability in inter-vehicle communication is very crucial. The contention-free and contention-based focuses on improving reliability and efficiency. Hybrid traffic scenarios involve Human-Driven Vehicles (HDVs) and Autonomous Platoons (APs) with hybrid modes of communication considering the different characteristics of these vehicle types [2].

The stability of vehicle platoons relies heavily on efficient inter-vehicle communication. In the United States, Wireless Access in Vehicular Environments (WAVE) is assigned 75 MHz in the 5.9 GHz spectrum. This spectrum is partitioned into six service channels (SCHs) and one control channel (CCH) [3]. The CCH primarily supports the exchange of platoon coordination signals, safety notifications, and other critical control

B. K. Mishra et al. (Eds.): ICT4GS 2024, ISEM 56, pp. 375–386, 2025.
https://doi.org/10.1007/978-3-031-94997-5_32

messages. In contrast, the SCHs handle non-safety-related communications. The IEEE 802.11p protocol employs contention-based CSMA/CA to manage channel access [4]. To facilitate channel coordination, the IEEE 1609.4 standard provides a framework that divides time into synchronization intervals (SIs). Each SI consists of a Service Channel Interval (SCHI) and a Control Channel Interval (CCHI), each spanning 50 ms. At the beginning of intervals, there is a guard interval (GI) of 4 ms. Vehicles can exchange timing information or use the Global Positioning System (GPS) for synchronization [5].

2 Related Works

The study [6] offers a comprehensive review of platooning vehicular networks. Meanwhile, [7] introduces a specialized intra-platoon communication framework tailored for Cooperative Adaptive Cruise Control (CACC) mode of communication between leader and follower. This system employs a contention-free MAC protocol combined with adaptive transmission power control. The work [8] proposes a transport layer retransmission scheme for reliability with bounded delay. Our protocol, presented in this paper, improves both Control Channel and Service Channel communications, offering a comprehensive solution. The research [9] explores communication mechanisms for heterogeneous platoons, while [10] presents a hybrid traffic communication scheme, dividing the Control Channel Interval for platoon beaconing and event-driven safety messages, despite [10]'s evaluation using packet ratios, its SCH reservation process lacks, leading to limited improvement in safety message reliability. In contrast, the proposed protocol enhances both CCH and SCH channel communications. In [11], focus on an infrastructure-aided framework that ensures timely and reliable message delivery but relies on costly roadside units. Our protocol stands out by eliminating the need for infrastructure and providing a cost-effective solution.

The research [12] introduces ADHOC MAC, a TDMA-based scheme ensuring reliable broadcast by synchronizing slot status. Extended in VeMAC [13] for collision reduction, it incurs a high synchronization cost. The research [14] presents a novel protocol emphasizing predictable communication with interference control. In our proposed protocol, a simplified version with an enhanced RTS/CTS mechanism, reducing communication overhead, is implemented. DMMAC [15] introduces a hybrid protocol, dividing CCHI into adaptive broadcast frames and contention periods. FMC-MAC [16] adjusts CCHI and SCHI lengths based on traffic densities, utilizing roadside units. Despite dynamic TDMA slots, frequent overlapping occurs due to unstable topology, requiring frequent reallocation.

3 System Model

3.1 Traffic Model

In a mixed traffic scenario, autonomous platoons and human-driven vehicles coexist on a multi-lane highway, utilizing the WAVE communication protocol. Each vehicle is equipped with GPS and sensors to communicate with other vehicles, sharing real-time data such as vehicle dynamics and AI-related features through regular beacons.

Within the platoon, APs maintain closer spacing through cooperative maneuvers, ensuring effective communication. To ensure communication stability, there are limits on the size of platoons, preventing too much communication and keeping direct connections within the platoon. This approach, known as the one-hop cluster concept, minimizes communication delays and synchronization issues, ensuring the reliability and performance of the platoon even with a fixed topology over short periods. Figure 1 presents the communication where the side vehicles in red represent the APs flowing a dedicated lane and the other vehicles represent the HDVs in blue. Information is exchanged between the leader of the following and the end vehicle of the leading platoon.

In the context of a multi-platoon system (MPS) sharing the same lane, multiple platoons are denoted as A_i, with A_{-1} representing the last platoon. Within each platoon, individual vehicles are designated as $V_{i,j}$, where i ranges from 0 to $N-1$, and $V_{i,-1}$ signifies the last vehicle of a platoon. Safety is ensured by maintaining a larger inter-platoon spacing compared to intra-platoon spacing. Specifically, $V_{i,-1}$ can communicate directly with $V_{i+1,0}$. However, vehicles like $V_{i,0}$ and $V_{i+1,0}$, or $V_{i,-1}$ and $V_{i+1,-1}$, cannot communicate directly, effectively creating a three-hop cluster between consecutive platoons. This structure eliminates interference between non-adjacent platoons. The platoon leading takes responsibility for inter-platoon coordination. In case a platoon leader doesn't receive beacons from platoons ahead, it identifies its own platoon as the leading one within the MPS.

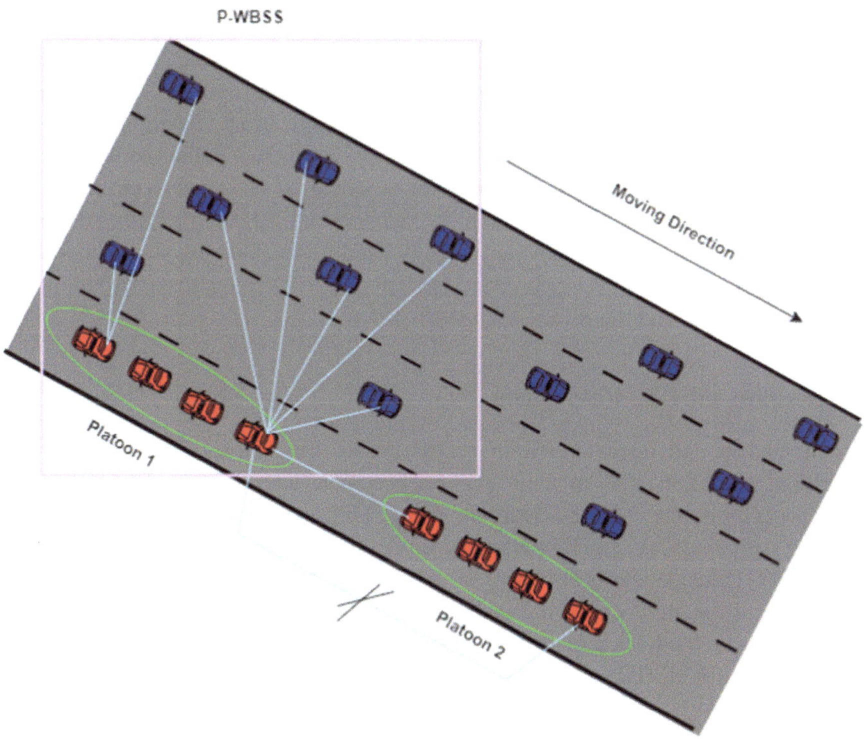

Fig. 1. Platoon Traffic model consisting of APs and HDVs.

3.2 Platoon-Centric Mechanism

The proposed communication scheme includes APs and its immediate neighboring HDVs collectively form a Platoon-based Wireless Service Set (P-WSS). HDVs can join a P-WSS upon receiving packets from the corresponding AP, exiting if they miss consecutive Synchronization Intervals. When HDVs simultaneously receive signals from two APs, they prioritize the one with direct communication to the AP leader or join behind if both platoons have leaders. Each P-WSS has a unique BSSID, while BSSID is reserved for HDVs and is not in any P-WSS.

The AP within a P-WSS serves various roles like scheduler and access point to Roadside Units (RSUs). The proposed scheme employs APs as schedulers for multi-channel access within related P-WSSs, with platoon leaders executing the algorithm. In a P-WSS, HDVs either directly communicate with the leader ($H_{i,k} \in \Theta_i$) or rely on relaying by the platoon tail ($H_{i,k} \in \Delta_i$), with a linear topology connecting those in Δ_i to $V_{i,-1}$. In a multi-platoon system, $\Delta_i \subset \Theta_{i+1}$ for $i = 1$ to $N-2$. The focus is specifically on the hybrid traffic system.

4 RTS/CTS-Based Broadcasting for Platoons

The broadcast of CCH lacks a retransmission mechanism, leading to potential issues. When collisions occur, vehicles attempting to transmit occupy the channel for extended periods, decreasing channel utilization. Prior solutions like UMB, SB, and BPAB have addressed this problem by revising RTS/CTS mechanisms. However, these traditional mechanisms introduce an extra contention process for determining CTS responders, which adds overhead and delays to multi-hop communications. In the proposed approach, APs act as CTS responders, and the process of channel contention differs from the RTS/CTS handshake. The mechanism for broadcast is designed to ensure the reliability of periodic safety-related message broadcasts on CCH, particularly cooperative awareness messages (CAMs). CAMs typically used for exchanging status information among neighboring vehicles, are broadcasted using a two-step process: determining the CTS responders set and the RTS/CTS handshake.

4.1 Deciding the CTS Responders

Platoon leaders and interlink within access points function as CTS responders by embedding a CTS header in their platoon beacons. The CTS header contains the sender's position, the P-WSS identifier, and a GPS-synchronized timestamp. Upon receiving such a beacon, an HDV adds the sender to its local list of the CTS responder. In case if a CTS responder is not heard from for two consecutive Synchronization Intervals (SIs) or if the HDV exits the P-WSS, it is removed from the list. Due to the small size of APs, which allows direct communication, the leaders and tails of the APs act as CTS providers for all HDVs within the P-WSS.

4.2 RTS/CTS Handshake

In a P-WSS, when an HDV intends to broadcast, it first consults its local CTS responder list and selects the nearest available responder. The HDV then transmits a short RTS packet containing the transmission duration, the CTS responder's address, and relevant control bits. An idle CTS responder replies with a CTS packet, marking itself as busy until the payload transmission is complete and updating the network allocation vector (NAV). If no CTS response is received within a specified timeframe, the HDV retries by sending a new RTS. Other vehicles that detect an RTS or CTS packet also update their NAVs to avoid collisions. In this single-hop broadcast process, acknowledgment packets (ACKs) are not used. If an HDV cannot locate a CTS responder, it broadcasts directly without utilizing the RTS mechanism. Separating CTS responder selection from the RTS/CTS handshake reduces contention overhead and improves channel efficiency.

5 Result Analysis

5.1 Simple Broadcast

The process by which HDVs access the CCH for broadcasting can be modeled as a 1-D Markov process, as illustrated in Fig. 2. The state variable in this model is the backoff counter, represented as $b(t)$. When a packet is ready for broadcasting, the backoff counter is initialized to a random value selected from the range $(0, W)$, where W represents the size of the contention window. The backoff counter decrements during idle slots but remains static during busy slots. Once the backoff counter reaches zero ($b(t) = 0$), the packet is forwarded to the physical layer for transmission.

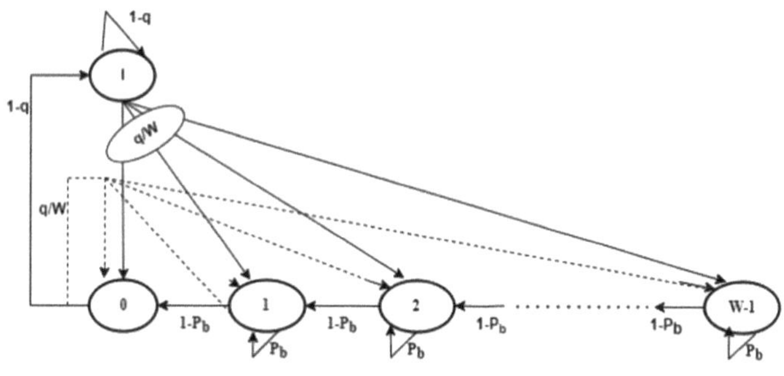

Fig. 2. One-dimension Markov Chain.

$$\begin{cases} P(I|I) = 1 - P_q, & \\ P(k|I) = \dfrac{P_q}{W}, & k \in [0, W), \\ P(k|k+1) = 1 - P_b, & k \in [0, W-1), \\ P(k|k) = P_b, & k \in [0, W), \\ P(I|0) = 1 - P_q, & \\ P(k|0) = \dfrac{P_q}{W}, & k \in [0, W), \end{cases} \tag{1}$$

Here, the probability of busy channel is P_b, and P_q is the probability of a packet being transmitted. Let b_k for $k \in [0,W)$ represent the backoff counter, while b_I denotes the idle channel probability.

The probability of transmission, τ, is given as:

$$\tau_s = \left(\frac{1}{P_q} + \frac{(W-1)}{2(1-P_b)}\right)^{-1} \tag{2}$$

where:

$$P_b = 1-(1-\tau)^n \tag{3}$$

The generation of HDVI safety packets is expressed as:

$$n_{\text{SCH}} = \lambda\,(T_{\text{SCHI}} + T_{\text{GI}} + T_{\text{API}}) \tag{4}$$

Here, $T_{\text{SCHI}} = 50\,\text{ms}$ represents the duration of the SCH Interval (SCHI), $T_{\text{GI}} = 4\,\text{ms}$ is the guard interval, and T_{API} depends on the platoon size and the TDMA time slot. The probability of a packet being ready for transmission is:

$$P_q = 1 - e^{-\lambda T_{\text{slot}}}\left(1 - \frac{T_{\text{S}} \cdot n_{\text{SCH}}}{T_{\text{CCHI}} - T_{\text{GI}}}\right) \tag{5}$$

where the slot duration, T_{slot}, is given by:

$$T_{\text{slot}} = (1-P_{\text{tr}})\sigma + P_{\text{tr}}P_{\text{s}}T_{\text{s}} + P_{\text{tr}}(1-P_{\text{s}})T_{\text{c}} \tag{6}$$

In this context, σ idle slot, T_s is time of successful transmission, and T_c is the time of packet collision. For basic broadcasting, $T_s = T_c = T$, and:

$$T = \frac{H+E[P]}{R} + \text{DIFS} + \delta \tag{7}$$

Here, H represents the combined size of the physical and MAC layer headers, δ denotes the delay in propagation. The bitrate is denoted as R.

The probability of at least one node transmitting in a given slot is:

$$P_{\text{tr}} = 1-(1-\tau_s)^n \tag{8}$$

The conditional success probability is:

$$P_s = \frac{n\tau_s(1-\tau_s)^{n-1}}{P_{\text{tr}}} \tag{9}$$

The normalized throughput can be expressed as:

$$S_s = \frac{n\tau_s(1-\tau_s)^{n-1}E[P]}{R \cdot T_{\text{slot}}} \tag{10}$$

5.2 Platoon-Centric Broadcast

This mechanism involves the retransmission of the RTS packet when the CTS is missed, accompanied by the doubling of the contention window (Fig. 3).

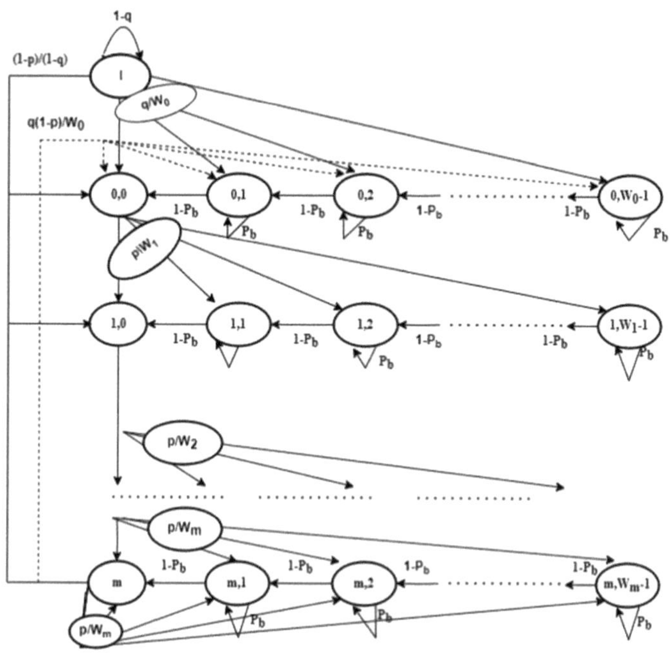

Fig. 3. Two-dimension Markov Model.

To accurately depict the access mechanism, a two-dimensional Markov model is essential. The state variable is $s(t)$, termed as the backoff stage, extends the model. This results in state variables represented as $\{s(t), b(t)\}$, where $b(t)$ is the backoff counter and $s(t)$ signifies the backoff stage for additional insights. Let $W = CW_{\min}$ represent the minimum contention window. If the maximum contention window is $CW_{\max} = 2^m W$, with m being the maximum backoff stage, the two-dimensional process $\{s(t), b(t)\}$ is established.

The one-step transition probabilities are:

$$\begin{cases} P(I|I) = 1 - P_q, & \\ P(i,k|i,k+1) = 1 - P_b, & i \in [0,m], k \in [0, W_i - 2], \\ P(i,k|i,k) = P_b, & i \in [0,m], k \in [1, W_i - 1], \\ P(0,k|I) = \dfrac{P_q}{W_0}, & k \in [0, W_0 - 1], \\ P(I|i,0) = (1-p)(1-P_q), & i \in [0,m], \\ P(0,k|i,0) = \dfrac{P_q(1-p)}{W_0}, & i \in [0,m], k \in [0, W_0 - 1], \\ P(i,k|i-1,0) = \dfrac{p}{W_i}, & i \in [1,m], k \in [0, W_i - 1], \\ P(m,k|m,0) = \dfrac{p}{W_m}, & k \in [0, W_m - 1]. \end{cases} \tag{11}$$

The transmission probability is given by:

$$\tau_{\text{pc}} = \frac{P_q}{(1-p)b_I}, \tag{12}$$

where $p = 1 - (1 - \tau_{\text{pc}})^{n-1}$. These equations can be solved numerically to determine τ_{pc} and p.

The stationary probability b_I satisfies:

$$1 = b_I \left[1 + \frac{P_q(1-2p)(W+1-2P_b) + P_q W(1-(2p)^m)}{2(1-P_b)(1-p)(1-2p)} \right]. \tag{13}$$

For the protocol involving the CCHI and SCHI reservation process, the transmission probability is:

$$P_q = 1 - e^{-\lambda T_S} \left[1 - \frac{T_S \cdot (n_{\text{SCH}} + 1)}{T_{\text{HDVI}}} \right]. \tag{14}$$

The average slot duration is:

$$T_{\text{slot}} = (1 - P_{\text{tr}})\sigma + P_{\text{tr}} P_{\text{s}} T_s^{\text{rts}} + P_{\text{tr}}(1 - P_{\text{s}}) T_c^{\text{rts}}, \tag{15}$$

where T_s^{rts} and T_c^{rts} are:

$$\begin{aligned} T_s^{\text{rts}} &= \text{RTS} + \text{SIFS} + \delta + \text{CTS} + \text{SIFS} + \delta + \frac{H + E[P]}{R} + \text{DIFS} + \delta, \\ T_c^{\text{rts}} &= \text{RTS} + \text{EIFS} + \delta. \end{aligned} \tag{16}$$

The normalized throughput is:

$$S_{pc} = \frac{P_{tr} \cdot P_s \cdot E[P]}{R \cdot T_{slot}}. \tag{17}$$

Proposition 1: For saturated communication conditions:

$$\tau_{pc} < \tau_s, \quad p_{pc} < p_s, \tag{18}$$

where for platoon centric broadcasts τ_{pc} and p_{pc} are the probability ofpacket transmission and conditional collision probability, and τ_s and p_s represent these metrics for simple broadcasts.
Proof: Under saturated conditions ($q = 1$):

$$\tau_s = \left(1 + \frac{W-1}{2(1-P_b)}\right)^{-1}, \tag{19}$$

and:

$$\tau_{pc} = \left(1 + \frac{W - 1 + 2(1-p)(1-P_b) + W\left(\frac{1-(2p)^m}{1-2p}\right)}{2(1-P_b)}\right)^{-1}. \tag{20}$$

Since τ decreases with P_b, and P_b increases with τ, it follows that $\tau_{pc} < \tau_s$ and $p_{pc} < p_s$.

Proposition 2: For dense networks, the platoon-centric broadcast throughput (S_{pc}) surpasses that of simple broadcasts (S_s).

Proof: Considering $T \approx T_s \gg T_c \gg \sigma$:

$$S_s = \frac{E[P] \cdot P_s}{T \cdot R}, \tag{21}$$

and:

$$S_{pc} = \frac{E[P]}{T_s \cdot R}. \tag{22}$$

5.3 Simulation Parameters

The simulation parameters are presented in Table 1.

Table 1. Parameter Settings

Parameter	Value
T_{SCHI}	50 ms
T_{GI}	4 ms
T_{API}	1.8 ms
T_{CCHI}	46 ms
σ	16 µs
δ	2 µs
PHY header	48 bits
MAC header	224 bits
RTS	160 bits + PHY header
CTS	112 bits + PHY header
R	6 Mbps
$E[P]$	1000 bytes
W	16, 32

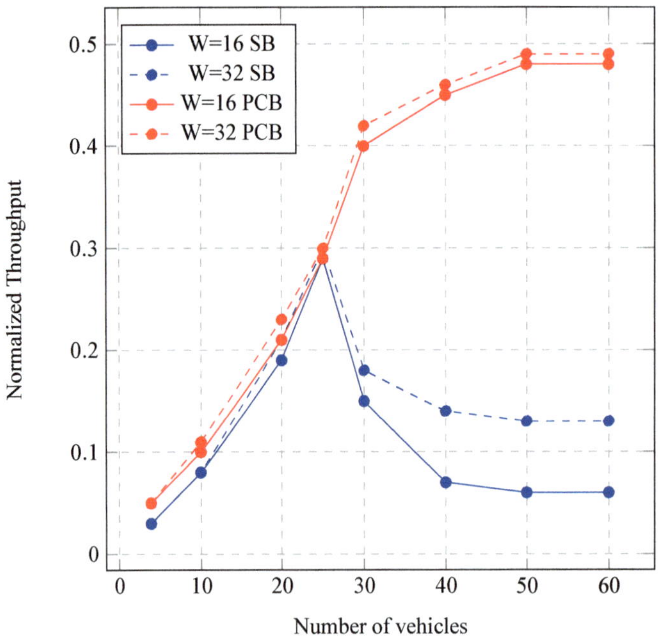

Fig. 4. Normalized throughput comparison for different traffic densities.

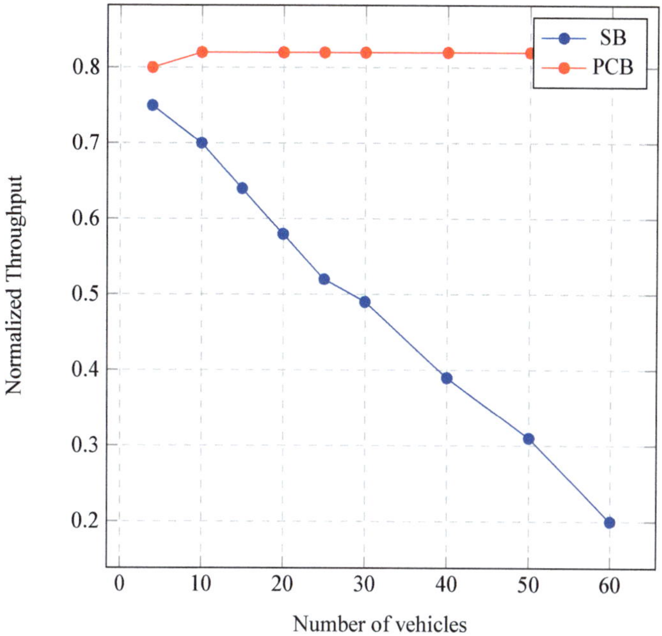

Fig. 5. Normalized throughput comparison for different traffic densities under saturation traffic.

The Fig. 4, depicts the normalised throughput at different traffic densities in a platoon-centric network, In this graph the comparison is made for both Simple Broadcasting (SB) and Platoon-Centric Broadcasting (PCB) for different contention window sizes W = 16, 32. Figure 5 illustrates the normalised throughput comparison for different traffic densities under saturation traffic.

6 Conclusion

The proposed mechanism considers hybrid traffic that includes APs and HDVs. The CCH Interval is divided into TDMA time intervals for APs CSMA/CA intervals for HDVs. Each autonomous vehicle (AV) is assigned dedicated slots to maintain quality of service. To enhance CCH throughput, an improved RTS/CTS platoon based broadcast scheme is employed. Numerical comparisons are made between simple broadcasting and platoon-centric broadcasting under various traffic densities. Future work will address scenarios involving multiple platoons in a single lane. Additionally, in real-world traffic, more than one lane may be allocated for APs, introducing a more complex situation. The P-WSS architecture can be expanded into a multi-AP P-WSS, where multiple APs within the same P-WSS collaboratively manage the allocation of channel resources for both APs as well as HDVs.

References

1. Huang, Y., Shen, Y., Wang, J., Zhang, X.: A platoon-centric multi-channel access scheme for hybrid traffic. IEEE Trans. Veh. Technol. **70**(6), 5404–5418 (2021)
2. Bianchi, G.: Performance analysis of the IEEE 802.11 distributed coordination function. IEEE J. Sel. Areas Commun. **18**(3), 535–547 (2000)
3. Lin, Z., Tang, Y.: Distributed multi-channel mac protocol for VANET: an adaptive frame structure scheme. IEEE Access **7**, 12868–12878 (2019)
4. Zhang, L., Liu, Z., Zou, R., Guo, J., Liu, Y.: A scalable CSMA and selforganizing TDMA MAC for IEEE 802.11 p/1609.x in VANETs. Wireless Pers. Commun. **74**(4), 1197–1212 (2014)
5. Wang, J.C.-P., Abolhasan, M., Franklin, D.R., Safaei, F.: Characterising the behaviour of IEEE 802.11 broadcast transmissions in ad hoc wireless LANs. In: Proceedings of IEEE International Conference on Communication, pp. 1–5 (2009)
6. Jia, D., Lu, K., Wang, J., Zhang, X., Shen, X.: A survey on platoon-based vehicular cyber-physical systems. IEEE Commun. Surv. Tutorialss **18**(1), 263–284 (2016)
7. Segata, M., et al.: Towards inter-vehicle communication strategies for platooning support. In: Proceedings of International Workshop on Communication Technologies for Vehicles (Nets4Cars-Fall), pp. 1–6 (2014)
8. Jonsson, M., Kunert, K., Böhm, A.: Increased communication reliability for delay-sensitive platooning applications on top of IEEE 802.11p. In: Berbineau, M., et al. (eds.) Nets4Cars/Nets4Trains 2013. LNCS, vol. 7865, pp. 121–135. Springer, Heidelberg (2013). https://doi.org/10.1007/978-3-642-37974-1_10
9. Jia, D., Ngoduy, D., Vu, H.L.: A multiclass microscopic model for heterogeneous platoon with vehicle-to-vehicle communication. Transportmetrica B: Transport Dyn. **7**(1), 448–472 (2019)
10. Liu, B., Jia, D., Lu, K., Ngoduy, D., Wang, J., Wu, L.: A joint control–communication design for reliable vehicle platooning in hybrid traffic. IEEE Trans. Veh. Technol. **66**(10), 9394–9409 (2017)
11. Liu, B., et al.: Infrastructure-assisted message dissemination for supporting heterogeneous driving patterns. IEEE Trans. Intell. Transp. Syst. **18**(10), 2865–2876 (2017)
12. Borgonovo, F., Capone, A., Cesana, M., Fratta, L.: ADHOC MAC: new mac architecture for ad hoc networks providing efficient and reliable point-to-point and broadcast services. Wireless Netw. **10**(4), 359–366 (2004)
13. Liu, Z., Han, Y., Fan, J., Zhang, L., Lin, Y.: Joint optimization of spectrum and energy efficiency considering the c-v2x security: a deep reinforcement learning approach (2020)
14. Li, C., Zhang, H., Zhang, T., Rao, J., Wang, L.Y., Yin, G.: Cyber-physical scheduling for predictable reliability of inter-vehicle communications. IEEE Trans. Veh. Technol. **69**(4), 4192–4206 (2020)
15. Lu, N., Ji, Y., Liu, F., Wang, X.: A dedicated multi-channel MAC protocol design for VANET with adaptive broadcasting. In: Proceedings of IEEE Wireless Communication Network Conference, pp. 1–6 (2010)
16. Yao, Y., Zhang, K., Zhou, X.: A flexible multi-channel coordination MAC protocol for vehicular ad hoc networks. IEEE Commun. Lett. **21**(6), 1305–1308 (2017)

A Comparative Analysis of Tesseract OCR and Amazon Textract for Handwritten Documents

D. Lakshmi Padmaja[1(✉)], S. Naga Prasad[2], G. K. Sri Harsha[3], and G. S. Deepak[4]

[1] Anurag University, Hyderabad, India
lakshmipadmajait@anurag.edu.in
[2] Department of Computer Science and Applications, Tara Government College (A), Osmania University, Hyderabad, India
[3] Arizona State University, Tempe, AZ, USA
[4] Cognizant Technology Solutions Corporation, Hyderabad, India

Abstract. Handwritten character recognition is a process of recognizing text from a handwritten image. This process is getting more and more attentive due to its wide range of applications. There are numerous Optical Character Recognition (OCR) instruments that identify and extract writing from pictures. These tools are utilized within the introductory steps of the investigation of scanned pictures to recognize and naturally prepare the data that the images contain. There are many OCR tools that are used for various purposes. Some of them are Google OCR, Azure OCR, Amazon Textract, and an open-source tool called Tesseract OCR. In this paper, our attempt is to compare Amazon Textract and Tesseract OCR with handwritten images as input, and we analyze the performance, accuracy, and execution time of various handwritten image datasets.

Keywords: Optical Character Recognition · Long Short-Term Memory · Convolution Neural Network · Support Vector Machine

1 Introduction

One technology that shows the most promise for the humanities and social sciences is optical character recognition (OCR) [1]. With the help of automated text extraction from digital photos, a significant number of hitherto unexplored historical writings can be computationally examined, possibly yielding important new understandings of human history. OCR tools, referenced as WinNT [2], can be employed to detect and extract material from records of varying complexity. In other cases, these tools are specifically used to extract content from preidentified regions of interest, such as paragraphs, tables, titles, and so on. Nevertheless, [1] OCR, or Optical Character Recognition, is a relatively new technology. The current software available for OCR varies in terms of accuracy. Typically, using a tailored method that involves unique pre-processing, preparation, or post-processing for a particular corpus produces the best outcomes even if these approaches may require a significant amount of time. The performance of pre-trained,

B. K. Mishra et al. (Eds.): ICT4GS 2024, ISEM 56, pp. 387–397, 2025.
https://doi.org/10.1007/978-3-031-94997-5_33

general OCR processors is of scientific interest due to their substantial potential for widespread adoption in the academic communsity [3].

2 Background

In this segment, we allow a brief diagram of the fundamental calculations of both algorithms used in Tesseract OCR and Amazon Textract [4]:

2.1 Background of Tesseract OCR

Tesseract is designed by various machine learning algorithms to extract handwritten text. Algorithms like Long Short-Term Memory (LSTM) [5], Convolutional Neural Network (CNN) 1 and Support Vector Machine(SVM). CNN, however, provides more precise results [6].

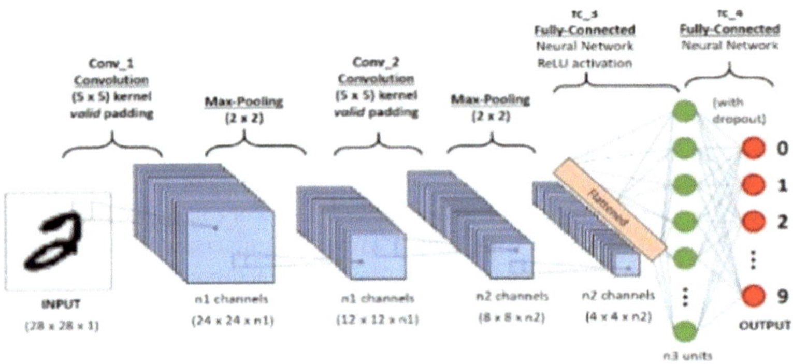

Fig. 1. Convolutional Neural Network (CNN)

Convolutional Neural Network (CNN): Figure 1 depicts a basic CNN model. The input layer is the top layer, and the input image is 28 by 28 pixels in size. The second layer is the convolution layer C2, and through convolution with the input picture, it can produce four separate feature map. The pooling layer P3 is the third layer. It calculates the input feature maps' local average or maximum [7, 8].

- Convolution: Convolution is mainly used for image manipulation, such as feature enhancement, camera smoothing and sharpening. It is not only used in CNNs, but also a core feature in several other algorithms for machine learning [8].
- Pooling: It is the moment's imperative component of CNN. Maximum pooling could be a well known approach for CNN method. Max-Pooling gives the most extreme esteem from its picture portion secured inside the kernel. But normal pooling gives way better clamor concealment [9, 10].
- Fully-connected neural network: At the final arrangement of CNN [11], completely connected neural systems are utilized. They are just completely as- sociated with

artificial neural networks. Amid the preparing session, weights that are associated with the arrangement are computed. Fully connected neural networks arrange computes the most-matched name to speak to the picture utilizing the convolution/pooling operation's yield. In this segment, the picture highlight vector and picture course are associated. The weights related to the organized association way are duplicated by the yields from the convolution/pooling operation. After that, the result is put through an actuation work [12].

2.2 Background of Amazon Textract

To extract and analyze handwritten text, Amazon Textract uses a variety of machine-learning methods. These algorithms include support vector machines(SVMs), neural networks such as artificial neural networks (ANNs) and deep learning techniques like recurrent neural networks (RNNs). Machine learning algorithms called ANNs are capable of learning to forecast the behavior of objects like neurons in a neural network [13]. Using a given data set, RNN, a particular kind of machine learning algorithm, can learn to predict the values in the data set's sequence. Using a set of training data, SVMs are a form of machine learning algorithm that can learn to predict the future.

With synchronous and asynchronous operations, Amazon Textract only returns the text it has found in a document. The following data is returned for both sets of operations in various Block objects [14, 15].

1. The text lines and words of the detected text.
2. The connections between the lines and words of the text that was detected
3. The webpage where the detected text is located
4. Where the lines and words of the text are located on the document page

A list of Block objects that contains detected text is returned by Amazon Tex tract operations. These objects stand for textual lines or words that have been found on a document page. Two lines of text comprised of numerous words are displayed in the text that follows: This is text.

In two separate lines. A Block object's Text field is where detected text is returned. The block type field establishes whether the text is a word or a line of text (word) [15]. A WORD is any combination of one or more non-spaced ISO basic Latin script characters. A line is a group of contiguous, tab-delimited words. Additionally, by using the Text type parameter, Amazon Textract can tell whether a passage of text was written by hand or printed. Both HAND- WRITTEN and painted are the results that are returned. For our datasets, we mostly employed the HANDWRITING Text type.

3 Literature Survey

Jay Amrutbhai Patel has connected the LSTM calculation and concluded that the framework they have used has great precision on printed and written content, but has direct precision when it comes to transcribed writings [16]. Future, So within the future the precision of manually written content can be expanded by training the tesseract at the next level. Hence, Tesseract 4. x and 5. x arrangements are valuable for discovering designs in writing that make recognizing content simple. In any case, they have not focused on manually written content.

Ray TeslaSmith, has expressed that tesseract is presently the driving commercial engine in terms of its exactness [17, 18]. Its primary quality is likely its abnormal choice of highlights. Its key shortcoming is likely its utilization of a polygonal estimation as input to the classifier rather than the raw outlines. And finally he concluded that the exactness may probably be moved forward essentially with the prudent addition of a Hidden-Markov-Model-based character n-gram and conceivably a made strides chopper [18].

Md. M. Hasan, Khandokar, utilized a profound learning procedure CNN is actualized for manually written character acknowledgement [19]. The center of this work is to explore CNN's capability to recognize the characters from NIST dataset preparation with a high degree of precision. It was found that the exact- ness obtained from 200 preparing pictures as 65.32% is moved forward steadily with expanding preparation pictures. The precision comes to 92.91% with prepared the 1000 preparing pictures. In this way, an advance increase in the preparation of pictures will proceed to increase the accuracy with respect to certain constraints [5]. Fathma Siddique, Md. Abu Bakr Siddique, and Shadman Sakib, had expressed the varieties of correctness for manually written digits that were watched for 15 years by changing the -covered up layers [20]. The exactness bends were created for the six cases for the diverse parameter using the CNN MNIST digit dataset. The six cases perform in an unexpected way because of the different combinations of covered-up layers. The layers were taken haphazardly in an occasional arrangement so that each case carries on in an unexpected way amid the try. The most extreme and least correctness were watched for distinctively covered-up layers with a bunch estimate of 100. Among all the perceptions, the greatest accuracy within the execution was found at 99.21% for 15 years and the least accuracy among all perceptions within the execution was found at 97.07% [21].

4 Experimentation

4.1 Dataset

We have collected and experimented with 100 handwritten samples randomly, among which 10 samples' results are presented in this paper.

Figure 2b shows the output of Textract for the input Fig. 2a. Figure 2c shows the output of tesseract for Fig. 2a. By comparing the input and output images, we can say Textract has more accuracy as most of the text is recognized correctly.

Figure 3b shows the output of Textract for Fig. 3a. Figure 3c shows the output of tesseract for Fig. 3a Here, the text is not recognized properly so it has less accuracy compared to Textract. By comparing with input and output images we can say Textract has more accuracy as most of the text is recognized correctly.

4.2 Additional Factors: Ease of Integration, Scalability, and Cost

1. Ease of Integration

(a) Tesseract OCR: Tesseract OCR is an open-source tool that is very adaptable and can be quick and easily added to many different systems. It works with many computer

(a) Handwritten Document Sample 1 (b) Textract (c) Tesseract

Fig. 2. Tesseract and Textract Output Comparison for Sample 1

(a) Handwritten Document Sample 2 (b) Textract (c) Tesseract

Fig. 3. Tesseract and Textract Output Comparison for Sample 2

languages and can be built into many different programs. Because Tesseract is open source, coders can change it and add to its features to suit their needs. Adding the Tesseract library to the project and using its API to process pictures and pull out text is a common way to integrate it.

(b) Textract from Amazon: Because Textract is part of the AWS environment, it works well with other AWS services like S3 for storage, Lambda for serverless computing, and SageMaker for machine learning. Users who are already familiar with the AWS services can easily connect Tex- tract to their account. Setting up the right permissions and using the Textract API to look at papers are part of the process. For businesses that already use AWS, this integration can make processes much easier.

2. Scalability

(a) Tesseract OCR: The ability of Tesseract to grow depends on how it is deployed. Since it is an open-source tool, it can be used on a number of cloud systems or servers that are located on-site. The ability to spawn multiple OCR processing jobs at the same time affects how scalable the system is. But managing and expanding the system takes a lot of work and knowledge.

(b) Amazon Textract is a program that can handle handling a lot of documents at once. When you use a managed service, it adjusts itself to handle the amount of work that comes in, so you don't have to do any- thing to process thousands of papers. This ability to grow is a big plus for businesses that need to process a lot of documents on a daily basis.

3. Cost

(a) Tesseract OCR is free to use because it is open source. Most of the costs of using Tesseract come from the hardware and software needed to run it. Businesses need to think about how much computers, storage, and upkeep cost. As a result, Tesseract is a good choice for small projects or those with limited funds.

(b) Amazon Textract: Users are charged based on how many pages they process with Amazon Textract when they use it. This might be a good deal for small to medium amounts of paper, but it can get more expensive when used on a large scale. Additionally, businesses must think about the prices of other AWS services they use with Textract. The managed nature and flexibility can make up for the higher costs for many businesses.

4.3 Experimental Environment

Windows 10 working operating system as test platform,

- CPU is Intel Center I7 6500u, which has double centers running on 2.4GHZ
- GPU is Nvidia GTX 970, 4 GB GDDR5
- Flash 16 GB

4.4 Experiment's Evaluation Factors:

The three calculations are assessed in understanding with the proposed demon- strate agreeing to the taking after Components [22].

- Accuracy
- Execution Time

5 Experimental Result and Discussion

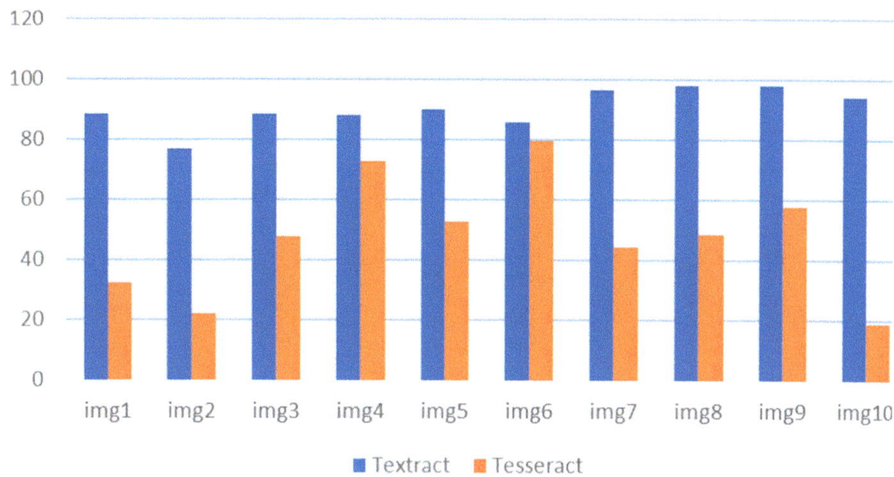

Fig. 4. Accuracy

Figure 4 shows the comparison of the accuracy of Textract and tesseract. Textract is more accurate than tesseract.

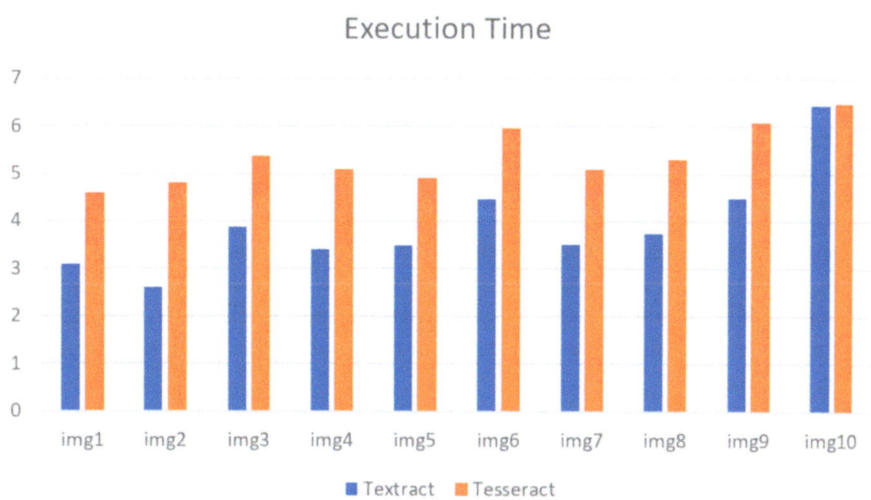

Fig. 5. Execution Time

Figure 5 shows the comparison of execution time for Textract and tesseract. Here the execution time of tesseract is less compared to Textract.

Sno.	File name	File size	Software tool	Accuracy	Execution Time
1.	Img1	45KB	Textract	88.54%	4.6 sec
			Tesseract	32.20%	3.1 sec
2.	Img2	120KB	Textract	76.92%	4.8 sec
			Tesseract	21.79%	2.6 sec
3.	Img3	95KB	Textract	88.60%	5.37 sec
			Tesseract	47.46%	3.87 sec
4.	Img4	430KB	Textract	87.87%	5.1 sec
			Tesseract	72.72%	3.42 sec
5.	Img5	46KB	Textract	90.13%	4.9 sec
			Tesseract	52.63%	3.49 sec
6.	Img6	103KB	Textract	85.57%	5.97 sec
			Tesseract	79.69%	4.47 sec
7.	Img7	109KB	Textract	96.49%	5.09 sec
			Tesseract	44.29%	3.51 sec
8.	Img8	107KB	Textract	98.18%	5.3 sec
			Tesseract	48.40%	3.74 sec
9.	Img9	90KB	Textract	98.10%	6.1 sec
			Tesseract	57.77%	4.48 sec
10.	Img10	172KB	Textract	94.33%	6.5 sec
			Tesseract	18.86%	6.45 sec

Fig. 6. Results of comparison

In the above Fig. 6 describes the comparison between accuracy and execution time of tesseract and Textract. Textract has high accuracy and tesseract has low execution time.

6 Practical Considerations and Challenges Encountered

The Challenges encountered during Testing phase are mentioned below.

1. **Accuracy with Complex Handwriting:** Tesseract and Textract both had trouble with complicated and varied handwriting styles. Even though Tesseract was taught on a lot of data, it had trouble with inconsistent hand- writing, which made it less accurate. Textract was generally more accurate, but it had trouble with handwriting styles that weren't common.

To deal with this problem, we added more pre-processing steps such as image enhancement and noise reduction. We also tried using custom training data for Tesseract to make handwriting detection more accurate for certain styles.

2. **Processing Speed and Execution Time:** Processing time was very important, especially for large data sets. Due mainly to its lightweight nature, Tesseract showed faster processing times than Textract. However, the time it took to upload papers to AWS and get results slowed down Textract's execution time.

To cut down on the time it took, we optimized the processing pipeline by uploading multiple documents at the same time and handling them all at the same time. Another thing we did was use AWS Lambda to run jobs in parallel, which helped us handle the load and boost throughput.

3. **Integration Complexity:** Setting up and configuring Tesseract needed more manual work, such as managing dependencies and setting up the environment. Textract, on the other hand, needed approval from AWS and management of API calls, which could be hard for people who aren't familiar with AWS.

To make the setup process go faster, we made interaction scripts and templates for both tools. Enclosing the environment in Docker containers made Tesseract's distribution easier. Textract used CloudFormation templates to set up all of their AWS services automatically.

4. **Scalability Issues:** extending Tesseract required a lot of manual work, like setting up extra servers and overseeing load balancing. AWS took care of Textract's ability to grow, but there were concerns about how to handle costs as the number of users increased.

To make the scaling process automatic for Tesseract, we set up auto-scaling groups and load balancers. Textract used budget alerts and cost optimization techniques as part of cost management methods to keep costs in check.

7 Conclusions

In this paper, we compared two OCR instruments for penmanship acknowledgment to decide the leading apparatus in terms of numerous components such as precision and execution. Other criteria such as execution time have been too taken in thought. Arbitrary and standard datasets of transcribed scripts have been utilized to assess the calculations. It appears that Textract is the most excellent device in terms of precision. Tesseract shows great in terms of execution time. Be that as it may, tesseract is open source program which can be sent on any framework though Textract may be a benefit advertised by AWS, it is as it were free for 1 year after making an account.

In sum, Tesseract OCR and Amazon Textract both have good and bad points. Tesseract is quick and doesn't cost much, but it takes a lot of work to grow and connect. Amazon Textract costs more than other options, but it has better accuracy and flexibility and works well with other AWS services. Which one an organization chooses will rest on its needs and limitations, such as its budget, existing infrastructure, and the number of documents it needs to process.

8 Future Work

There are many OCR tools available which can be done by using the existing tools. If you are looking for a tool that gives more accurate results go with Amazon Textract. If you are looking for a tool that performs fast in execution, go with Tesseract OCR. It has to be noted that these results are applicable to handwritten scripts only.

References

1. Gomez, L., Karatzas, D.: Object proposals for text extraction in the wild. In: 2015 13th International Conference on Document Analysis and Recognition (ICDAR), IEEE, pp. 206–210 (2015)
2. : Capital business systems inc: https:www.capitalbusinessinc.com/. Accessed: 17 June 2024
3. Dengel, A., Hoch, R., Hönes, F., Jäger, T., Malburg, M., Weigel, A.: Techniques for improving ocr results. In: Handbook of Character Recognition and Document Image Analysis. World Scientific. pp. 227–258 (1997)
4. Wang, J.: A study of the ocr development history and directions of development. Highlights Sci. Eng. Technol. **72**, 409–415 (2023)
5. Breuel, T.M., Ul-Hasan, A., Al-Azawi, M.A., Shafait, F.: High-performance ocr for printed english and fraktur using lstm networks. In: 2013 12th International Conference on Document Analysis and Recognition, IEEE, pp. 683–687 (2013)
6. : Tesseract-ocr: (2019). https://github.com/tesseract-ocr/tesseract/. Accessed: 17 June 2024
7. LeCun, Y., et al.: Handwritten digit recognition with a back-propagation network. Adv. Neural Inf. Process. Syst. (1989)
8. Yin, Y., Wu, J., Zheng, H.: Ncfm: accurate handwritten digits recognition using convolutional neural networks. In: 2016 International Joint Conference on Neural Networks (IJCNN), IEEE, pp. 525–531 (2016)
9. Zelic, F., Sable, A.: A comprehensive guide to ocr with tesseract, opencv and python. Automate Manual Data Entry Using AI (2021)
10. Pasi, K.G., Naik, S.R.: Effect of parameter variations on accuracy of convolutional neural network. In: 2016 International Conference on Computing, Analytics and Security Trends (Cast), IEEE, pp. 398–403 (2016)
11. Arif, R.B., Siddique, M.A.B., Khan, M.M.R., Oishe, M.R.: Study and observation of the variations of accuracies for handwritten digits recognition with various hidden layers and epochs using convolutional neural network. In: 2018 4th International Conference on Electrical Engineering and Information & Communication Technology (iCEEiCT), IEEE, pp. 112–117 (2018)
12. Finkelstein, B., Kuncan, K.: The study of handwriting recognition algorithms based on neural networks. Int. J. Hybrid Inf. Technol **14**(1), 69–80 (2021)
13. website described about working of textract: https://www.websitebuilderinsider.com/how-does-aws-textract-work. Accessed: 17 June 2024
14. Used as reference from amazon textract developer guide provided by aws. https://docs.aws.amazon.com/pdfs/textract/latest/dg/textract-dg.pdf Accessed: 17 June 2024
15. Used as reference from amazon textract developer guide provided by aws. https://github.com/aws-samples/amazon-textract-serverless-large-scale-document-processing. Accessed: 17 June 2024
16. Joshi, K., Arolkar, H.: Comparative analysis of outcomes of tesseract ocr for different languages. In: 2024 5th International Conference on Intelligent Communication Technologies and Virtual Mobile Networks (ICICV), IEEE pp. 95–100 (2024)
17. Smith, R.: Building a multilingual ocr engine: training lstm networks on 100 languages and test results (google inc., june 20, 2016), **16**, 17
18. Hegghammer, T.: Ocr with tesseract, amazon textract, and google document AI: a benchmarking experiment. J. Comput. Soc. Sci. **5**(1), 861–882 (2022)
19. Khandokar, I., Hasan, M., Ernawan, F., Islam, S., Kabir, M.N.: Handwritten character recognition using convolutional neural network. In: Journal of Physics: Conference Series. Volume 1918., IOP Publishing (2021) 042152

20. Siddique, F., Sakib, S., Siddique, M.A.B.: Recognition of handwritten digit using convolutional neural network in python with tensorflow and comparison of performance for various hidden layers. In: 2019 5th International Conference on Advances in Electrical Engineering (ICAEE), IEEE, pp. 541–546 (2019)
21. Wu, M., Zhang, Z.: Handwritten digit classification using the mnist data set. Course project CSE802: pattern classification & analysis (2010)
22. Bieniecki, W., Grabowski, S., Rozenberg, W.: Image preprocessing for improving ocr accuracy. In: 2007 International Conference on Perspective Technologies and Methods in MEMS Design, IEEE, pp. 75–80 (2007)

Development of a Restricted Access and Energy-Efficient IoT-Based Greenhouse System for Tropical Climates

Tolulope C. Erinosho[1(✉)], Kamoli A. Amusa[1], Abdultaofeek Abayomi[2,3], and Azeez Lawal[1]

[1] Department of Electrical and Electronic Engineering, Federal University of Agriculture, Abeokuta, Ogun State, Nigeria
erinoshotc@funaab.edu.ng

[2] Faculty of Engineering, Built Environment and Information Technology, Water Sisulu University, Buffalo City Campus, East London 5200, South Africa

[3] Innovation and Advanced Science Research Group (IASRG), Summit University, PMB 4412, Offa 250101, Kwara, Nigeria

Abstract. Agriculture as a cornerstone of global food production, faces sustainability and environmental impact issues. Conventional greenhouses while aiding crop production, often contribute to high energy consumption and substantial greenhouse gas (GHG) emissions. They are also vulnerable to the introduction of diseases due to uncontrolled access. Theoretical frameworks highlight the greenhouse effect and the impact of climate change on traditional agricultural methods, emphasizing the urgency of alternative sustainable practices. This paper introduces and implements a Radio Frequency Identification (RFID) and Internet of Things (IoT)-based energy-efficient greenhouse system for tropical climates applications. The system employs a solar energy source, an Arduino microcontroller, and various units for soil moisture, temperature monitoring, and RFID-based access control. The Arduino system enhances security by managing a list of approved RFID card (UIDs), granting access only to authorized users, and activating a buzzer for unauthorized access attempts. Performance tests of the prototype demonstrated superior results in temperature and soil resistance management compared to open-field farming, thus highlighting its potential for increased yield, resource efficiency, and enhanced farm security. This eco-friendly system is particularly suitable for deployment in energy-deprived tropical regions, offering improved farm yield and robust protection for greenhouse operations.

Keywords: energy · agriculture · greenhouse · IoT · radio frequency identifier

1 Introduction

One of the most pressing issues confronting the world today is the provision of adequate nourishment to a population exceeding eight billion individuals worldwide [1]. Despite its favorable agro-ecological endowments of a total landmass of 92.4 million hectares

B. K. Mishra et al. (Eds.): ICT4GS 2024, ISEM 56, pp. 398–409, 2025.
https://doi.org/10.1007/978-3-031-94997-5_34

of land, food and nutrition insecurity is prevalent in Nigeria, as only approximately 32 million hectares or 34.63% of the available landmass are under cultivation. As such, food insecurity and the prevalence of undernutrition in Nigeria are among the worst globally [2].

Climate change is one of the numerous contributing factors to the prevalence of food insecurity [3, 4]. A number of serious threats resulting from climate change, including harsh weather, unexpected temperature changes, and rainfall variations, drastically affect food production, which in turn affects food security [5, 6]. According to [7], developing countries such as Nigeria and most other African countries are mostly affected by climate change due to their warm climates coupled with an increasing and widespread poverty, which makes it very difficult for them to implement policies that can reduce the effects of climate change [8]. As reported by [9], 10–14% of total global greenhouse emissions are produced by agricultural processes, such as rice padding fields, nitrous-oxide emissions from fertilizers, deforestation, mechanized agriculture, soil degradation, methane waste emissions from livestock, and food waste emissions [10–12]. Controlled-environment agricultural production technologies, such as greenhouse farming, have been found to be suitable solutions to solve food insecurity and decrease greenhouse emissions caused by agriculture [13–16]. However, it has been observed that the adoption of greenhouse technology has not been efficiently explored, especially in developing countries such as Nigeria, occasioned by lack of funds for optimal operation, unstable power supply, and inadequate technical know-how [17]. Traditional greenhouse practices depend heavily on non-renewable energy sources such as fossil fuels for heating, cooling, and lighting, which consequently contribute to carbon emissions and global warming [18, 19]. In addition, by integrating IoT technologies into greenhouse systems, it is possible to monitor and control growing factors, such as light, water, and temperature, and reduce energy consumption as well [20]. Energy-efficient systems aim to achieve a scheme with reduced energy dissipation and consumption. A sustainable greenhouse involves incorporating renewable energy sources [21–24], such as solar panels, and implementing energy management systems to optimize energy usage [15, 25]. Hence, this current approach employed energy efficiency technique in the choice of lighting fixtures and their operations. To enhance the security of the system, an RFID based control system is incorporated into the design which deny access to unauthorized individuals whose information are not available in the preloaded database of the unit.

This paper is organized as follows: Sect. 2 highlights some of the previous studies carried out on the subject, Sect. 3 focused on the materials and method used in the design while the numerical results and discussion are presented in Sect. 4, with the concluding remarks treated in Sect. 5.

2 Literature Review

Several studies have been conducted with respect to the application of IoT in the development of greenhouse farms. Chodisetti et al. [31] used Raspberry Pi as the main component and LabVIEW to display the values of major parameters like the temperature, Humidity, pH, soil moisture, and light intensity. These parameters were continuously measured and when readings fall beyond the predetermined levels, it triggers the fan, humidifier, pH-balancing chemicals, water pump, and light respectively.

Subahi et al. [32] developed a complex closed loop temperature regulation system to regulate the temperature of the greenhouse specifically designed for Saudi Arabia which has mostly very hot weather. It comprises advanced PID controllers that depend on complex algorithms that constantly perform computations on the data received from the temperature sensor. It also has a lightning system that can be opened, half-closed, or closed depending on the direction of the sun and the temperature of the greenhouse.

In another study, Ardiansah et al. [33] examined different sensors that could be used to detect various factors that affect plant growth: temperature sensor using LM35, temperature and humidity sensor using DHT11, and soil moisture sensor. It also compares the different types of wireless technology that can be used, namely Bluetooth, Zigbee, and IoT. Bluetooth and Zigbee are used for short-range applications while the IOT cover long-distance applications.

However, Kodali et al. [35] developed an irrigation system using drip irrigation. The water tank in the study was equipped with an ultrasonic sensor that continuously measured the level of water and alerted the user through a short message service (SMS) whenever the level was low. The moisture sensors deployed at the site measured the soil moisture and hence controlled the valve of the water pump. Temperature and humidity sensors were used to measure the temperature and humidity, and when the temperature rose above a preset threshold, the relay triggered a fogger that sprinkled tiny water droplets in the air. A Peltier fan was used to extract heat from the greenhouse until the temperature was normalized. A special light emitting diode (LEDS), which emits light in a very specific spectrum that serves as growing light for the plant, is placed in the greenhouse. The light affects plant primary and secondary metabolism, and significantly improves plant growth. A marketing system was set up to connect the farm storage house with consumers only and to bypass middlemen. Thus, whenever farm produce is stored in the containers, the farmer will swipe an RFID tag and the data from the containers will update a spreadsheet and an email will be sent to an e-commerce website for the intending consumers to be aware of the new stock for possible purchase.

Ahmad et al. [34] utilized Raspberry-Pi microprocessor to control temperature and humidity in order to evaluate the impact on field crop production, soil management and quality of tomatoes. The variable dataset for tomato growth is set to different conditions and times, which is then used to determine which is best for the tomato plant. The final system involved the use of humidity sensors, temperature sensors, and timers which changed the settings of the temperature and humidity sensors to maintain microclimate conditions.

In tropical climate region like Nigeria, though there have been several studies and advocacies on the benefits of smart agriculture over the traditional approaches [26–28] in farming activities, the system of reducing the carbon footprint to minimal level and using an access control in Greenhouse design and energy efficiency has not been extensively reported in the literature [29, 30]. These gaps are incorporated in the current study for an optimal design of the system towards an improved food production and security.

3 Materials and Methods

In this current study, an IoT enabled and net zero greenhouse using Arduino microcontroller, RFID and hydroponics sensors such as temperature and humidity sensor DHT11, and soil humidity sensor is developed as depicted in Fig. 1. The system is designed to provide optimum growing environment for plants especially vegetables to grow using solar energy as the main source of energy for the farm and, also ensure that the net-zero carbon balance is achieved. The system is to maintain the temperature, soil moisture of the soil and control access in and out of the farm to ensure only authorized personnel access. The system made use of IoT which will put all the system data on ThingSpeak server for data collection and further analysis. Displayed in Figs. 1 and 2 is the design procedure and the flow chart of the developed system respectively. When the device is powered, the RFID security system allows or deny access into the farm. If it detects an abnormality, the buzzer system comes on. At that same moment, the Arduino uno which is the central control system is constantly checking the temperature and humidity sensor and the soil moisture sensor to determine if the temperature and water level are still within the statutory limits and if not, will trigger the cooling system and the water irrigation system respectively. During these processes, the ThingSpeak platform collects relevant data from the greenhouse via a Wi-Fi module sensor where it can be integrated into a mobile application.

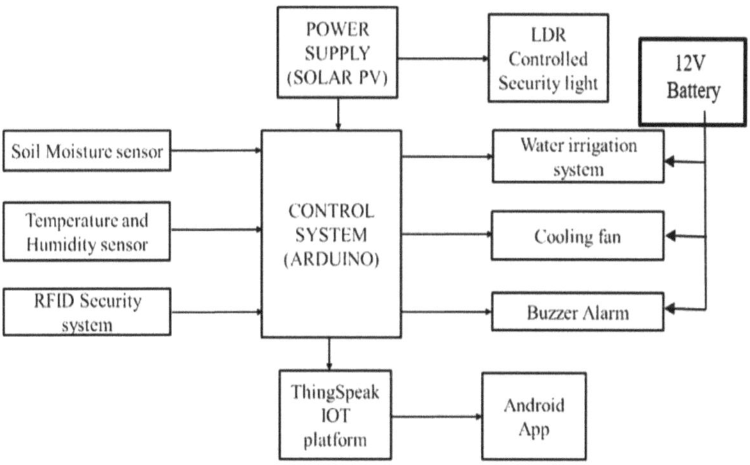

Fig. 1. Design of the IoT system for Net Zero Greenhouse.

The System Design - Presented are the design consideration and calculation for each component unit of the developed system.

3.1 Soil Moisture Unit

This unit is responsible for triggering the water irrigation system when the water level falls below the appropriate threshold and vice versa. When the moisture level is below

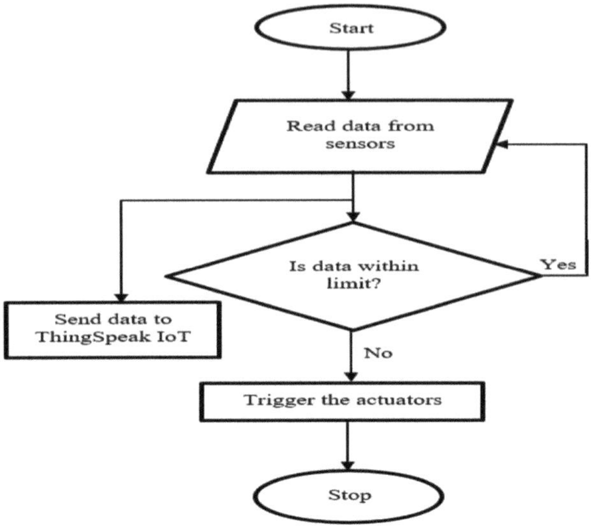

Fig. 2. Flowchart of the proposed IoT-based Greenhouse system.

a predefined threshold, it triggers the Arduino to activate the water pump and start the drip irrigation process. Arduino reads the analog voltage output from the soil moisture sensor, which corresponds to the soil moisture level. The analog output of the soil moisture sensor ranges from 0 to 1023. These analog values are then converted to digital values using the sensor's inbuilt Analog to Digital Converter. The moisture content (in percentage) is displayed on a serial monitor.

This is represented as the moisture value in terms of percentage using Eqs. (1) and (2).

$$Analog\ Output = \frac{ADC\ Value}{1023} \tag{1}$$

$$Moisture\ in\ percentage = 100 - (Analog\ output * 100) \tag{2}$$

For zero moisture, we obtained the maximum value of the 10-bit ADC, that is, 1023. This in turn yielded 0% moisture content. The digital value was then compared to a predefined low-threshold value and if the soil moisture is below the low threshold, the Arduino turns on the relay module, which energizes the water pump through a transistor circuit. The drip irrigation system, powered by 5V water pumps, delivers water to the plants through the water pump and drip line to maintain the appropriate soil moisture level.

3.2 Temperature and Humidity Unit

This unit is responsible for triggering cooling when the heat level falls below the appropriate threshold. The DHT11 sensor senses temperature and humidity in the greenhouse. When the temperature exceeded a predefined high threshold (23 °C), it triggered the Arduino to activate the cooling fan and maintain the appropriate temperature by dissipating excess heat and maintaining a suitable environment for plant growth.

The equation for calculating relative humidity is given by Ogonori et al. (2023) as:

$$\mathrm{RH} = \frac{\rho_{\mathrm{w}}}{\rho_{\mathrm{s}}} \times 100\% \tag{3}$$

where

RH = Relative Humidity.

ρ_w = Density of water vapour.

ρ_s = Density of water vapour at saturation.

3.3 Farm Access Control Unit

This unit controls access in and out of the greenhouse farm so that only authorized personnel can access the farm. RFID cards are assigned to authorized personnel, and the Arduino checks whether the presented RFID card matches the authorized list before granting access. The RFID reader module communicates with the Arduino to read the unique identifier (UID) from the RFID cards when brought into proximity with the reader (Fig. 3).

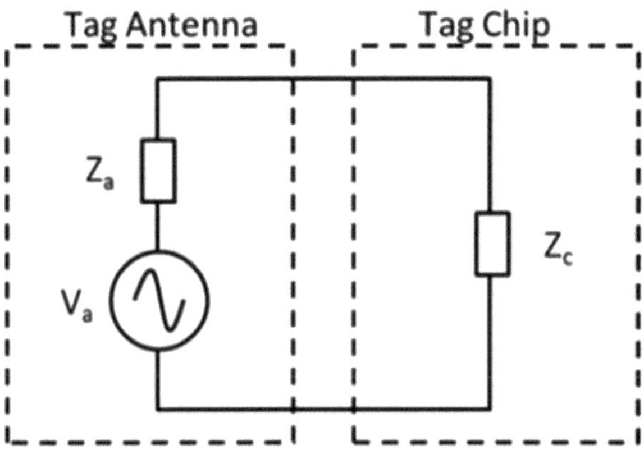

Fig. 3. Equivalent circuit of a RFID Tag.

where:

Z_a = Antenna Impedance.

Z_c = Chip Impedance.

V_a = Input voltage.

Arduino stores a list of authorized RFID card UIDs in its memory, which can be stored in an array or another data structure in the program. When an RFID card is presented to the reader, the RFID reader module reads the UID and sends it to Arduino. The Arduino then compares the received UID with a list of authorized UIDs. If there is a match, it grants access and otherwise, access is denied. When access or unauthorized card usage is detected, Arduino activates a digital output connected to the buzzer.

3.4 Power Supply Unit

This unit is a 12 V DC general power supply to the system, from the Arduino to the sensors and the actuators. The DC supply system comprises a solar panel, charge control unit, voltage regulator (LM7812), and lead acid battery. The Arduino board is powered directly by the solar PV system because it operates within the 7-12 V voltage range. Therefore, the output voltage of the voltage regulator (12 V) can be used to power the Arduino board by connecting it to the Arduino's Vin pin or DC power jack.

Additionally, low-power components such as the soil moisture sensor, DHT11 sensor, and RFID Reader module are powered directly from the Arduino's 5 V pin since they usually require low current consumption within the capabilities of the Arduino's onboard voltage regulator. Components with higher power requirements, such as the cooling fan, buzzer, and water pump, are powered by the battery rather than directly from the Arduino to prevent overloading of the Arduino's voltage regulator.

3.5 IoT Integration Unit

This unit collects the soil parameters (temperature, humidity and soil moisture) inside the greenhouse from the Arduino and sends it to the ThingSpeak IoT cloud platform in real-time. ThingSpeak is an IoT platform to collect, analyze, and visualize sensor data from connected devices like Arduino. The process is divided into two parts namely:

1. ThingSpeak Platform Setup for Data Logging: The ThingSpeak was set up for temperature, humidity, and soil moisture data logging with the following steps:
   - ThingSpeak account was created for sign in.
   - A new channel on ThingSpeak (a channel represents a collection of data fields for specific sensor readings) was created.
   - Three fields in the channel for temperature, humidity, and soil moisture data were defined.
   - API Key was obtained for the newly created channel which was used by the Arduino to send data to ThingSpeak.
2. Wi-Fi Module (ESP8266) and Arduino connection: The Wi-Fi module (ESP8266) connects to the Arduino through Universal Asynchronous Receiver/Transmitter (UART) communication using the serial interface. The Tx (Transmit) pin of the Wi-Fi module was connected to the Rx (Receive) pin of the Arduino (for example, D1). The Rx (Receive) pin of the Wi-Fi module was connected to the Tx (Transmit) pin of the Arduino (for example, D0). Then the Ground (GND) pin of the Wi-Fi module to the GND pin of the Arduino. The 3.3V pin of the Wi-Fi module was connected to a 3.3V power supply pin on the Arduino or use an external 3.3V power supply. The circuit diagram of the IoT integration unit is depicted in Fig. 4.

3.6 Construction of the Greenhouse Prototype

The greenhouse structure is constructed from aluminum frame and clear transparent plastic measuring 5m in length, 5m in height and 7m in width as depicted in Fig. 5.

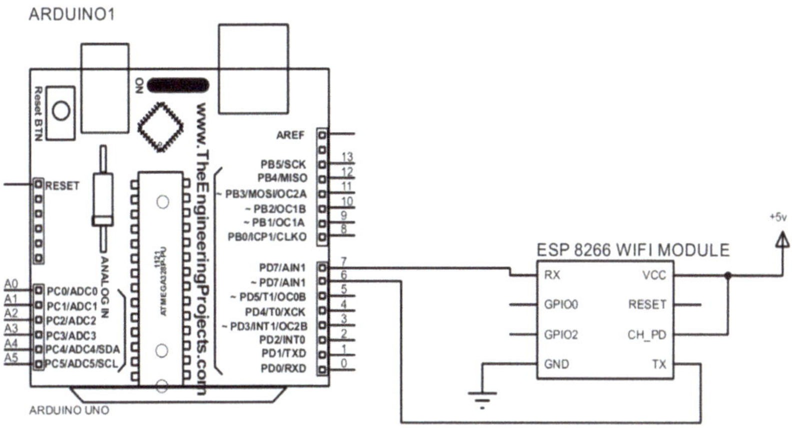

Fig. 4. Wi-Fi connection to Arduino circuit.

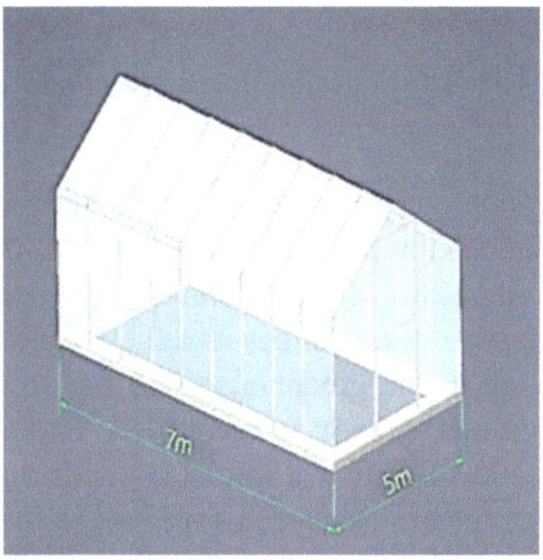

Fig. 5. Greenhouse CAD design.

4 Results and Discussion

The developed system was tested under different conditions and the results obtained are presented. The plate of the developed greenhouse system is shown in Fig. 6.

When the device is powered from the battery, the LCD came on and prompts that the access cards be scanned. This card scanning is to be done on the mrfc522 RFID reader on the controller. If the right card is scanned, it opens the door for 5 s and then closes the door. This is to simulate opening and closing of the security door. At that same moment, the Arduino is constantly checking the DHT11 temperature and humidity sensor and the soil moisture sensor respectively to determine if the temperature and water level are still within the statutory limits and if not, will trigger the relays of the cooling fan and the water pumps to blow air and supply water to the farm respectively. During these processes, the ThingSpeak platform collects relevant data from the greenhouse via a Wi-Fi module sensor where it can be integrated into a mobile application or web application.

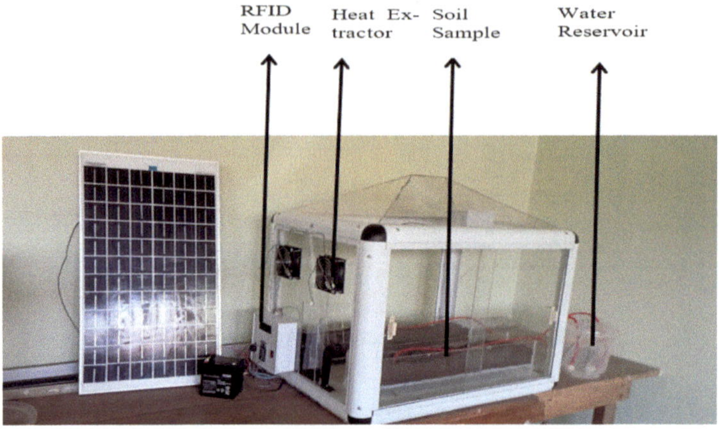

Fig. 6. The plate of the developed greenhouse system.

Figure 7 shows the temperature variations inside the greenhouse and outside the greenhouse (open field), a mean temperature of 23.4 °C was obtained for the entire 24 h in the greenhouse compared to 28.2 °C mean value of the open field. This is important because certain vegetables such as cucumber have better yield when it grows at 23 °C degree. As a result, growing in the greenhouse will have more yield than outside in the open where the temperature cannot be controlled. Furthermore, Fig. 8 shows the soil resistance variations in the greenhouse soil and outside the greenhouse (open field). As shown in the Figure, the mean resistance of the soil in the greenhouse is 500 ohms and that of the open field as 794.5 ohms over a 24-h period. Soil moisture however is inversely proportional to soil resistance as a wet soil conducts more than a dry soil. This means that there is constant supply of water to the soil in the greenhouse which will help the plants to grow much faster with better yield.

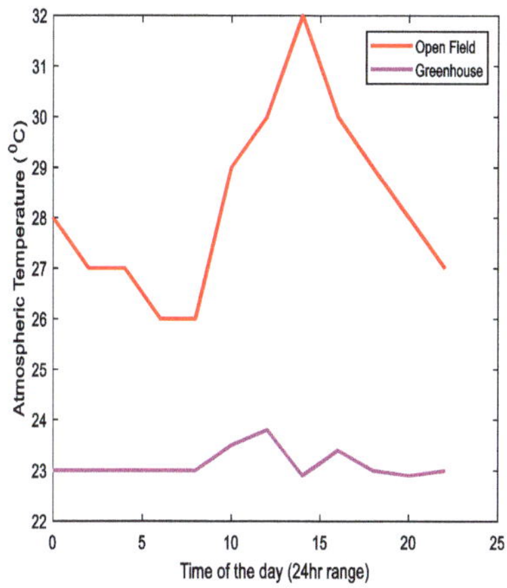

Fig. 7. Plot of temperature variations

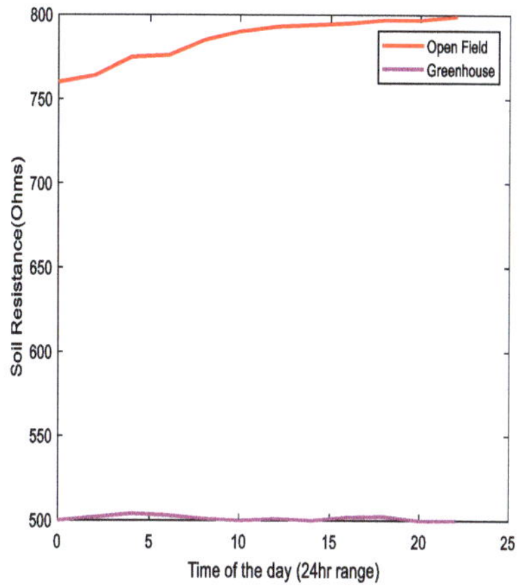

Fig. 8. Plot of soil resistance variations

5 Conclusion

The development of an IoT-based access-controlled energy greenhouse for tropical applications represents a significant advancement in sustainable agriculture. As proposed in this current study, the Arduino system enhances security by managing a list of approved RFID card UIDs, granting access only to authorized users, and activating a buzzer for unauthorized access attempts. This system, coupled with precise environmental monitoring and automation, optimizes conditions for plant growth and resource utilization. Additionally, the integration of IoT technologies and energy-efficient principles provides farmers with real-time insights and improves overall greenhouse management.

References

1. Norrman, K.: World population growth: a once and future global concern. World **4**(4), 684–697 (2023). https://doi.org/10.3390/world4040043
2. Nwozor, A., Olanrewaju, J.S., Ake, M.: National insecurity and the challenges of food security in Nigeria. Acad. J. Interdisc. Stud. (2019). https://doi.org/10.36941/ajis-2019-0032
3. Anabaraonye, B., Okonkwo, U.C., Ewa, B.O.: The impact of climate change on crop production in Nigeria. Int. J. Res. Civ. Eng. Technol. **3**(1), 2707–8272 (2019)
4. Malhi, G.S., Kaur, M., Kaushik, P.: Impact of climate change on agriculture and its mitigation strategies: a review. Sustainability **13**(3), 1318 (2021). https://doi.org/10.3390/su13031318
5. Gopalakrishnan, T., Hasan, M.K., Haque, A.K.M.A., Jayasinghe, S.L., Kumar, L.: Sustainability of coastal agriculture under climate change. Sustainability **11**(24), 7200 (2019). https://doi.org/10.3390/su11247200
6. Ayinde, I.A., Akinbode, S.O., Otekunrin, O.A.: Food security in Nigeria: impetus for growth and development. J. Agric. Econ. Rural Dev. **6**(2), 808–820 (2020)
7. Ojo, T., Baiyegunhi, L.J.S.: Climate change perception and its impact on net farm income of smallholder rice farmers in South-West Nigeria. J. Cleaner Prod. **310**, 127373 (2021). https://doi.org/10.1016/j.jclepro.2021.127373

8. Auffhammer, M., Kahn, M.E.: The farmer's climate change adaptation challenge in least developed countries. In: Elsevier eBooks Vol. 4, pp. 193–229 (2018). https://doi.org/10.1016/bs.hesenv.2018.04.001
9. Hussain, S., et al.: Rice production under climate change: adaptations and mitigating strategies. In: Springer eBooks, pp. 659–686 (ss2020). https://doi.org/10.1007/978-3-030-49732-3_26
10. Driga, A.M., Drigas, A.: Climate change 101: how everyday activities contribute to the ever-growing issue. Int. J. Recent Contrib. Eng. Sci. IT **7**(1), 22 (2019). https://doi.org/10.3991/ijes.v7i1.10031
11. Bhattacharya, A.: Global climate change and its impact on agriculture. Elsevier eBooks **4**(2), 1–50 (2019). https://doi.org/10.1016/b978-0-12-816209-5.00001-5
12. Changqing, C., et al.: Quantifying the impact of global warming on china's rice production. Jiangsu Collaborative Innovation Center for Modern Crop Production (2019)
13. Latake, P.T., Pawar, P., Ranveer, A.: The greenhouse effect and its impacts on environment. Int. J. Innovative Res. Creative Technol. (2015). https://www.ijirct.org/viewPaper.php?paperId=IJIRCT1201068
14. Kweku, D.W., et al.: Greenhouse effect: greenhouse gases and their impact on global warming. J. Sci. Res. Rep. **17**(6), 1–9 (2018). https://doi.org/10.9734/jsrr/2017/39630
15. Gorjian, S., Ebadi, H., Najafi, G., Chandel, S.S., Yildizhan, H.: Recent advances in net-zero energy greenhouses and adapted thermal energy storage systems. Sustain. Energy Technol. Assess. **43**, 100940 (2021). https://doi.org/10.1016/j.seta.2020.100940
16. Stan, C. 2022. Carbon dioxide, the Greenhouse effect and the warming of the Earth's atmosphere. In: Springer eBooks pp. 36–48. https://doi.org/10.1007/978-3-662-64162-0_6
17. Belay, A., et al.: Knowledge of climate change and adaptation by smallholder farmers: evidence from Southern Ethiopia. Heliyon **8**(12), e12089 (2022). https://doi.org/10.1016/j.heliyon.2022.e12089
18. Navarro-Pedreño, J., Almendro-Candel, M.B., Zorpas, A.A.: The increase of soil organic matter reduces global warming, myth or reality? Sci **3**(1), 18 (2021). https://doi.org/10.3390/sci3010018
19. Pata, U.K.: Linking renewable energy, globalization, agriculture, CO2 emissions and ecological footprint in BRIC countries: a sustainability perspective. Renew. Energy **173**, 197–208 (2021). https://doi.org/10.1016/j.renene.2021.03.125
20. Malik, P., et al.: Industrial internet of things and its applications in industry 4.0: state of the art. Comput. Commun. **166**, 125–139 (2021). https://doi.org/10.1016/j.comcom.2020.11.016
21. Chaturvedi, V.: A vision for a Net-zero energy system for India. Energy Clim. Change **2**, 100056 (2021). https://doi.org/10.1016/j.egycc.2021.100056
22. Elavarasan, R.M., et al.: A Comprehensive review on renewable energy development, challenges, and policies of leading Indian States with an international perspective. IEEE Access **8**, 74432–74457 (2020)
23. Amit, K.J., Tripathi, N.T., Katitar, V.: climate change impacts on agriculture. IJRAR Int. J. Res. Anal. Rev. (2019)
24. Bistline, J., Blanford, G.J.: The role of the power sector in Net-zero energy systems. Energy Clim. Change **2**, 100045 (2021)
25. Akhator, P., Obanor, A., Sadjere, E.: Electricity situation and potential development in Nigeria using off-grid green energy solutions. J. Appl. Sci. Environ. Manag. **23**(3), 527 (2019)
26. Gassner, A., et al.: Poverty Eradication and food security through agriculture in Africa: rethinking objectives and entry points. Outlook Agric. **48**(4), 309–315 (2019). https://doi.org/10.1177/0030727019888513
27. Gavrilova, N.G.: Impediments to the development of Nigeria's agricultural exports. IOP Conf. Ser. Earth Environ. Sci. **548**(2), 022085 (2020)

28. Dickson, M.S., Amannah, C.I.: Augmented IoT model for smart agriculture and farm irrigation water conservation. Int. J. Intell. Sci. **13**(4), 131–163 (2023)
29. Oguntosin, V., Okeke, C., Adetiba, E., Abdulkareem, A., Olowoleni, J.: Int. J. Comput. Digital Syst. **14**(1), 469–483 (2023)
30. Folorunso, T.A., Oshiga, O., Bala, J.A., Ojogunwa, D.A.: Development of an internet of things based smart greenhouse. Int. J. Comput. Trends Technol. **72**(5), 96–101 (2024)
31. Chodisetti, L.S., Sindhu, K.M., Manikanta, M.N., Satya, V.S., Balaji. M.M., Venkateswara, R.C.: Smart watering system for terrace garden using LabVIEW. In: 5th International Conference on Advances in Computing, Communication Control and Networking, Greater Noida, India, 2023, pp. 979–982 (2023). https://doi.org/10.1109/ICAC3N60023.2023.10541339
32. Subahi, A.F., Bouazza, K.E.: An intelligent IoT-based system design for controlling and monitoring greenhouse temperature. IEEE Access **8**, 125488–125500 (2020). https://doi.org/10.1109/ACCESS.2020.3007955
33. Ardiansah, I., Bafdal, N. Suryadi, E., Bono, A.: Greenhouse monitoring and automation using arduino: a review on precision farming and internet of things (IoT). Int. J. Adv. Sci. Eng. Inf. Technol. **10**, 703–709 (2020). https://doi.org/10.18517/ijaseit.10.2.10249
34. Ahmad, B., et al.: Evaluation of smart greenhouse monitoring system using raspberry-pi microcontroller for the production of tomato crop. . J. Appl. Res. Plant Sci. **4**(01), 452–458 (2023). https://doi.org/10.38211/joarps.2023.04.01.54
35. Kodali, R.K.: A low-cost smart irrigation system using MQTT protocol. IEEE Region 10 Symposium, India (2017). https://doi.org/10.1109/TENCONSpring.2017.8070095.

Smart Air Quality Monitoring and Conditioning System (SAQMCS)

Arunya Paul(✉), Sasmita Pahadsingh, and Tejaswini Kar

School of Electronics Engineering, KIIT Deemed to be University, Bhubaneswar, India
arunyapaul@gmail.com, {spahadsinghfet,tkarfet}@kiit.ac.in

Abstract. Air pollution poses significant health risks globally, especially with increasing urbanization and industrialization. The Smart Air Quality Monitoring and Conditioning System (SAQMCS) addresses this challenge by integrating real-time monitoring with active air conditioning to ensure cleaner and healthier indoor environments. Unlike traditional systems that merely report air quality, SAQMCS actively controls it through a combination of sensors and actuators. Key components include the ESP32 microcontroller, MQ-135 air quality sensor, BMP280 pressure sensor, and DHT11 temperature and humidity sensor, all managed via Arduino IDE developed code. The system dynamically adjusts inlet and outlet fans and a humidifier based on sensor data, ensuring optimal air quality and comfort. Data is visualized on an OLED display and transmitted to ThingSpeak for remote monitoring and analysis. The user-friendly design, affordability, and scalability makes SAQMCS accessible for residential and small-scale commercial use. Extensive testing demonstrated its effectiveness and reliability. Future advancements will focus on integrating advanced air purification technologies, enhancing data analytics through machine learning, and improving compatibility with smart home ecosystems, furthering its capabilities in autonomous air quality management.

Keywords: Air quality monitoring · IoT · ESP32 · ThingSpeak · Humidity control · Indoor air conditioning

1 Introduction

Air pollution has become a critical global concern, posing significant health risks. With urbanization and industrialization on the rise, air pollutant concentrations have reached hazardous levels, necessitating innovative solutions for air quality monitoring and improvement [1, 2]. The "Smart Air Quality Monitoring and Conditioning System (SAQMCS)" addresses this need by integrating real-time monitoring with active air conditioning to ensure cleaner and healthier environments in the indoors.

Traditional air quality monitoring systems primarily focus on passive detection and reporting of pollutants, often lacking the capability to actively improve air quality. [3, 4]. Many existing solutions are also too expensive or complex for widespread residential and small-scale commercial use [5]. Our work distinguishes itself by combining affordability,

B. K. Mishra et al. (Eds.): ICT4GS 2024, ISEM 56, pp. 410–421, 2025.
https://doi.org/10.1007/978-3-031-94997-5_35

simplicity, and functionality in a compact system that not only monitors air quality but also actively control and conditions it based on detected environmental parameters and level of pollution.

SAQMCS's uniqueness lies in its integrated conditioning mechanism. Unlike conventional systems that merely monitor air quality, our system also employs a system of fans along with air filters, coupled with a humidifier to regulate air quality actively. By adjusting airflow and humidity in real-time, the system ensures a more comfortable and healthier indoor environment. The system's cost-effective and scalable design is another key innovation. By using readily available components such as ESP32 microcontroller, MQ-135 air quality sensor, BMP280 pressure sensor, and DHT11 humidity sensor, we ensure that the SAQMCS is both affordable and easy to reproduce. The system's selective working mechanism is not only cost-effective but also energy efficient at the same time making it suitable for widespread adoption.

The SAQMCS also features enhanced data visualization and IoT integration through the ThingSpeak platform. Our system includes an OLED display for immediate on-site feedback and integrates with ThingSpeak for remote monitoring and data analysis. This dual approach ensures access to both real-time and historical data, enabling better decision-making regarding indoor air quality management. The plug-and-play design of the system also allows for straightforward installation and minimal maintenance, while the intuitive control logic ensures that users can easily understand and operate the system without requiring technical expertise. This user-centric design ensures that the SAQMCS can even be effectively used by individuals without specialized knowledge, broadening its applicability.

2 Literature Review

The growing concern about air pollution and its adverse health effects has led to significant research and development in air quality monitoring and purification systems. Over the years, several approaches and technologies have been proposed to address this issue. One of the early comprehensive reviews on air quality monitoring systems was conducted by Kumar et al. (2015), who discussed various sensing technologies and their applications in monitoring air pollutants. They highlighted the importance of low-cost sensors in expanding the deployment of air quality monitoring networks, particularly in urban areas where pollution levels are critical [6]. The integration of Internet of Things (IoT) in air quality monitoring has also gained considerable attention. An IoT-based approach was explored by Pavani et al. (2017), who developed a system that combined wireless sensor networks with cloud computing to provide real-time air quality monitoring. Their work demonstrated the feasibility of using wireless sensor networks for large-scale air quality monitoring and the benefits of remote data access and management [7].

Recent advancements have focused on enhancing the accuracy and functionality of air quality monitoring systems. A study by Kang et al. (2022) evaluated the performance of low-cost air quality sensors in comparison with traditional monitoring methods. They found that while low-cost sensors had some limitations in terms of accuracy, they were valuable for providing high-resolution spatial data when used in dense networks [8].

Active air purification mechanisms have also been investigated. Pitarma et al. (2017) designed an indoor air quality monitoring system that included a purification unit. Their system used various sensors to detect pollutants and automatically activated purification processes to maintain a healthy indoor environment. This study underscored the importance of integrating monitoring and purification in a single system for enhanced effectiveness [9].

Furthermore, the use of machine learning in air quality prediction has shown promising results. Liu et al. (2024) developed a predictive model using historical air quality data and meteorological factors. Their model achieved high accuracy in predicting air quality indices, which could be used to trigger air purification mechanisms proactively [10].

Several researchers have explored the use of specific sensors for air quality monitoring. For example, Popoola et al. (2018) evaluated the performance of the MQ-135 sensor for detecting ammonia, benzene, and smoke, highlighting its effectiveness in low-cost air quality monitoring applications [11]. Similarly, the DHT11 sensor, widely used for humidity and temperature monitoring, was assessed by Malik et al. (2023) for its reliability in various environmental conditions [12]. The BMP280 sensor, known for its accuracy in measuring atmospheric pressure, has been integrated into several air quality monitoring systems. A study by Kortoçi et al. (2022) demonstrated its application in a portable air quality monitoring device, which provided real-time data and remote access via a mobile app [13]. IoT platforms such as ThingSpeak have facilitated the remote monitoring and analysis of air quality data. ThingSpeak was utilized by Guvvala et al. (2023) in their study, where they developed a low-cost air quality monitoring system that transmitted data to the cloud for real-time visualization and analysis [14].

In summary, the literature reveals a clear trend towards the integration of low-cost sensors, IoT, and machine learning in air quality monitoring and purification systems [15]. These advancements have made air quality monitoring more accessible, accurate, and proactive, paving the way for the development of comprehensive systems like the Smart Air Quality Monitoring and Conditioning System (SAQMCS).

3 Methodology

3.1 System Overview

The Smart Air Quality Monitoring and Conditioning System (SAQMCS) integrates advanced sensors, actuators, and IoT technology to monitor and regulate indoor air quality in real-time. The system comprises hardware components including an ESP32 microcontroller, various sensors, actuators such as fans and a humidifier, and an OLED display for local feedback. Additionally, the system utilizes ThingSpeak, a cloud-based IoT platform, for remote data logging, visualization, and analysis.

3.2 Hardware Components

ESP32 Microcontroller: The ESP32 serves as the central processing unit of the SAQMCS, responsible for interfacing with sensors, controlling actuators, and managing data communication with ThingSpeak.

Sensors

- *MQ-135 Air Quality Sensor,* measures volatile organic compounds (VOCs), ammonia, carbon monoxide, and other pollutants.
- *BMP280 Pressure Sensor,* provides atmospheric pressure data.
- *DHT11 Humidity and Temperature Sensor,* monitors indoor humidity and temperature levels.

Actuators

- *Inlet Fan,* draws polluted air into the system for purification.
- *Outlet Fan,* expels purified air back into the environment with greater speed.
- *Humidifier,* adjusts indoor humidity levels based on sensor inputs.
- *OLED Display,* provides real-time feedback on air quality parameters such as pollutant levels, temperature, humidity,pressure and system status.

3.3 System Operation

The SAQMCS operates in a closed-loop system architecture where sensor data continuously informs the control block to adjust the operation of actuators:

Data Acquisition Sensors (MQ-135, BMP280, DHT11) collect air quality metrics including VOC levels, atmospheric pressure, temperature, and humidity.

Data Processing The ESP32 microcontroller processes sensor data in real-time, using predefined thresholds and algorithms to assess air quality conditions.

Control Logic
Pollution Detection: The system utilizes data from the MQ-135 sensor to detect elevated levels of pollutants in the indoor environment. The filtration system is activated in case of increased pollution levels. In this case the suction (inlet) fan is activated.

Temperature and Humidity Regulation: DHT11 data guide the operation of the outlet fan and humidifier to maintain optimal indoor conditions. In case of higher temperatures, the outlet fan also gets activated allowing increased air throw thus providing better atmospheric cooling; humidifier gets activated in dry atmospheric conditions.

Fan Control: Inlet and outlet fans are activated based on pollutant concentrations and predefined thresholds. Inlet fan draws polluted air into the system which is forced through the filtration mechanism back into the environment. Outlet fan provides greater expulsion to the air leaving, during hotter environmental conditions.

This control logic (see Fig. 1) ensures that the Smart Air Quality Monitoring and Conditioning System (SAQMCS) responds dynamically to environmental changes, enhancing indoor air quality by adjusting fan speeds and humidification levels as needed. The thresholds can however be adjusted according to the atmospheric needs.

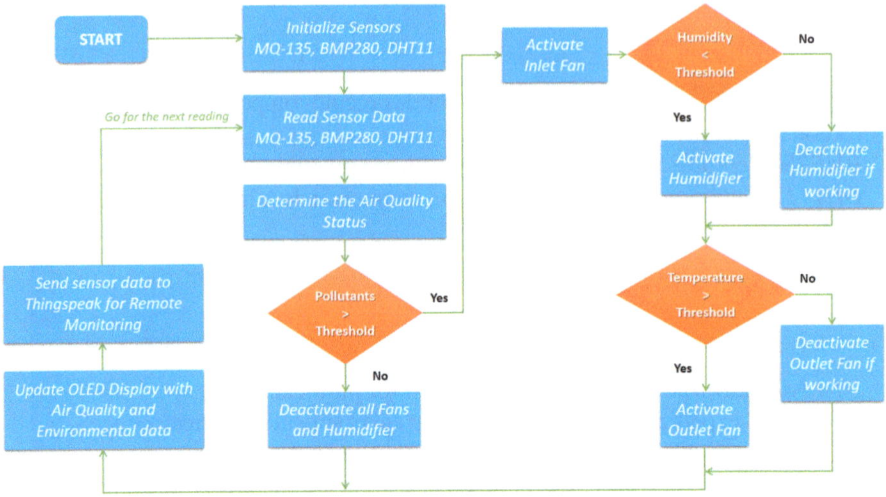

Fig. 1. Flowchart depicting the control logic of SAQMCS

User Interface:

Local Display: The OLED display shows real-time air quality parameters and system status on the device itself.

Remote Monitoring: ThingSpeak allows users to monitor historical data trends and receive alerts remotely. This is useful for IoT applications, utilizing the Wi-Fi (IEEE 802.11) capabilities of ESP32 for seamless remote data transfer and observation. The system architecture block diagram can be referred to in Fig. 2.

3.4 Structural Model

The SAQMCS device is designed with a comprehensive structure to effectively monitor and purify air. The device consists of several key components: a net cover, an inlet fan, a central chamber with a refillable humidifier reservoir, an outlet fan, a filter mesh, and a chamber termination. The process begins when polluted air enters the system through the inlet fan. This incoming air first passes through the central chamber where occasional humidification occurs, followed by forceful exertion through the filter mesh, which traps dust particles and other contaminants. The structural model has been well illustrated in Fig. 3.

In the central chamber, the air is subjected to thorough monitoring using a suite of sensors. The MQ-135 sensor evaluates air quality, the BMP280 sensor measures pressure, and the DHT11 sensor gauges humidity and temperature. Based on these readings, the system makes decisions by comparing the data to predefined thresholds. Inlet fan works to drive the polluted air in, outlet fan improves air current in hotter temperatures, whereas the humidifier works during drier climates. This structured approach not only purifies the air but also maintains a comfortable indoor environment, making SAQMCS an effective solution for air quality management.

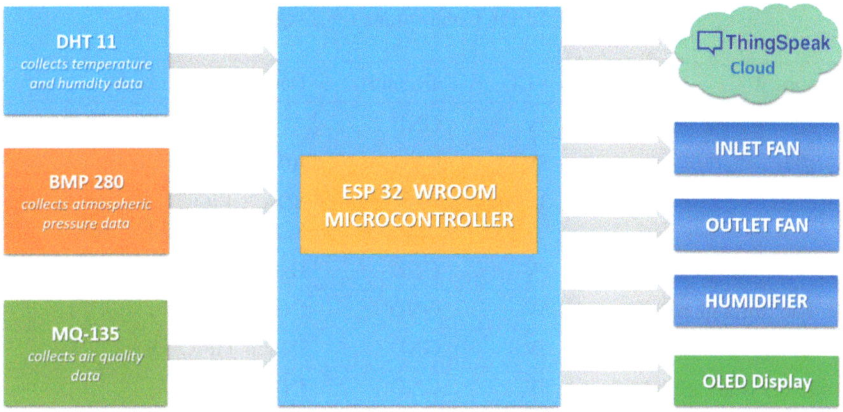

Fig. 2. Block Diagram of SAQMCS

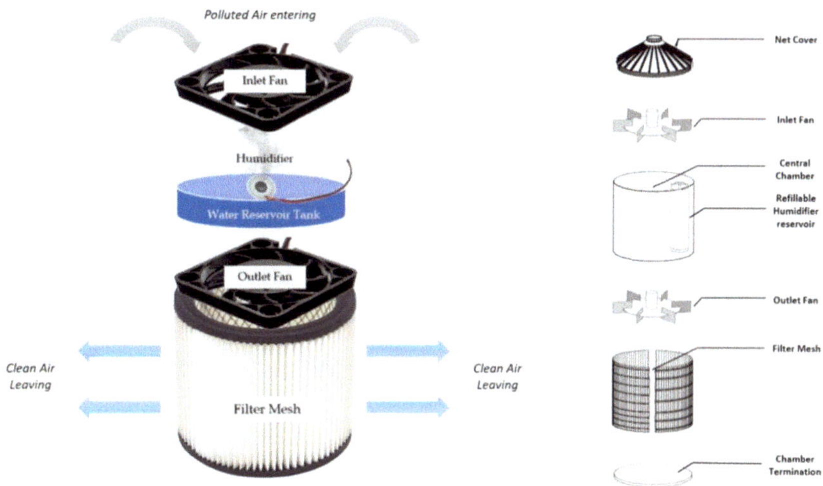

Fig. 3. Structural Model of SAQMCS

3.5 Circuit Design

The following Table 1 outlines the pin connections for the sensors, actuators, and OLED display to the ESP32 microcontroller for the SAQMCS System. The pin connection diagram has also been depicted in Fig. 4.

3.6 Firmware Development

The firmware development for the Smart Air Quality Monitoring and Conditioning System involves programming the ESP32 microcontroller to interact with various sensors and actuators, process data, and implement control logic. The platform utilized here is Arduino IDE. The following steps outline the firmware development process:

Table 1. Pin Connections for SAQMCS

Component	VCC	Ground	Data/Signal
ESP 32	5V	GND	-
MQ-135	5V	GND	A0 (Analog Output): GPIO 34
BMP 280	3.3V	GND	SDA: GPIO 21 (I2C SDA) SCL: GPIO 22 (I2C SCL)
DHT 11	5V	GND	Data: GPIO 4
Inlet Fan (through relay)	5V	GND	IN Pin: GPIO 5
Outlet Fan (through relay)	5V	GND	IN Pin: GPIO 13
Humidifier (through relay)	5V	GND	IN Pin: GPIO 32
OLED Display	3.3V	GND	SDA: GPIO 21 (I2C SDA) SCL: GPIO 22 (I2C SCL)

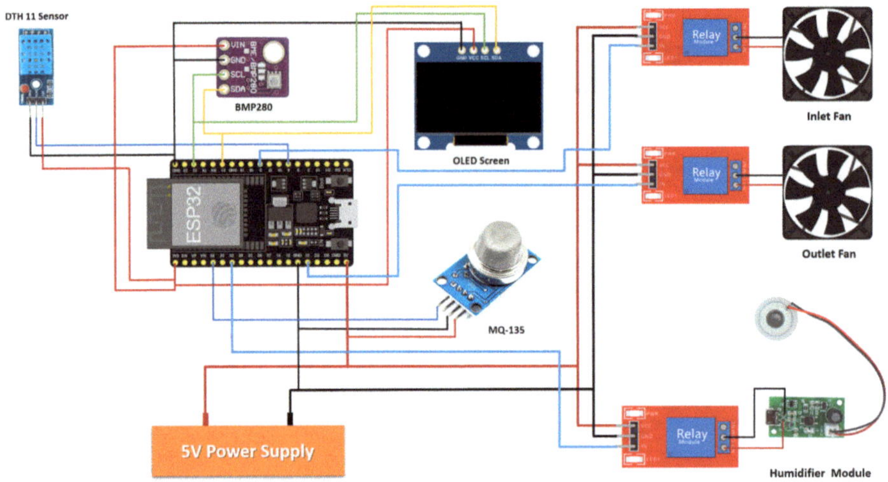

Fig. 4. Circuit diagram of the SAQMCS

Setup and Initialization:

Arduino IDE: The Arduino IDE software was installed into the PC and was used to write and upload the firmware code to the ESP32 microcontroller.

Required Libraries: The following libraries were included for the various sensors and the display integration:

"Adafruit_Sensor.h"; Interfacing of sensor libraries for consistent data retrieval.
"Adafruit_BMP280.h"; Provides functions to interface with the BMP280 sensor.
"DHT.h",Communication with DHT11 for humidity and temperature measurements.
"Adafruit_GFX.h"; Graphics library for drawing shapes and text on OLED display.
"Adafruit_SSD1306.h"; Manages communication with the SSD1306 OLED display.
"WiFi.h"; Enables WiFi connectivity for the ESP32 to connect to the network.

"ThingSpeak.h"; Facilitates communication of data with ThingSpeak IoT platform.

Sensor and Display Initialization: The MQ-135, BMP280, and DHT11 sensors, along with the OLED display, were initialized to start collecting and displaying data.

Control Logic Implementation: The control logic for the Smart Air Quality Monitoring and Conditioning System (as discussed in the Subsect. 3.3) is designed to monitor air quality parameters in real-time and take necessary actions based on the sensor readings. This involves defining thresholds for different pollutants, temperature, and humidity, and implementing conditional statements to trigger actuators such as fans. The control logic also involves sending notifications or alerts through the IoT platform when certain conditions are met.

Data Logging and IoT Integration: Sensor data is sent to the ThingSpeak IoT platform for remote monitoring and analysis. This allows for long-term data logging and visualization.

Testing and Debugging: The testing and debugging phase of the Smart Air Quality Monitoring and Conditioning System (SAQMCS) involved a multi-stage process to ensure accuracy and reliability. Initially, the system was tested in a controlled environment to verify sensor readings and actuator responses. The Arduino IDE's serial monitor was then used to print sensor values for real-time debugging. Finally, the system was deployed in a real-world environment to validate its performance under various conditions, ensuring it could maintain optimal air quality and respond accurately to environmental changes. This comprehensive approach ensured the robustness and effectiveness of the SAQMCS.

3.7 ThingSpeak Platform Configuration

The Smart Air Quality Monitoring and Conditioning System (SAQMCS) leverages advanced IoT technologies to provide real-time monitoring and control of indoor air quality. Central to our approach is the integration of ThingSpeak, an open-source IoT application and API that offers a user-friendly platform for storing, analyzing, and visualizing data from sensors and devices. ThingSpeak is particularly well-suited for systems like SAQMCS due to its robust real-time data handling capabilities, ease of integration, and compatibility with MATLAB for advanced data analysis. The following steps outline the configuration process for the model:

ThingSpeak Channel Creation and Configuration: A new channel is created on ThingSpeak with the name "SAQMCS" to log sensor data. The fields are configured for temperature, humidity, pressure, and air quality.

Data Transmission: The generated API key is copied and passed into the firmware code. Data posting to ThingSpeak is done by setting up a Wi-Fi connection and making HTTP requests to the ThingSpeak API.

Data Visualization: ThingSpeak's built-in tools are used to create graphs and widgets for visualizing the sensor data in real-time. Configuration of alerts and notifications can also be done based on threshold values to monitor air quality and environmental conditions remotely.

4 Result Analysis and Discussion

The Smart Air Quality Monitoring and Conditioning System (SAQMCS) was rigorously tested to evaluate its performance in real-time environmental monitoring and control. The results were analyzed based on data collected over several weeks, focusing on air quality, temperature, humidity, and the effectiveness of the control mechanisms. The following sections summarize the key findings:

4.1 Air Quality Monitoring

The MQ-135 sensor enabled continuous monitoring of air quality, detecting various pollutants and logging data on the ThingSpeak platform for real-time and historical analysis. Key findings include:

Pollutant Levels: The system effectively detected variations in air quality, with pollutant levels showing expected daily fluctuations. Peaks in pollution were observed during high-traffic hours and cooking activities.

Threshold Exceedance: When pollutant levels exceeded predefined thresholds, the inlet fan was activated, demonstrating the system's responsiveness to poor air quality.

4.2 Temperature and Humidity Monitoring

The DHT11 sensor monitored temperature and humidity levels efficiently, and the system's control logic regulated these factors effectively:

Temperature Control: The outlet fan was activated to maintain optimal indoor temperatures when they exceeded comfort thresholds, particularly to prevent overheating in hot weather.

Humidity Control: The humidifier was activated when humidity levels dropped below a set threshold, ensuring a comfortable indoor environment during drier climates.

4.3 Local Monitoring

The OLED display (see Fig. 5) provided immediate, on-site feedback on air quality and environmental conditions. It displayed real-time data on temperature, humidity, and air quality index, enabling users to monitor the system's status at a glance.

4.4 Remote Monitoring

The integration with ThingSpeak significantly enhanced the SAQMCS by providing robust data visualization and remote monitoring capabilities:

Data Visualization: Real-time graphs and widgets facilitated smooth, continuous monitoring of air quality, temperature, pressure, and humidity remotely over network. A dedicated dashboard for the SAQMCS on ThingSpeak is depicted in Fig. 6.

Fig. 5. OLED Display Showing Real-time Environmental Data and System Status of SAQMCS

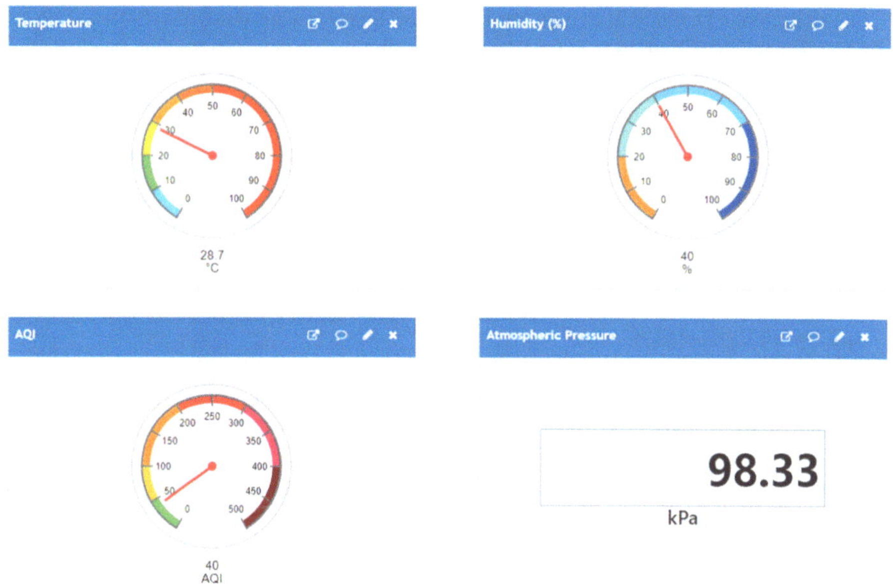

Fig. 6. Remote Air Quality Monitoring Dashboard for SAQMCS on ThingSpeak Platform

Data Analysis: Historical data analysis tools helped identify patterns and trends over time. MATLAB facilitates advanced data analysis, which can also be used to further improve the system's utility.

5 Conclusion

Thus the Smart Air Quality Monitoring and Conditioning System (SAQMCS) provides an effective solution for real-time indoor air quality monitoring and control. Using advanced IoT technologies, sensor networks, and automated controls, the system maintains optimal air quality and comfort. ThingSpeak enhances data visualization and remote monitoring, allowing users to manage air quality from anywhere. Testing confirmed the system's reliability and responsiveness in various conditions, detecting and responding to changes in air quality, temperature, and humidity with minimal user intervention.

Future improvements will include advanced air purification technologies, modular designs, and enhanced data analytics with machine learning. We aim to integrate SAQMCS with smart home systems, optimize energy efficiency, and improve user interfaces, setting new standards in IoT-driven environmental control.

References

1. Kumar, A., Kumar, C., Kumar, A., Dhanda, A.K., Kumar, H.: Innovative solutions for urban air pollution in Delhi/NCR region. Int. J. Sci. Res. Eng. Manage. (IJSREM) **8**(4), 1–7 (2024)
2. Manusmare, A., Gajarlawar, P.J.: Advance pollution monitoring system with health critical gas indexing using IoT. Int. J. Eng. Appl. Sci. Technol. (IJEAST) **8**(3), 53–63 (2023)
3. De Capua, C., Fulco, G., Lugarà, M., Ruffa, F.: An improvement strategy for indoor air quality monitoring systems. Sensors **23**, 3999 (2023)
4. AbdulWahhab, R.S., Jetly, K., Shakir, S.: Indoor air-quality monitoring systems: a comprehensive review of different IAQM systems. In: Husain, M.S., Adnan, M.H.B.M., Khan, M.Z., Shukla, S., Khan, F.U. (eds) Pervasive Healthcare. EAI/Springer Innovations in Communication and Computing. Springer, Cham (2022)
5. European Environment Agency (EEA): Air quality in Europe - 2020 report. (2020)
6. Kumar, P., et al.: The rise of low-cost sensing for managing air pollution in cities. Environ. Int. **75**, 199–205 (2015)
7. Pavani, M., Rao, P.: Urban air pollution monitoring using wireless sensor networks: a comprehensive review. Int. J. Commun. Netw. Inf. Secur. **9**(3), 439–449 (2017)
8. Kang, Y., Aye, L., Ngo, T.D., Zhou, J.: Performance evaluation of low-cost air quality sensors: a review. Sci. Total Environ. **818**, 151769 (2022)
9. Pitarma, R., Marques, G., Ferreira, B.R.: Monitoring indoor air quality for enhanced occupational health. J. Med. Syst. **41**(2), 23 (2017)
10. Liu, Q., Cui, B., Liu, Z.: Air quality class prediction using machine learning methods based on monitoring data and secondary modeling. Atmosphere **15**(5), 553–564 (2024)
11. Popoola, O.A.M., Carruthers, D., Lad, C., Bright, V.B., Mead, M.I., Stettler, M.E.J.: Use of networks of low-cost air quality monitoring sensors to quantify air quality in urban settings. Atmos. Environ. **194**, 58–70 (2018)
12. Malik, P. K., Duggal, A.S., Aluvala, S., Sahithi, R., Geetanjali, Gehlot, A.: Development of a low-cost IoT device using ESP8266 and Atmega328 for real-time monitoring of Outdoor Air Quality with Alert. 2023 3rd International Conference on Advancement in Electronics & Communication Engineering (AECE), GHAZIABAD, India, pp. 125–129 (2023)
13. Kortoçi, P., et al.: Air pollution exposure monitoring using portable low-cost air quality sensors. Smart Health **23**, 100241 (2022)

14. Guvvala, M.V., Ch, D.: ThingSpeak based air pollution monitoring system using raspberry Pi. In: 3rd Asian Conference on Innovation in Technology (ASIANCON), Ravet IN, India, 2023, pp. 1–6 (2023)
15. Taştan, M.: A low-cost air quality monitoring system based on Internet of things for smart homes. J. Ambient Intell. Smart Environ. **14**, 1–24 (2022)

Toxicity and Efficacy of ITK Based Chili-Neem Biopesticide on Sucking Pests in Vegetable Crops

Rinny Swain[1](✉), Mamata Behera[1], Farin Bano[2], Tasmia Nausad[2], Suvam Roy[2], Yugesh Kumar Negi[2], and Subhankar Panda[2]

[1] Crop Improvement Division, School of Agriculture, GIET University, Gunupur, Rayagada, Odisha 765022, India
rinnyswain@giet.edu

[2] EYUVA Fellow, School of Agriculture, GIET University, Gunupur, Odisha 765022, India

Abstract. This research paper explores the efficacy of an innovative ITK (Indigenous Technology Knowledge) based chili-neem biopesticide in targeting common sucking pests infesting vegetable crops. The biopesticide's strategic focus on sucking pests necessitates a targeted approach to pest management, utilizing potent ingredients such as green chili extract, red chili extract, neem leaf extract, molasses, garlic, and cow urine to enhance efficacy. The effect was evaluation firstly by brine shrimp lethality assays to assess the lethal dose. The most effective lethal dose of 7–10% were further evaluated through laboratory trials on pests such as aphids, mealy bugs, and hadda beetles demonstrates the concentration-dependent effectiveness of biopesticides in controlling pest populations. Overall, ITK-based chili-neem biopesticide is a novel innovative Agro product promising a sustainable alternative for managing pest infestations in vegetable crops, aligning with the principles of eco-friendly and healthy agriculture.

Keywords: Chili · Neem · Biopesticide · Bioefficacy

1 Introduction

In the ever-evolving realm of agriculture, ITK (Indigenous Technology Knowledge) based biopesticide can emerges as a beacon in blending centuries-old wisdom with cutting-edge formulations. The various organic mixtures crafted with meticulous precision, not only redefines pest management but also addresses the imperative shift from synthetic to organic solutions. As per the US (United States) Environmental Protection Agency, biopesticides, derived from natural materials like animals, plants, bacteria, and certain minerals, are utilized to prevent crop damage through products such as genes or metabolites from these biocontrol agents. Biopesticides offer significant advantages over traditional chemical pesticides, being both eco-friendly and host-specific. Enhancing the use and application, biopesticides an Agro-based product in agriculture proves to be a more sustainable product protecting crop plants from invading and infecting pests.

Since ancient times, the utilization of synthetic (chemical) pesticides for enhancing crop production by controlling pests has been recognized [1]. Formulated from chemicals and carriers, such as polymers synthetic pesticides are tailored for specific pests,

B. K. Mishra et al. (Eds.): ICT4GS 2024, ISEM 56, pp. 422–429, 2025.
https://doi.org/10.1007/978-3-031-94997-5_36

ranging from herbicides for weed control to insecticides for insect management [2]. This classification encompasses various types, including algicides for algae, fungicides for fungi, miticides/acaricides for mites or ticks, bactericides for bacteria, rodenticides for rodents, termiticides for termites, molluscicides for molluscs, and nematicides for nematodes [3]. An alternative method of classifying pesticides is based on their active ingredients, such as organochlorines, dichlorvos, diazinon, diamide, chlorpyrifos, and more. Despite the positive impact of synthetic pesticides on crop yield and productivity, they impose adverse effects on soil biodiversity, animals, aquatic life, and humans. Typically, synthetic pesticides make the soil more fragile, decrease soil respiration, and diminish the activities of certain microorganisms like earthworms. They also diminish the traits of animal offspring, animal disease immunity, vitality, and the success of mating in animals [4]. Additionally, they have detrimental effects on soil microorganisms by restricting their biological contributions to the production of specific plant growth-promoting traits, such as siderophores, nitrogen, indole-3-acetic acid, etc. [5].

Biopesticides, known for being cost-effective, eco-friendly, specific in action, sustainable, residue-free, and devoid of greenhouse gas emissions, come in various forms, including phytopesticides (plant origin), microbial pesticides (microbial origin) and nanobiopesticides (nanoparticles created from biological agents) [6]. In contrast to synthetic pesticides, microbial pesticides exhibit specificity in action, are readily obtainable without the requirement for costly chemicals, and are environmentally sustainable with no lingering effects. The majority of plant-derived pesticides are applied for controlling insect pests and mitigating plant parasites such as nematodes, fungi, bacteria, and viruses [7–12]. Botanical pesticides could serve as alternative tools for integrated pest management, given their positive effects on environmental preservation, low toxicity to mammals, and minimal risk of developing resistance in target pests.

2 Materials and Methods

2.1 Preparation of ITK Based Chili-Neem Biopesticide Formulation

250 g each of green and red chili were cut longitudinally, submerged in 400 mL of water for 24 h. Later, 250 g garlic and neem alongwith chili was grinded to obtain extract. The aqueous extract is mixed with 10 mL of cow urine, 5 mL of molasses, 5 mL oil stirred rigorously to produce an emulsion. Post-formulation, the biopesticide was carefully stored in suitable containers safeguarded from direct sunlight to preserve its efficacy.

2.2 Bioassay

Brine shrimp lethality assay- Brine shrimp eggs were incubated in water for 48 h to facilitate hatching of the larvae (nauplii). The newly hatched nauplii were then subjected to cytoplasmic toxicity tests by exposing them to different concentrations of a plant extract dissolved in the appropriate solvent. The concentrations tested ranged from 1% to 5% of the plant extract. The number of live nauplii was recorded at 24-h intervals over a period of two days to monitor the mortality rate induced by the plant extract.

2.3 Statistical Analysis

Mortality data were expressed as percentage and analyzed using one-way analysis of variance (ANOVA).

3 Results and Discussion

3.1 Brine Shrimp Lethality Assay

The evaluation of cytotoxicity is crucial in the initial screening of potential bioactive compounds from natural sources such as plants. The brine shrimp lethality assay has gained prominence due to its simplicity, cost-effectiveness, and ability to provide rapid insights into the cytotoxic effects of test substances. This assay relies on the principle that toxic compounds will induce mortality in brine shrimp larvae upon exposure. In this study, we utilized laboratory-cultured brine shrimp larvae (nauplii) as a model organism to evaluate the cytotoxic effects of various concentrations of a plant extract over a 24-h exposure period. The results obtained from the brine shrimp lethality assay revealed a concentration-dependent increase in mortality rates following exposure to the plant extract. At higher concentrations (4% and 5%), the mortality rates reached 70% and 86% respectively after 24 h, and further increased to 90% and 93% by the end of the 48-h period. Conversely, lower concentrations (1% to 3%) exhibited relatively lower mortality rates, with 24-h mortality ranging from 10% to 63%, and 48-h mortality reaching up to 84%. These findings suggest that the plant extract possesses cytotoxic properties against brine shrimp larvae in a concentration-dependent manner (Tables 1, 2, 3 and 4).

Table 1. Records of number of motile nauplii within 48 h (hrs) of experiment.

Timing 12:30 pm	Control (0 ml NBF)	T1 (1%)	T2 (2%)	T3 (3%)	T4 (4%)	T5 (5%)
No. of live nauplii	30	30	30	30	30	30
Day 1	27	19	13	11	9	4
Day 2	25	9	6	3	3	2
% Death after 24 h	10	36	56	63	70	86
% Death after 48 h	16	70	84	90	90	93

3.2 Bio-efficacy Test on Pests at Laboratory Level

This study aimed to assess the effectiveness of novel ITK-based chili-neem biopesticide, an Agro on various prevalent sucking pests known to cause significant damage to vegetables. Various commonly occurring sucking pests like aphids, mealy bugs, and hudda beetle were targeted for evaluation of chili-neem biopesticide were applied in different concentration. Pest populations were meticulously monitored pre- and post-treatment

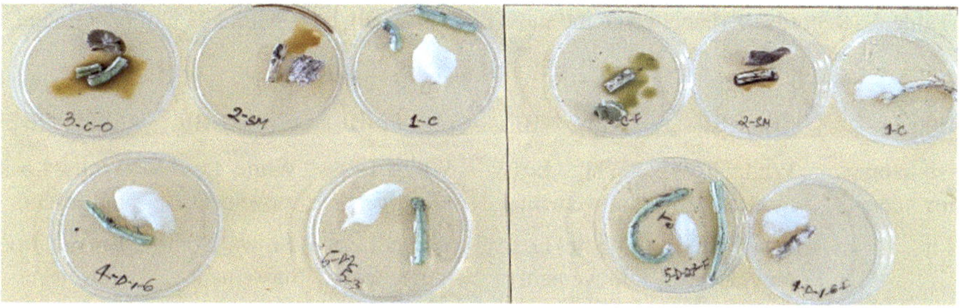

Fig. 1. Aphids from infested Cowpea bean collected for bio efficacy trial.

Table 2. Records of number of motile aphids within 48 h of experiment.

Timing 10:30 am	Control (0 ml NBF)	T1 (50% concentrate)	T2 (100% formulation)	T3 (2.5%)	T4 (5%)
No. of live aphids	40	40	40	40	40
24 h	40	6	3	11	7
48 h	35	0	0	2	0
% Death after 24 h	0	90	92	72	70
% Death after 48 h	1.3	100	100	95	100

using standardized sampling techniques. Data on pest mortality were systematically recorded (Figs. 1, 2 and 3).

Cowpea bean plants infested with aphids were collected and divided into 4 to 5 segments. Each segment was treated with varying concentrations of the biopesticide solution using cotton balls for application. The concentrations tested included a 50% concentrate, 100% formulation, 2.5% diluted formulation solution, and 5% diluted formulation solution. The population of live aphids was monitored at 24- and 48-h post-treatment, with observations recorded for each concentration group. At the 24-h mark, both the 50% and 100% concentrations demonstrated remarkable efficacy, with approximately 90% and 92% population reduction, respectively. By the 48-h assessment, complete eradication of aphids was achieved across all concentration groups, indicating 100% mortality.

Coleus stems infected with mealy bugs were divided into segments, and varying concentrations of Nature Bloom Pesticides were applied directly onto the infected areas using cotton balls. Concentrations tested ranged from 1.0% to 2.5%. The movement and coloration of the mealy bug-infested segments were closely monitored at 24- and 48-h post-treatment. Observations were recorded, and mortality rates were calculated based on the reduction in mealy bug activity and color change indicative of response to the biopesticide treatment. At lower concentrations (1.0% and 1.5%), a gradual reduction in mealy bug movement and color change from white to brown were observed within 48 h post-application, resulting in mortality rates of approximately 70% and 82%, respectively.

Table 3. Records of visible layer and movement of mealy bugs within 48 h of experiment.

Timing 11:00 am	Control (0% NBF)	T1 (1.0% NBF)	T2 (1.5% NBF)	T3 (2.0% NBF)	T4 (2.5% NBF)
Visible layer/ Movement	White/ Less movement	White/ Less movement	White/ Less movement	White/ Less movement	White/ Less movement
24 h	White/ Less movement	White/ Less movement	White/ Less movement	Brown/ Some movement	Brown/ No movement
48 h	White/ Some movement	Brown/ Some movement	Brown/ No movement	Brown/ No movement	Brown/ No movement
% Death after 48 h	50	70	82	100	100

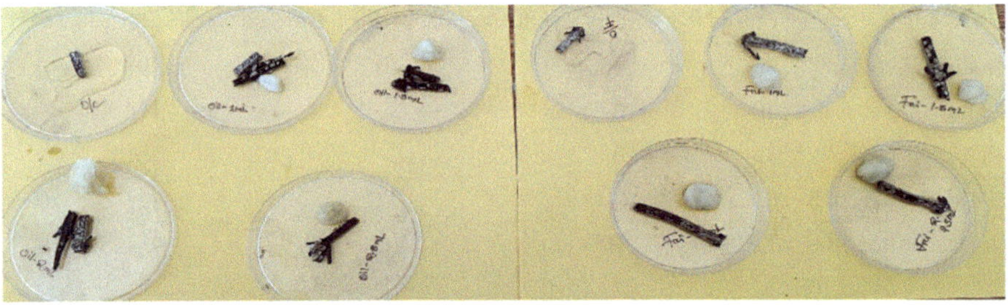

Fig. 2. Mealy bugs from infested Coleus stems collected for bio efficacy trial.

Table 4. Records of number of motile beetles within 48 h of experiment.

Timing 10:00 am	Control (0% NBF)	T1 (2.5% NBF)	T2 (5.0% NBF)	T3 (7.5% NBF)	T4 (10.0% NBF)
No. of live beetles	5	5	5	5	5
24 h	5	5	4	2	1
48 h	5	4	3	0	0
% Death after 24 h	0	0	20	60	80
% Death after 48 h	0	20	40	100	100

Hadda beetles were collected from infested Brinjal plants, and Brinjal leaves were immersed in different concentrations of the biopesticide solution. The concentrations tested ranged from 2.5% to 10.0%. Petri plates were used to cover the treated leaves with beetles, allowing them to feed on the treated foliage. Observations were made at 24- and 48-h post-treatment to record mortality rates among the Hadda beetles. At lower concentrations (2.5% and 5.0%), moderate mortality rates of 20% and 40% were observed among the Hadda beetles, respectively, within 48 h post-treatment. Higher

Fig. 3. Hadda beetles from infested Brinjal plants collected for bio efficacy trial.

concentrations (7.5% and 10.0%) resulted in complete eradication of the Hadda beetle population within the same timeframe, achieving mortality rates of 100%.

Sucking pests pose a substantial threat to agricultural productivity, particularly in vegetable cultivation. Traditional pesticides have shown limited effectiveness and pose environmental risks. Evaluating the effectiveness of ITK-based chili-neem biopesticide and Agro product on common sucking pests in laboratory conditions provided useful insights into their potential as sustainable pest management tools. The observed cytotoxic effects of the plant extract on brine shrimp larvae indicate its potential bioactivity against living organisms. The concentration-dependent increase in mortality rates suggests that higher concentrations of the extract exert more pronounced cytotoxic effects. The combination of chili and neem shows significant efficacy against common sucking pests, such as aphids, jassids, and whiteflies. Studies indicate that neem oil, especially when combined with other botanical extracts like chili, effectively reduces pest populations by disrupting feeding and reproductive cycles. This aligns with the results observed and suggested by Halder et al. and Hezakiel et al. in pest populations reduction and rapid onset of mortality in laboratory conditions [13, 14]. The observed pest population reductions in laboratory trials underscore the efficacy of biopesticides in controlling pest infesting in plants. The biopesticide's mode of action likely disrupts feeding and reproduction, leading to population decline. The rapid onset of mortality within 48 h suggests a potent effect, indicative of the biopesticide's potential as a promising alternative to conventional pesticides. The rapid eradication achieved with higher concentrations suggests a potent and fast-acting mode of action against Hadda beetles. Increasing the concentration of the biopesticide led to higher mortality rates among the beetles, with concentrations of 7.5% and 10.0% achieving complete eradication within 48 h. Analogous research highlights that neem-based biopesticides, when applied at higher concentrations, show significant pest control within a short period, supporting the rapid eradication observed within 48 h at concentrations of 7.5% and 10.0% [13]. The complete eradication of all evaluated pest within this 48-h timeframe is particularly noteworthy, emphasizing the efficacy of biopesticides in achieving pest management goals. Biopesticides like neem and chili extracts are generally considered safe for non-target organisms, including beneficial insects and natural predators. Research has shown that these biopesticides are less harmful compared to synthetic pesticides, making them suitable for integrated pest management strategies that aim to preserve ecological balance. These findings highlight the potential of based chili-neem biopesticide as an effective and rapid-acting solution for managing sucking pest infestations different vegetable crops, warranting further exploration and field validation.

4 Conclusion

This study employs laboratory trials to assess the efficacy of biopesticide against common sucking pests having destructive impact on crops, as they feed on plant sap, causing stunted growth, decreased yield, and the spread of plant diseases. Traditional pesticides, frequently used to combat these pests, often necessitate repeated applications, which can lead to the development of pesticide-resistant strains and environmental contamination. The novel introduced ITK-based chili-neem biopesticide Agro-product can be presented as a sustainable alternative, harnessing naturally occurring compounds sourced from plants to effectively target pests while minimizing harm to the environment. Preliminary findings from laboratory trials reveal promising results, with significant reductions in pest populations observed following the application biopesticides. Further research should focus on isolating and characterizing the specific bioactive compounds responsible for the observed cytotoxicity and pest control efficacy. Laboratory results should be validated through field trials to confirm the effectiveness of the chili-neem biopesticide under different environmental conditions and against a broader range of pests. Moreover, further field trials are necessary to demonstrate its effectiveness in crop health and yield compared to conventional pesticide treatments.

Acknowledgement. The authors wish to acknowledge the BIRAC (Biotechnology Industry Research Assistance Council), Govt. of India for providing financial assistance under EYUVA (Empowering Youth for Undertaking Value Added Innovative Translational Research) Scheme to authors as fellowship and research grant through EYUVA GIET Centre (Gandhi Institute of Engineering and Technology).

References

1. Ngegba, P.M., Cui, G., Khalid, M.Z., Zhong, G.: Use of botanical pesticides in agriculture as an alternative to synthetic pesticides. Agriculture **12**, 600 (2020). https://doi.org/10.3390/agriculture12050600
2. Pavela, R.: Advances in plant biopesticides-limitation of plant biopesticides. Singh, D., ed. Springer: New Delhi, India, pp. 347–359 (2014)
3. Tijjani, A., Bashir, K.A., Mohammed, I., Muhammad, A., Gambo, A., Musa, H.: Biopesticides for pests control: a review. Biopesticides Agric. **3**(1), 2465–7115 (2016)
4. Bag, D.: Pesticides and health risks. Econ. Polit. **35**(Wkly), 3381–3383 (2000). https://doi.org/10.1111/j.1552-6909.2009.01092.x
5. Kumar, N., Pathera, A.K., Saini, P., Kumar, M.: Harmful effects of pesticides on human health. Ann. Agri. Bio. Res. **17**, 125–127 (2012)
6. Chowdhury, S.K., Banerje, M., Basnett, D., Mazumdar, T.: Natural pesticides for pest control in agricultural crops: an alternative and eco-friendly method. Plant Sci. Today ISSN 2348-1900 (2023)

7. Sirohi, A., Tandon, P.: Insecticidal effects of various parts of Azedarach indica on adults of Aulacophora foveicollis (Lucas) (Coleoptera: Chrysomelidae). Trends Biosci. **7**, 3947–3949 (2014)
8. Kumar, R., Kaushik. J.K., Mohant, A.K., Kumar, S. Cow urine: Identification of bioactive components behind the antimicrobial activity of Cow urine by Peptide and metabolite profiling. Anim. Biosci. 1–13 (2020). https://doi.org/10.5713/ab.22.0249 pISSN 2765–0189 eISSN 2765–0235
9. Ramjan, M.d.: Extraction of plant based botanical pesticides for management of vegetable pests. Just Agric. **1**(015), Issue-5 (2021)
10. Ray, K.R.: Ginger garlic extract as a bio-pesticide for organic farming. Just Agric. (e-ISSN: 2582–8223), **2**(102), Issue-6 (2022)
11. Kamaliyah, S.N., Wahyun, D.R.: Effect of EM4 and molasses levels on quality of organic liquid fertilizer of cow urine based. Jurnal Ilmiah Peternakan Terpadu, **11**(3), 190–200 (2023)
12. Djiwanti, S.R., Rismayani, H.R., Aunillah, A.: Nematicidal effectivity of neem oil, castor oil, and pyrethroids toward foliar nematode (Aphelenchoides fragariae) mortality. IOP Conf. Ser. Earth Environ. Sci. **1297**(1), 012058 (2024). https://doi.org/10.1088/1755-1315/1297/1/012058
13. Halder, J., Divekar, P.A., Rani, A.T.: Compatibility of entomopathogenic fungi and botanicals against sucking pests of okra: an ecofriendly approach. Egypt. J. Biol. Pest Control **31**, 30 (2021). https://doi.org/10.1186/s41938-021-00378-6
14. Hezakiel, H.E., Thampi, M., Rebello, S., et al.: Biopesticides: a green approach towards agricultural pests. Appl. Biochem. Biotechnol. (2023). https://doi.org/10.1007/s12010-023-04765-7

Author Index

B. K. Mishra et al. (Eds.): ICT4GS 2024, ISEM 56, pp. 431–432, 2025.
https://doi.org/10.1007/978-3-031-94997-5

MIX
Papier aus verantwortungsvollen Quellen
Paper from responsible sources
FSC® C105338

If you have any concerns about our products,
you can contact us on
ProductSafety@springernature.com

In case Publisher is established outside the EU,
the EU authorized representative is:
Springer Nature Customer Service Center GmbH
Europaplatz 3, 69115 Heidelberg, Germany

Printed by Libri Plureos GmbH
in Hamburg, Germany